Materials Modeling for High-Performance Solid-State Batteries

Junhua Hu, Hongjie Xu, Jinjin Ban, Fanfan Liu, and Shilin Zhang

WILEY VCH

Authors

Prof. Junhua Hu
Zhengzhou University
Zhengzhou, 450001
China

Dr. Hongjie Xu
North China University of Water Resources
and Electric Power
Zhengzhou, 450045
China

Dr. Jinjin Ban
Zhengzhou University
Zhengzhou, 450001
China

Dr. Fanfan Liu
Zhengzhou University
Zhengzhou, 450001
China

Dr. Shilin Zhang
Zhengzhou University
Zhengzhou, 450001
China

Cover Design: Wiley
Cover Image: Just_Super/Getty Images

Typesetting: Lumina Datamatics, Madurai, India

Library of Congress Card No.: applied for

British Library Cataloguing-in-Publication Data
A catalogue record for this book is available from the British Library.

Bibliographic information published by the Deutsche Nationalbibliothek
The Deutsche Nationalbibliothek lists this publication in the Deutsche Nationalbibliografie; detailed bibliographic data are available on the Internet at <http://dnb.d-nb.de>.

Print ISBN: 9783527354771
ePDF ISBN: 9783527850723
ePub ISBN: 9783527850716
oBook ISBN: 9783527850730

Printed and bound by CPI Group (UK) Ltd, Croydon, CR0 4YY

C9783527354771_070726

Contents

Preface

With the in-depth advancement of the global energy transition and the rapid expansion of new energy industries—including electric vehicles and stationary energy storage systems—lithium-ion batteries have emerged as the core energy storage devices underpinning society's low-carbon transformation. Despite their widespread application, traditional lithium-ion batteries with liquid electrolytes still suffer from inherent bottlenecks in energy density, safety, cycle life, and environmental adaptability, which significantly hinder their further penetration into high-end fields. As a revolutionary solution to these constraints, all-solid-state batteries have garnered extensive attention from both academic and industrial communities, owing to their superior energy density, exceptional safety, long cycle life, and broad operating temperature range. Consequently, they have become a pivotal research direction in the field of new energy storage materials and devices.

In recent years, significant strides have been made in solid-state battery research; however, numerous core technical challenges remain unresolved. These include the theoretical design and fabrication of high-performance electrode and electrolyte materials, the optimization of solid-solid interface compatibility, the development of advanced characterization and testing technologies, and the refinement of assembly processes. Against this backdrop, compiling a monograph that integrates the academic fundamentals and engineering practices of new energy lithium batteries—with a specific focus on solid-state batteries—holds profound significance. Such a work not only systematically sorts out the latest research progress and summarizes key technical bottlenecks but also provides guidance for the in-depth development of subsequent research and industrialization efforts.

Centered on solid-state batteries, this book focuses on the frontier research and engineering applications of new energy lithium batteries, systematically organizing the research progress, theoretical basis, technical difficulties, and development trends in related fields. The content is closely integrated with academic research and engineering practice, spanning multiple interdisciplinary fields such as materials science, electrochemistry, energy engineering, and artificial intelligence. As such, it is well-suited for researchers, engineers, and graduate students engaged in the research and development of lithium batteries and related areas.

Comprising nine chapters, the book forms a coherent logical framework that progresses from basic principles to practical applications and from material design to process optimization. Chapter 1 lays the theoretical foundation for the entire book by elaborating on the applications and challenges of solid-state batteries, covering research background, electrochemical principles, working mechanisms, development history, core challenges, and future prospects. Building on this foundation, Chapter 2 focuses on the application of high-throughput technology in lithium batteries, with an emphasis on the role of first-principles and machine learning in battery material design—reflecting the cross-integration of emerging technologies with battery research. Chapter 3 then delves into the material and structural design of high-performance all-solid-state batteries, focusing on the research progress of cathode materials, the development status and applications of common commercial cathode materials (e.g., $LiCoO_2$, $LiFePO_4$, and ternary cathodes), as well as their preparation methods, thereby providing critical technical support for the development of high-performance solid-state batteries.

As the core components of solid-state batteries, oxide and sulfide solid electrolytes are the focus of Chapters 4 and 5, respectively. These chapters systematically introduce their research status, preparation technologies, performance characteristics, existing challenges, and future development prospects, enabling readers to grasp the key technical points of different types of solid electrolytes. Complementing the research on electrode materials, Chapter 6 focuses on the research progress of anode materials for lithium-ion batteries, including material types, existing problems, and future development trends. Chapter 7 expands the scope to the research progress and current status of magnesium-ion batteries, offering readers a broader research perspective. Chapter 8 then focuses on the electrochemical performance testing and advanced characterization methods essential for battery research, including crystal structure characterization, morphology characterization, and in-situ electrochemical performance testing. Finally, Chapter 9 summarizes the assembly technology of solid-state batteries and provides an overall outlook on their development, forming a comprehensive conclusion and prospect for the entire book.

In compiling this book, the authors have incorporated the latest domestic and international research achievements, emphasizing the integration of academic rigor and engineering practicability to ensure the accuracy, comprehensiveness, and forward-looking nature of the content. Additionally, the book highlights the interdisciplinary characteristics of lithium battery research, showcasing the application of new theories, methods, and technologies, with the aim of providing valuable references for relevant researchers and engineers.

It is important to note that the field of new energy lithium batteries is evolving rapidly, with new technologies and achievements emerging continuously. Due to the limitations of the authors' academic level and research experience, there may be deficiencies in this book. We sincerely welcome comments and corrections from readers and peers to further improve the quality of the work.

Finally, we thank all the researchers who have made important contributions in the field of new energy lithium batteries, and the colleagues and friends who have provided support and help in the compilation of this book. At the same time, we are very grateful to Mr. Cao Guoqin, Mr. Li Dabing and Mr. Zheng Ningchong for their guidance in the compilation of the manuscript, and to all the teachers involved in the format Finally, we thank all the researchers who have made important contributions in the field of new energy lithium batteries, and the colleagues and friends who have provided support and help in the compilation of this book. At the same time, we are very grateful to Mr. Cao Guoqin, Mr. Li Dabing and Mr. Zheng Ningchong for their guidance in the compilation of the manuscript, and to all the teachers involved in the format adjustment and revision of this book.

1

Solid-state Lithium-ion Batteries: Applications and Challenges

1.1 Background and Introduction

Organic liquid lithium-ion batteries have been widely used in many sectors of the national economy (such as portable electronic devices, electric vehicles, and large-scale energy storage, among others) and are gradually expanding into specialized application fields like deep-sea, deep-space, deep-earth, polar regions, individual soldier equipment, drones (military and civilian), and outer space exploration. The introduction of the national "Dual Carbon" strategy has further provided an excellent development opportunity for secondary electrochemical energy storage technologies, typically represented by lithium-ion batteries.

However, organic liquid lithium-ion batteries pose safety hazards, primarily due to the following:

1. Mainstream carbonate solvents are generally volatile, have low flash points, are flammable and explosive at elevated temperatures, and possess certain toxicity, which limit their operating temperature range.
2. Dendrites readily form on the anode surface, potentially piercing the separator and causing a short circuit between the cathode and anode, leading to battery fires. When the anode surface is uneven, repeated charge–discharge cycles can lead to excess lithium accumulation and deposition on the anode surface, forming lithium dendrites. These dendrites may penetrate the microporous separator, connect to the cathode electrode, and cause an internal short circuit.

Meanwhile, as the demands for energy density and safety performance of lithium-ion batteries in mobile electronic devices like smartphones and notebooks, as well as in electric vehicles and large-scale energy storage, continue to rise, traditional liquid lithium-ion batteries inevitably undergo side reactions during charging and discharging. These reactions not only affect battery lifespan but also raise safety concerns due to the flammable organic electrolyte and multiple industry incidents of vehicle spontaneous combustion, causing public apprehension about lithium battery safety.

From January to July 2023, several spontaneous combustion incidents involving new energy vehicles attracted the attention of national regulators. On July 28, 2023, the Equipment Industry Development Center of the Ministry of Industry and Information Technology issued the "Notice on Carrying Out the 2023 New Energy Vehicle Safety Hazard Investigation Work," which deployed tasks for investigating safety hazards in new energy vehicles and specified clear deadlines and content requirements for reporting fire incidents.

Analyzing the root causes of battery safety hazards and revealing the underlying failure mechanisms are prerequisites for constructing highly safe battery systems. In commercial liquid lithium-ion batteries, polyolefin separators with poor dimensional stability and flammable, leak-prone organic electrolytes are considered the primary factors leading to thermal runaway. However, research indicated that improving separator thermal stability and employing flame-retardant electrolytes can only mitigate the battery's exothermic behavior to a certain extent and cannot fundamentally prevent thermal runaway [1, 2]. Tracing back to the origin, a profound understanding of the evolution path of the chain exothermic reactions during thermal runaway is crucial from a macroscopic perspective [3]. Based on an in-depth analysis of battery exothermic characteristics, Cui and coworkers from the Qingdao Institute of Bioenergy and Bioprocess Technology, Chinese Academy of Sciences, proposed that the thermal compatibility between battery materials (electrode materials, electrolytes, additives, etc.) is crucial for battery safety. The team explored the failure mechanism of high-nickel NMC batteries through coupled in situ/ex situ methods at multiple levels, from materials to the battery. Using an isotope titration–mass spectrometry online gas detection device, they identified the presence of lithium hydride (LiH) on the graphite anode of NMC batteries and confirmed that the poor thermal stability between this component and the electrolyte is the main factor inducing the chain self-exothermic behavior of the battery. Furthermore, using a self-developed double-reaction sphere test device for in situ detection of gas shuttling behavior during thermal runaway, they demonstrated that H_2 generated by LiH on the anode side can shuttle to the cathode side, accelerating intense exothermic reactions and causing thermal runaway [4–6]. Based on this research, it was found that the chemical composition of the anode interface layer and its derivatives significantly impact battery thermal runaway. Therefore, effectively suppressing the accumulation of LiH and the generation of H_2 on the anode side is key to intrinsically solving battery safety issues from a materials perspective.

The intrinsic properties of traditional liquid lithium-ion battery materials indicate that they cannot meet the stringent future requirements for energy density, cycle life, and safety of high-performance devices. Consequently, replacing flammable and explosive liquid electrolytes with non-hydrogen-containing solid-state electrolytes is expected to fundamentally solve the hydrogen generation problem in lithium batteries, cut off the initiation source of thermal runaway, fabricate lithium battery systems with inherently high safety and reliability, and enable the safe design of high-energy-density power batteries. Thus, solid-state lithium batteries have become a research and development hotspot (Figure 1.1).

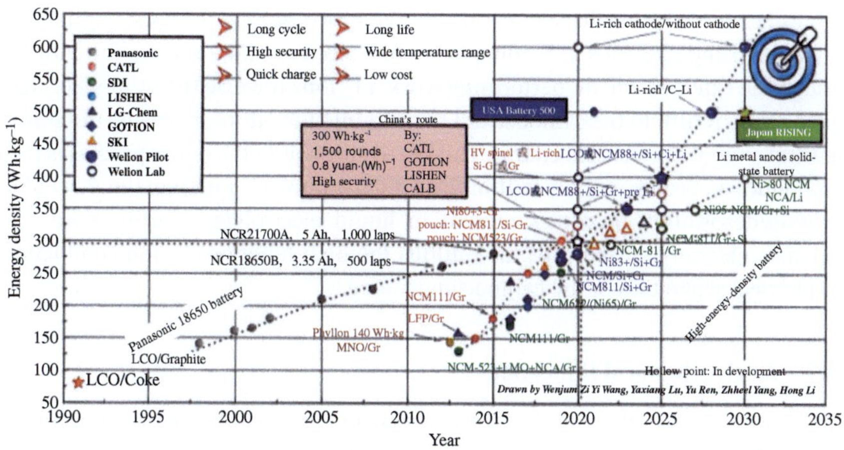

Figure 1.1 Solid-state lithium batteries enhance the safety, energy density, cycle life, and other characteristics of the battery system [7].

1.2 Electrochemical Principles

1.2.1 Working Principle of Solid-state Batteries

As a revolutionary energy storage system, the basic structure of a solid-state battery primarily consists of four core components: the cathode, solid-state electrolyte, anode, and current collectors. The cathode typically employs lithium transition-metal oxides (e.g., NCM, NCA), lithium iron phosphate (LFP), or high-capacity lithium-rich manganese-based materials. The solid-state electrolyte, acting as the "soul" of the entire battery, fulfills the dual mission of conducting lithium ions while blocking electrons. The anode can be materials like graphite or silicon–carbon, but the most revolutionary prospect lies in directly using a metallic lithium anode. The current collectors are similar to conventional batteries, with aluminum foil for the cathode side and copper foil for the anode side. The core feature of this structure is that the ion conduction medium is entirely solid-state. This fundamental change not only greatly simplifies the battery structure but also makes possible high-energy-density system architectures like bipolar stacking (integrating multiple cells in series within a single package). In terms of working principle, solid-state batteries macroscopically still follow a "rocking-chair" mechanism similar to liquid lithium-ion batteries, where lithium ions shuttle back and forth between the cathode and anode via intercalation and deintercalation. Specifically, during charging, an external power source applies a voltage, driving electrons (e^-) from the cathode through the external circuit to the anode. Simultaneously, lithium ions (Li^+) deintercalate from the cathode material lattice, enter and migrate through the solid-state electrolyte toward the anode side, finally reaching the anode where they combine with electrons and are stored as atoms embedded in the anode material or deposited on the metallic lithium surface, thereby converting electrical energy into chemical energy. During discharge, this process reverses: Li at the anode undergoes oxidation, losing electrons to become Li^+, while electrons flow back to the cathode through the external circuit, performing work. Li^+ ions traverse the solid-state electrolyte again, return to the cathode, and reinsert into the lattice, releasing the stored chemical energy as electrical energy. The key distinction in the entire process is that ion conduction occurs entirely within the solid medium, eliminating the inevitable solvation/desolvation processes present in liquid electrolytes. Furthermore, the solid-state electrolyte itself integrally fulfills the dual functions of ion conduction and physical separation of the cathode and anode.

1.2.2 Electrochemical Principles

If the working principle describes the macroscopic process of "where lithium ions come from and where they go," the electrochemical principles profoundly reveal the microscopic mechanisms and core scientific questions of "how they migrate and interact" within the solid-state system. First, the ion conduction mechanism of the solid-state electrolyte is the foundation for all this, and its ionic conductivity is a core performance metric. Conduction mechanisms vary across material systems: oxide electrolytes (e.g., LLZO) rely on lattice interstitial or vacancy mechanisms, where Li^+ "hops" through predefined crystal channels—they are chemically stable but rigid. Sulfide electrolytes (e.g., $Li_{10}GeP_2S_{12}$ [LGPS]) utilize the weak binding effect due to the large radius of sulfur ions, enabling migration via vacancy mechanisms or even a "push–pull" coordinated motion between Li^+ and the lattice framework, achieving ultrahigh ionic conductivity comparable to liquids, but they are extremely sensitive to the environment. Polymer electrolytes (e.g., polyethylene oxide [PEO]-based) conduct ions relying on chain segment motion; Li^+ migrates within the flexible polymer through coordination and de-coordination with polar groups, offering advantages like good interfacial contact but typically low room-temperature ionic conductivity. The fundamental differences in the conduction mechanisms of these three types of electrolytes directly determine their distinct application prospects and technical challenges.

However, the true "Achilles' heel" of solid-state batteries lies not in the bulk electrolyte itself, but in the solid–solid interfacial electrochemistry formed between the electrolyte and the electrodes. This issue manifests at three levels. First, interfacial contact: point contact between rigid solids cannot achieve perfect wetting like liquids, resulting in small effective contact area and high interfacial impedance, which further deteriorates under the volume changes of electrode materials (e.g., silicon, lithium) during cycling. Second, interfacial chemical stability: most solid-state electrolytes, when in contact with high-potential cathodes or low-potential metallic lithium anodes, undergo thermodynamically driven decomposition reactions, forming an interfacial layer. This interphase plays a dual role: if it is ionically conductive but electronically insulating, it can kinetically stabilize the interface (similar to a good solid electrolyte interface [SEI] film); conversely, if its ionic conductivity is poor, it becomes a barrier for ion transport, drastically increasing impedance and consuming active lithium. Third, lithium dendrite growth: this is not exclusive to liquid systems. In solid electrolytes, dendrites can grow along grain boundaries, cracks, or other defects, or be caused by electrons penetrating into the electrolyte reducing Li^+, or even use the immense mechanical stress generated by deposition to "wedge open" the electrolyte, ultimately leading to short circuits.

It is these complex interfacial problems that bring both enormous opportunities and challenges for the application of the metallic lithium anode—the ultimate choice. The theoretical appeal of solid-state batteries largely stems from their combination with the metallic lithium anode, which possesses an extremely high theoretical specific capacity and the most negative electrochemical potential. Ideally, a robust solid-state electrolyte should suppress dendrites and provide a uniform Li^+ flux, enabling uniform lithium deposition/stripping. However, in reality, the immense volume change associated with the "hostless" nature of metallic lithium poses a devastating test for solid–solid interfacial contact; simultaneously, its strong reducibility causes reactions with almost all solid-state electrolytes, making interfacial stability extremely difficult to maintain. Therefore, realizing the application of metallic lithium anodes shifts the core scientific problem beyond the electrolyte material itself to the synergistic regulation of interfacial conduction, mechanical contact, and electrochemical stability.

1.3 Development History

1.3.1 Development History of Solid-state Batteries

The development history of solid-state batteries is not a brand-new field that has suddenly emerged in recent years to address the bottlenecks of liquid lithium-ion batteries, as people usually imagine. Instead, it is a magnificent scientific and technological epic that spans nearly two centuries, full of the spirit of scientific exploration, evolves together with materials science and basic electrochemical theories, and experiences ups and downs and revivals. The germination of this idea can be traced back to the era of Physicist Michael Faraday in the early 19th century. In the 1830s, his discovery of ion conduction phenomena in solid materials (such as silver sulfide and lead fluoride) first revealed that the carriers of electric current are not limited to electrons, and ions can also migrate in the solid lattice. This laid the most original theoretical foundation for the birth of all solid-state ionics in later generations. However, throughout the 19th century and even the first half of the 20th century, due to the shallow understanding of the microscopic mechanism of ion transport in solids and the extremely low ionic conductivity of known solid ionic conductors, this field remained in the stage of exploring basic physical phenomena for a long time and failed to form a meaningful connection with practical chemical power sources. The real turning point occurred in the 1950s and 1960s. With the advancement of solid-state physics, crystal chemistry, and electrochemical testing technologies, scientists began to systematically search for fast ionic conductors with practical value. During this period, high-temperature oxygen ion conductors such as zirconia doped with calcium oxide or yttrium oxide, silver ion conductors (such as silver iodide–silver sulfide systems), and sodium ion conductors (such as β-alumina) were successively discovered and in-depth studied [8]. They showed that in specific crystal structures, ions can achieve migration capabilities close to or even exceeding those of some liquid electrolytes. This greatly inspired the research community, proved the feasibility of developing high-performance solid electrolytes at room temperature, and gave birth to the formal establishment of "solid-state ionics" as an independent discipline. At the same time, high-temperature solid-state battery systems based on β-alumina electrolytes, such as sodium–sulfur batteries and ZEBRA batteries, were widely developed in engineering in the 1970s and 1980s due to their potential application prospects in large-scale energy storage fields. Although they were difficult to apply in consumer electronics due to their high operating temperatures (usually 300–350 °C), these practices accumulated valuable early experience for the design, packaging, and interface treatment of solid-state batteries.

Despite the progress made in high-temperature solid-state batteries, the breakthrough in lithium-ion conduction technology in solids is what truly connects solid-state battery research with today's mainstream lithium battery applications. In the 1970s, Dr. Stanley Whittingham from Exxon Corporation constructed the first rechargeable lithium–titanium disulfide battery [9]. Although its initial version still used a liquid electrolyte, it triggered a global research upsurge in lithium-based batteries and naturally inspired scientists to search for solid electrolytes that can match metallic lithium negative electrodes. The entire period from the 1980s to the 1990s was a golden age for the exploration of solid electrolyte materials. Research on the three mainstream systems—polymers, oxides, and sulfides—was launched one after another and showed different development trajectories. The pioneer of polymer electrolytes was the French Scientist Michel Armand. In the late 1970s, he systematically put forward the concept of "polymer electrolytes" and predicted that the composite of PEO and lithium salts could conduct lithium ions. The mechanism relies on the segmental movement of the amorphous region of the polymer, enabling lithium ions to migrate through continuous coordination–de-coordination with ether oxygen bonds. This pioneering work laid the foundation for flexible and easy-to-process all-solid-state batteries. However, the inherently low room-temperature ionic conductivity and narrow electrochemical window of PEO-based electrolytes have long limited their application in high-temperature or low-power scenarios. In terms of oxide electrolytes, researchers have developed from early amorphous lithium phosphorus oxynitride (LIPON) thin films and perovskite-type (such as LLTO) structures to the discovery and optimization of garnet-type (such as LLZO) structures from the late 1990s to the early 21st century. LIPON was prepared by Oak Ridge National Laboratory in the early 1990s using radio frequency sputtering. Although its bulk conductivity is not high (about 10^{-6} S cm^{-1}), it has been successfully applied in micro-thin-film solid-state batteries due to its excellent film-forming property, good interface stability with high-voltage positive electrodes, and certain ability to inhibit lithium dendrites. It still has a place in the field of microelectronics today, which proves that even solid electrolytes with moderate performance can achieve commercial application as long as the interface is stable and the process is mature. Later, garnet-type LLZO electrolytes became star materials in the oxide system due to their high ionic conductivity (especially after being stabilized in the cubic phase) and good stability to metallic lithium [10]. The most remarkable breakthrough occurred in the field of sulfide electrolytes. Japanese scientists first discovered the thio-LISICON structure (such as Li_4GeS_4–Li_3PS_4 system) in the 1980s, but its conductivity was still not ideal. At around 2000, the team of Professor Ryoji Kanno from Tokyo Institute of Technology in Japan successively discovered argyrodite-type (such as LGPS) and its derivatives (such as Li_6PS_5Cl) with "superionic conductor" characteristics [11]. Their room-temperature ionic conductivity reached the order of 10^{-2} S cm^{-1} for the first time, which is comparable to or even better than that of liquid organic electrolytes. This milestone discovery completely subverted the traditional cognition that "the ionic conductivity of solid electrolytes must be lower than that of liquids," greatly boosted the research confidence of the entire field, and marked the formal transition of solid-state battery technology from "possible" to "feasible."

Entering the 21st century, especially after the 2010s, the driving force for the development of solid-state batteries has undergone a fundamental change. On the one hand, with the rapid increase in the requirements for energy density and safety in industries such as smartphones and electric vehicles, traditional liquid lithium-ion batteries are approaching their theoretical limits in terms of material systems (especially graphite negative electrodes), and safety accidents such as combustion and explosion caused by organic electrolytes occur frequently. As a result, the industry and investment community have turned their attention to solid-state batteries, which are regarded as the "ultimate solution." On the other hand, the accumulation of basic research over the previous decades has finally reached a critical point of qualitative change from quantitative change in material synthesis, characterization technology, and interface understanding. The research focus in this stage has gradually deepened and concentrated on a core scientific issue—the regulation of solid-solid interfaces, shifting from the initial pursuit of high ionic conductivity of the bulk solid electrolyte alone. Researchers have clearly realized that even with a bulk electrolyte with high conductivity, if the huge interface impedance and chemical/electrochemical instability problems between it and the porous positive electrode particles, as well as between it and the metallic lithium negative electrode, cannot be solved, any commitment to high performance will become empty talk in the laboratory. On the positive electrode side, the point contact between rigid solids leads to a small effective ion transport area, and positive electrode materials with high potential (>4 V vs. Li/Li^+; such as lithium cobalt oxide and high-nickel NCM) are thermodynamically unstable with most sulfide and oxide electrolytes, and side reactions will occur at the interface, forming a high-impedance interface layer. On the negative electrode side, although the infinite volume change and dendrite growth problems of metallic lithium during cycling are physically inhibited to a certain extent in the solid-state system, they have not been eradicated. Dendrites can still penetrate the electrolyte along the grain boundaries, pores of the electrolyte, or through electrochemical–mechanical coupling, leading to short circuits. These challenges have prompted global research efforts to turn to precise interface engineering: for example, by constructing an ultrathin (nanoscale) coating layer (including $LiNbO_3$, $LiTaO_3$, LLZO, etc.) on the surface of positive electrode active particles, it can not only physically isolate direct contact with the electrolyte and inhibit side reactions but also serve as a bridge for ion transport, significantly reducing interface impedance. On the negative electrode side, strategies such as introducing a flexible polymer buffer layer, constructing an artificial SEI (ASEI) film, or using alloyed negative electrodes (such as Li–In) are explored to improve interface contact and stability. At the same time, the concept of composite electrolytes has become increasingly popular, aiming to combine electrolytes with different characteristics (such as incorporating inorganic fillers into polymer matrices or constructing polymer–inorganic multilayer structures) to achieve complementary advantages. For example, the flexibility of polymers is used to improve contact, and inorganic fillers are used to enhance ionic conductivity and mechanical modulus.

In terms of industrialization, a global competitive pattern of multi-route parallel development and a hundred flowers blooming has been formed. Japan and South Korea have obvious leading advantages in sulfide electrolytes relying on their profound accumulation in materials science and consumer electronics. Toyota Motor has invested heavily since 2010, accumulated more than 1,000 patents, and launched prototype vehicles equipped with sulfide solid-state batteries for road testing. Enterprises such as Murata Manufacturing focus on the commercialization of small polymer-based solid-state batteries. As the world's largest producer and application market of lithium-ion batteries, China has followed up on all technical routes under the dual drive of national scientific research plans and corporate venture capital, with layouts in oxide (such as LLZO), sulfide, and even polymer routes. A number of start-ups and industry–university–research cooperation platforms have emerged, and forward-looking attempts have begun in capacity planning and vehicle matching. Europe and the United States rely more on their strong basic scientific research capabilities and national-level R&D alliances, while start-ups (such as QuantumScape focusing on the oxide separator route and Solid Power focusing on the sulfide route) are promoting technological breakthroughs and have carried out close cooperation with mainstream automakers. Looking back at this history that is still being written rapidly, we can obviously see a clear path from basic scientific discovery (Faraday effect) to material system innovation (exploration of fast ionic conductors), then to focusing on core issues (interface science), and finally moving toward engineering integration and commercial verification. The development of solid-state batteries is no longer a simple laboratory technology breakthrough, but a grand system engineering involving materials science, electrochemistry, mechanics, thermal management, intelligent manufacturing, and even supply-chain reconstruction. The final barrier it faces is not only the breakthrough of a single material performance index but also the huge challenge of achieving cost control and large-scale, consistent production of the entire manufacturing chain under the premise of ensuring comprehensive performance (energy density, power density, cycle life, and safety). This history reveals to us that the next revolution in energy storage technology is likely to be nurtured in the in-depth understanding and precise regulation of the microscopic world of these solid–solid interfaces.

1.3.2 Domestic Policies on Solid-state Lithium Batteries

In November 2020, the National Development and Reform Commission issued the "New Energy Vehicle Industry Development Plan (2021–2035)," which mentioned the implementation of battery technology breakthrough action, to accelerate the solid-state power battery technology research and development and industrialization. For the first time, the research and development of solid-state batteries has risen to the national level. In addition, the national key projects have further strengthened the support for solid-state batteries. For example, among the 18 key projects released by the High-tech Department of the Ministry of Science and Technology in February 2021, there are three projects (key projects of high-end functions and smart materials, key projects of new energy vehicles, and key projects of energy storage and smart grid technology) that are closely related to solid-state batteries.

In January 2023, the Ministry of Industry and Information Technology, the Ministry of Education, the Ministry of Science and Technology, the People's Bank of China, the China Banking and Insurance Regulatory Commission, and the National Energy Administration jointly issued the "Guiding Opinions on Promoting the Development of the Energy Electronics Industry." Its core content includes "developing safe and economical new energy storage batteries, strengthening the industrialization technology research of new energy storage batteries, promoting the large-scale application of advanced energy storage technologies and products; researching and developing solid-state batteries, and strengthening the research on the standard system of solid-state batteries." It can be seen that various ministries and commissions of the state attach great importance to solid-state batteries. To implement the relevant arrangements for national scientific and technological innovation during the "14th Five-Year Plan" period, the National Key R&D Program launched the implementation of the key special project "Key Technologies and Equipment for Deep Sea and Polar Regions." In June 2023, the "2023 Project Application Guidelines for the Key Special Project of Key Technologies and Equipment for Deep Sea and Polar Regions" continued to focus on the "Ultra-High Energy Density All-Solid-State Pouch Batteries (General Key Technology Category)." It aims to break through the bottlenecks of solid-state batteries in interface construction from particles to cells under 110 MPa environment, structural evolution stress distribution, life attenuation, safety boundaries, etc., and build a disruptive ultrahigh-energy-density all-solid-state battery system to meet various deep-sea applications.

To sum up, the support of national policies will greatly promote the rapid development of solid-state lithium batteries.

1.3.3 Foreign Policies on Solid-state Lithium Batteries

Overseas countries have taken the lead in R&D and layout of all-solid-state batteries, implemented financial subsidies, and vigorously promoted technology implementation. For example, Japan has built a joint R&D system between automakers and battery factories, with government funding support exceeding 200 billion yen (approximately 10 billion yuan), and strives to achieve the commercialization of all-solid-state batteries by 2030, with an energy density target of 500 Wh kg^{-1}. The South Korean government provides tax credits to support the R&D of solid-state batteries, and with the joint promotion of leading power battery companies, the goal is to develop commercial technology with an energy density of 400 Wh kg^{-1} by 2025–2028 and complete vehicle installation by 2030. Among the European countries, Germany has invested the most in R&D and layout. The United States is funded by the Department of Energy, with R&D led by start-ups, and has reached cooperation with many automakers, aiming to achieve an energy density of 500 Wh kg^{-1} by 2030.

1.4 Research Status and Challenges

As a new type of electrochemical energy storage system that replaces liquid electrolytes and separators with solid electrolytes, solid-state batteries, relying on their theoretically high safety and potential to break through the existing energy density bottleneck, have become the recognized next-generation power battery technology direction in the global academic and industrial circles. Their development is at a critical node from laboratory R&D to industrial-scale manufacturing. As the "soul" component of solid-state batteries, the technical maturity of solid electrolytes directly determines the commercialization process of solid-state batteries. Currently, the entire field presents a complex picture of frequent good news in basic research, in-depth advancement of engineering challenges, and accelerated competition and cooperation in industrial layout.

1.4.1 Overall Development Status of Solid-state Batteries

Under the grand narrative of global energy transformation and carbon neutrality goals, major economies have elevated solid-state battery technology to the national strategic level. In 2025, the Ministry of Industry and Information Technology of China included all-solid-state batteries in the core of new industry standard construction for the first time and began to establish a standard system. This is not only a technical specification but also a top-level design of the industrial ecosystem, aiming to set a rule framework for the global solid-state battery industry. At the same time, local governments such as Beijing and Shanghai have also included it in local industrial plans, highlighting its huge application potential in fields such as new energy vehicles, energy storage, consumer electronics, and low-altitude economy. This series of policy measures has injected strong impetus into the R&D and industrialization of solid-state batteries. In terms of market size prediction, industry data shows that the global solid-state battery market is expected to grow by more than five times in the next five years. By 2030, its proportion in China's lithium battery market is expected to reach about 10%, forming a segmented market with a scale of more than 250 billion yuan. The capital market has also responded enthusiastically. As of October 2025, the market value of leading battery companies such as

CATL has doubled, the financing amount in the solid-state battery field has surged by three times year-on-year, and leading automakers such as Tesla and BYD have signed long-term supply agreements with material suppliers, showing the market's strong confidence in technological iteration.

In terms of industrialization, a global competitive pattern of multi-technical-route parallel development and a hundred flowers blooming has been formed, and the mass production schedule has become increasingly clear. Chinese enterprises are showing a collective acceleration trend. After building the first semi-solid-state battery production line in 2024, CATL plans to realize small-batch production of all-solid-state batteries by 2026, and the energy density of its developed sulfide all-solid-state battery cells has exceeded 500 Wh kg^{-1}. The Chengdu mass production base of the Solid-State Battery Research Institute of EVE Energy was officially unveiled, and its "Longquan No. 2" all-solid-state battery successfully rolled off the production line, marking a solid step from the laboratory to industrialization. Gotion High-Tech successfully developed the vehicle-grade sulfide all-solid-state battery "Jinshi Battery" with an energy density exceeding 400 Wh kg^{-1}, and plans to start on-vehicle verification in 2025. In addition, Qingtao Energy, Funeng Technology, and other companies are also actively promoting technology implementation through material innovation or industrial chain cooperation models. From a global perspective, Toyota of Japan plans to realize on-vehicle testing of all-solid-state batteries in 2027 and large-scale mass production after 2030; Samsung SDI of South Korea is expected to start mass production in 2027; a number of American start-ups are seeking breakthroughs in energy density and charge–discharge speed. Based on the plans of all parties, 2026–2027 is regarded as a critical period for mass production and on-vehicle application, and all-solid-state batteries are expected to achieve comprehensive commercialization after 2030. Currently, the application of solid-state batteries will show a gradient characteristic. They will not replace liquid lithium-ion batteries immediately but will first be applied in scenarios with extreme requirements for performance and safety, such as high-end electric vehicles, electric vertical take-off and landing aircraft (eVTOL), humanoid robots, and aerospace. These scenarios can also bear the relatively high initial cost.

1.4.2 R&D Status of Solid-state Electrolytes

Solid electrolytes are the core of solid-state batteries, and their R&D mainly focuses on three major systems: polymers, oxides, and sulfides. Each route has its distinct advantages, disadvantages, and applicable scenarios. Polymer electrolytes, represented by PEO, have the greatest advantage of good flexibility and mature film-forming technology, and high compatibility with existing liquid battery production lines. Therefore, they are often used as a transitional solution for semi-solid-state batteries. However, their inherently low room-temperature ionic conductivity (usually $<10^{-4}$ S cm^{-1}) and narrow electrochemical window limit their application in all-solid-state and high-voltage systems. To overcome this problem, the team of Professor Qiang Zhang from Tsinghua University [12] has made a breakthrough. They successfully developed a new type of fluorine-containing polyether electrolyte by designing a "fluoride-rich solvation structure." By introducing strong electron-withdrawing fluorine-containing groups, this electrolyte improves its resistance to high-voltage oxidation, enabling it to match 4.7 V high-voltage lithium-rich manganese-based positive electrodes. At the same time, it induces the formation of a stable interface layer rich in fluorides. The 8.96 Ah large-capacity polymer pouch all-solid-state battery assembled with this electrolyte has an energy density of up to 604 Wh kg^{-1}, and the capacity retention rate reaches 72.1% after 500 cycles at 0.5C rate. It has successfully passed the needle puncture and 120 °C hot box tests, demonstrating high safety characteristics. This achievement has pushed the performance of polymer solid-state batteries to a new height.

Oxide electrolytes, typical representatives of which are garnet-type LLZO and perovskite-type LLTO, have the greatest advantages of good chemical stability, wide electrochemical window (up to more than 5 V), and stability to air, thus having high intrinsic safety. However, they are hard in texture, resulting in poor solid–solid contact with electrodes, leading to large interface impedance. Their room-temperature ionic conductivity is usually in the order of 10^{-4}–10^{-3} S cm^{-1}, which can meet some applications but still has room for improvement. Enterprises such as Gotion High-Tech have chosen the oxide electrolyte as the core technical route and have built pilot lines. Sulfide electrolytes, such as LGPS and its derivative Li_6PS_5Cl, are currently the solid electrolyte systems with the highest ionic conductivity, reaching the order of 10^{-2} S cm^{-1} at room temperature, which is even comparable to that of liquid electrolytes. Its conduction mechanism is derived from the weak binding effect of the large radius of sulfide ions on Li^+, and the "push–pull" coordinated movement between Li^+ and the framework sulfide ions, forming a "superionic conductor" structure. This also makes it a key layout direction for giants such as CATL and Toyota in the field of power batteries. However, the fatal weakness of sulfide electrolytes is their extreme sensitivity to humidity. They react with water to produce toxic hydrogen sulfide (H_2S) gas, which puts extremely strict requirements on the production environment (such as dry room requirements) and packaging processes, directly increasing the manufacturing cost.

In addition to the above three traditional systems, the exploration of emerging electrolyte materials has never stopped. Halide electrolytes have attracted much attention in recent years. In October 2025, the team of Academician Xueliang Sun from Ningbo Oriental Institute for Advanced Materials and the team of Professor Yifei Mo from the University of Maryland, United States [13], jointly developed a new type of superionic conductor $Li_3Ta_3O_4Cl_{10}$. This material has refreshed the record of room-temperature ionic conductivity of halide-based solid electrolytes, reaching 13.7 mS cm^{-1}, and proposed a fast ion percolation network theory based on "tetrahedron-to-tetrahedron." The all-solid-state battery prepared based on this electrolyte shows excellent cycle stability under extremely low-temperature conditions of −50 °C and high room-temperature rate (3C) conditions. At the same time, the material also has significantly improved air stability, providing a new technical path for all-solid-state batteries. In addition, to take into account the advantages of different materials, the strategy of composite electrolytes has become increasingly popular. For example, inorganic fillers (oxides or sulfides) are incorporated into polymer matrices, or polymer–inorganic multilayer structures are constructed, aiming to use the flexibility of polymers to improve interface contact and at the same time use inorganic fillers to enhance the overall ionic conductivity and mechanical modulus, achieving the effect of "combining rigidity and flexibility."

1.4.3 Core Challenges and Bottlenecks

Despite the broad prospects, the mass production of solid-state batteries, especially all-solid-state batteries, still faces a series of severe challenges. Among them, the solid–solid interface problem is generally recognized as the "Achilles' heel." The interface challenges are specifically reflected in three aspects. First is the problem of physical contact. Unlike liquid electrolytes that can perfectly wet the uneven surface of the electrode, the contact between rigid solid electrolytes and solid electrodes is mechanical point contact, resulting in a small effective contact area and a large interface impedance. During the charge–discharge process, the electrode materials (especially metallic lithium and silicon) will undergo volume expansion and contraction, which will further deteriorate the physical contact, leading to uneven local current density and even complete failure of some active materials, greatly affecting the cycle life and rate performance of the battery. Wang Fang, Chief Scientist of China Automotive Technology and Research Center Co., Ltd. [14] pointed out that the stability problem of the conductive path of high-specific-capacity positive electrode materials in a low-pressure environment has not been completely solved, which will affect the further improvement of energy density.

Second, there is the problem of interface (electro)chemical stability. Most solid electrolytes have a limited electrochemical stability window. When in direct contact with high-potential positive electrodes (such as > 4 V vs. Li/Li^+) or low-potential negative electrodes (metallic lithium, 0 V vs. Li/Li^+), thermodynamically driven decomposition reactions will occur at the interface, forming an interface phase. This interface phase plays a dual role: if it has low ionic conductivity and high electronic conductivity (such as Li_2S and Li_2O), it will become a barrier for ion transport, leading to large interface impedance and continuous consumption of active lithium, reducing the Coulomb efficiency; on the contrary, if a stable interface phase that is ion-conducting and electron-insulating (similar to an excellent SEI film in liquid batteries) can be formed through artificial regulation, it can prevent further decomposition of the electrolyte and achieve dynamic stability. The traditional solution relies on continuous external pressure application (up to 5 MPa) to maintain interface bonding, but this method significantly increases the volume, weight, and complexity of the battery system, making it difficult to meet commercial needs.

Third problem is that of lithium dendrite growth. The traditional view holds that solid electrolytes can physically block lithium dendrites, but studies have found that lithium dendrites can also grow inside solid electrolytes. The mechanisms include: growth along defects such as grain boundaries (in polycrystalline electrolytes) or cracks; electrons may penetrate into the electrolyte to reduce Li^+, forming metallic lithium deposition; during cycling, excessive local current density leads to the concentrated deposition of Li^+ at certain weak points, generating huge mechanical stress, thereby "wedging open" the electrolyte to produce cracks, and lithium dendrites extend accordingly, eventually causing a short circuit. Wang Fang [15] also emphasized that the safety boundary of solid-state batteries is indeed wider than that of liquid batteries, but once this boundary is broken, the consequences may be more serious. This means that its safety risk characteristics are different from those of liquid batteries, requiring a brand-new evaluation system.

In addition to interface problems, the high cost is another major obstacle restricting its large-scale popularization. According to research institute reports, the unit cost of solid-state batteries is as high as 1,200 yuan $(kWh)^{-1}$, which is more than three times the cost of traditional liquid lithium-ion batteries. This is mainly due to two aspects: one is the high material cost. The solid electrolytes themselves, as well as the high-performance positive and negative electrode materials (such as lithium-rich manganese-based positive electrodes and metallic lithium negative electrodes) needed to adapt to them, are expensive. The other is the complex and immature manufacturing process. For example, sulfide electrolytes have extremely strict requirements on the production environment, requiring a full-process dry room, resulting in huge equipment investment. At the same time, the yield and efficiency of process links such as the preparation of electrodes and electrolyte films and multilayer stacking need to be improved. Although Gotion High-Tech announced that the battery yield rate of its pilot line has reached 90%, the stability and consistency control from laboratory pilot to large-scale mass production is still a huge challenge.

1.4.4 Cutting-edge Breakthroughs and Solutions

Facing severe interface challenges, the global scientific research community is constantly proposing innovative solutions. In October 2025, the team of Researcher Huang Xuejie from the Institute of Physics, Chinese Academy of Sciences [16], jointly with multiple units, published a breakthrough research result—anion regulation technology—in the international top journal *Nature Sustainability*. By introducing iodide ions into the sulfide electrolyte, this technology utilizes the interface migration characteristics of iodide ions under the action of an electric field to in situ form an LiI-rich dynamically adaptive interface (DAI) layer at the negative electrode/electrolyte interface. This interface layer can automatically fill all gaps and holes between the electrode and the electrolyte, keeping them in close contact at all times. Its mechanism of action is vividly compared to the "octopus tentacle" behavior. The test results showed that the all-solid-state battery prototype prepared based on this technology achieves stable cycling of the metallic lithium negative electrode with high areal capacity under zero external pressure or low external pressure, and the performance remains stable after hundreds of cycles. Professor Chunsheng Wang from the University of Maryland [17] commented: "This research solves the key bottleneck problem restricting the commercialization of all-solid-state batteries and takes a decisive step towards realizing their practical application." In addition, interface coating is also a widely used and effective strategy. By constructing an ultrathin (nanoscale) coating layer (such as $LiNbO_3$ and $LiTaO_3$) on the surface of positive electrode active particles, it can physically isolate direct contact with the electrolyte, inhibit side reactions, and at the same time serve as a bridge for ion transport, significantly reducing interface impedance.

To promote the industrialization of solid-state batteries, vertical integration and collaborative innovation of the industrial chain are crucial. Qingtao Energy has adopted a unique "material + battery" bundled cooperation model, jointly building production lines with upstream and downstream enterprises in the industrial chain. This model can solve the material adaptation problem of solid-state batteries from the source. National policies also encourage automakers and battery enterprises to jointly build laboratories, promote the vertical integration of "materials–cells–systems," and provide support for the localization of core equipment such as solid-state electrolyte coating machines. In terms of standard and safety system construction, the China Society of Automotive Engineers released the "Determination Method for All-Solid-State Batteries" in May 2025, which clearly defines "all-solid-state batteries" for the first time. It requires that ion transfer be completely realized through solid electrolytes, which forms a strict technical boundary with mixed solid–liquid electrolyte batteries (often referred to as "semi-solid-state batteries"), helping to standardize the market and avoid concept speculation. At the same time, it is urgent to establish a safety standard for all-solid-state batteries. The test results in the laboratory are not equal to the actual performance of mass-produced on-vehicle batteries. A unified and strict standard is needed to ensure product reliability and consumer confidence in safety.

1.4.5 Future Development Directions

In Lux's [18] biennial market forecast report (Figure 1.2), we found that by 2035, the annual revenue of the entire energy storage market will grow to $546 billion, and the annual deployment will reach 304.6 billion kWh. By 2035, population mobility will

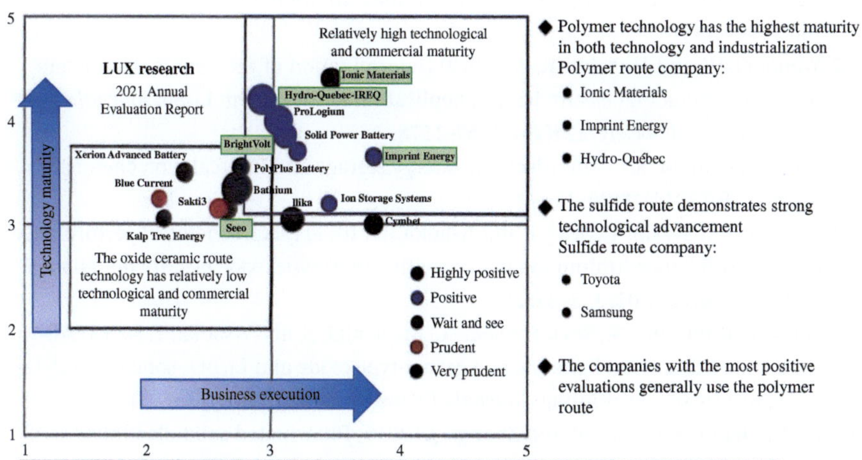

Figure 1.2 Analysis of the commercial maturity and technical advancement of solid-state electrolyte routes [7].

still be the long-term driving factor for the annual revenue and demand of energy storage, accounting for 74% of the total market share in terms of annual revenue and 91% in terms of demand. At the same time, the stationary storage market will surpass the electronic device market in 2035, and Lux expects it to become an industry worth $30.4 billion with an installed capacity of 52.5 GWh by then.

References

1 Liu, X., Ren, D., Hsu, H., Feng, X., Xu, G.-L., Zhuang, M., Gao, H., Lu, L., Han, X., Chu, Z., Li, J., He, X., Amine, K. and Ouyang, M. (2018) Thermal runaway of lithium-ion batteries without internal short circuit. *Joule*, 2(10), 2047–2064.

2 Lu, D., Zhang, S., Li, J., Huang, L., Zhang, X., Xie, B., Zhuang, X., Cui, Z., Fan, X., Xu, G., Du, X. and Cui, G. (2023) Transformed solvation structure of noncoordinating flame-retardant assisted propylene carbonate enabling high voltage Li-ion batteries with high safety and long cyclability (Adv. Energy Mater. 28/2023). *Advanced Energy Materials*, 13(28), 2300684.

3 Xu, G., Huang, L., Lu, C., Zhou, X. and Cui, G. (2020) Revealing the multilevel thermal safety of lithium batteries. *Energy Storage Materials*, 31, 72–86.

4 Huang, L., Xu, G., Du, X., Li, J., Xie, B., Liu, H., Han, P., Dong, S., Cui, G. and Chen, L. (2021) Uncovering LiH triggered thermal runaway mechanism of a high-energy $LiNi_{0.5}Co_{0.2}Mn_{0.3}O_2$/graphite pouch cell. *Advanced Science*, 8(14), e2100676.

5 Huang, L., Lu, T., Xu, G., Zhang, X., Jiang, Z., Zhang, Z., Wang, Y., Han, P., Cui, G. and Chen, L. (2022) Thermal runaway routes of large-format lithium-sulfur pouch cell batteries. *Joule*, 6(4), 906–922.

6 Zhang, X., Huang, L., Xie, B., Zhang, S., Jiang, Z., Xu, G., Li, J. and Cui, G. (2023) Deciphering the thermal failure mechanism of anode-free lithium metal pouch batteries. *Advanced Energy Materials*, 13(8), 2203648.

7 Cui, G. (2024) *High Energy Density Solid-State Lithium Batteries*. Beijing: Chemical Industry Press.

8 Weppner, W. and Huggins, R. A. (2019) Determination of the kinetic parameters of mixed-conducting electrodes and application to the system Li_3Sb. *Journal of The Electrochemical Society*, 124(10), 1569–1578.

9 Whittingham, M. S. (1976) Electrical energy storage and intercalation chemistry. *Science*, 192(4244), 1126–1127.

10 Ohta, S., Kobayashi, T., Seki, J. and Asaoka, T. (2011) Electrochemical performance of an all-solid-state lithium ion battery with garnet-type oxide electrolyte. *Journal of Power Sources*, 202, 332–335.

11 Ohta, S., Komagata, S., Seki, J., Saeki, T., Morishita, S. and Asaoka, T. (2013) All-solid-state lithium ion battery using garnet-type oxide and Li_3BO_3 solid electrolytes fabricated by screen-printing. *Journal of Power Sources*, 238, 53–56.

12 Li, T., Zhang, X. Q., Shi, P. and Zhang, Q. (2019) Fluorinated solid-electrolyte interphase in high-voltage lithium metal batteries. *Joule*, 3, 2647–2661.

13 Zhao, F., Zhang, S., Wang, S., Reid, J. W., Xia, W., Liu, J., King, G., Kaduk, J. A., Liang, J., Luo, J., Gao, Y., Yang, F., Zhao, Y., Li, W., Alahakoon, S. H., Guo, J.,

Huang, Y., Sham, T.-K., Mo, Y. and Sun, X. (2025) Anion sublattice design enables superionic conductivity in crystalline oxyhalides. *Science*, 390, 199–204.

14 Wang, F., Han, C., Yan, P. F., Wang, W. N., Ma, X. L. and Liu, S. Q. (2025) Review on technological innovation and typical safety evaluation of traction battery. *Science and Technology Review*, 43, 87–96.

15 Wang, F., Han, C., Wang, W. N., Liu, S. Q. and Li, P. (2025) Construction and challenges of testing and evaluation systems for solid-state batteries. *Journal of the Chinese Ceramic Society*, 53, 1478–1491.

16 Cen, G., Yu, H., Xiao, R., Ben, L., Qiao, R., Zhu, J., Zhang, X., Liu, G., Jiang, K., Yao, X., Zhang, H. and Huang, X. (2025) Adaptive interphase enabled pressure-free all-solid-state lithium metal batteries. *Nature Sustainability*, 8, 1360–1370.

17 Deng, T., Wang, C., Wan, H., Li, A.-M., He, X., Wang, Z., Cao, L., Fan, X. and Wang, C. (2025) A universal design of lithium anode via dynamic stability strategy for practical all-solid-state batteries. *Angewandte Chemie International Edition*, 64, e202418811.

18 Lux Research. Global Energy Storage Market Forecast 2019. *Boston: Lux Research*, 2019.

2

Application of High-throughput Computing in Lithium Batteries

2.1 First-principles Calculation Method

2.1.1 Introduction to First-principles

First-principles is a calculation method based on the basic principles of quantum mechanics. It is usually based on density functional theory (DFT) and is used to numerically calculate the properties of materials and substances. It is very important in materials science, chemistry, physics, and other fields. This method relies not on empirical data, but directly on the basic laws of physics to calculate the electronic structure and properties of materials. The core of the first-principles is to use the principles of quantum mechanics, starting from the basic laws of physics, to predict and explain the properties of matter through mathematical derivation and calculation. This method does not depend on empirical parameters or semiempirical models, but is directly based on the principles of interaction between nucleus and electrons, by solving the Schrödinger equation to obtain the ground-state properties of the system [1].

When first-principles calculations are performed, first, it is necessary to construct an atomic model of the studied material, including the arrangement of atoms and the lattice structure. Hamiltonian in quantum mechanics is used to describe the total energy of the system, including kinetic energy term and potential energy term. The Schrödinger equation is used to describe the wave function of the electron. Bloch's law is a commonly used method in solid-state physics, which can effectively simplify the representation of wave functions. In the numerical simulation, it is necessary to select the appropriate basis set, such as plane basis set and atomic orbital basis set, among others. On this basis, we can use the iterative method to solve the Schrödinger equation until the wave function and energy of the ground state are obtained. This technique is called self-consistent field (SCF) method. Based on this method, the band structure, density of states, electron density distribution, and optical properties of the system can be studied. In order to test the accuracy of the method, it is also necessary to compare with the experimental data.

2.1.2 Calculation Formula

2.1.2.1 The Schrödinger Equation

The structure and properties of matter can be described by analyzing the interaction between atoms and molecules. The interactions between atoms and molecules can be expressed by the Schrödinger equation, which helps in obtaining the structure and properties of matter, which greatly facilitates our calculations.

The Schrödinger equation is a differential equation describing the quantum deductive change, which contains the wave number of the object and the Hamiltonian operator. By solving the Schrödinger equation, we can obtain the energy, structure, parameters, and other information of the material, which help us to understand the properties and behaviors of the material.

As a basic equation in quantum mechanics, the Schrödinger equation is of great significance. It was proposed by the Austrian Physicist Erwin Schrödinger in 1926. The Schrödinger equation describes that the wave function in the quantum system changes with time. It is a core equation and constitutes the theoretical basis of quantum mechanics together with other basic principles [2].

The specific form of Schrödinger equation is as follows:

$$H\psi(r,R) = E^{\mathrm{H}}\Psi(r,R) \tag{2.1}$$

One of the core operators in the Schrödinger equation is the Hamiltonian operator, which is used to describe the overall energy of the entire system. In the first-principles calculations, the Hamiltonian operator obtains the sum of the kinetic energy and the potential energy of the material system. The kinetic energy of the whole system is related to the movement of electrons in the material, and the potential energy of the whole system is mainly related to the geometric structure of the crystal and the interaction between the nuclei [3].

2.1.2.2 Born–Oppenheimer Approximation

Born–Oppenheimer approximation is also a commonly used method to solve problems and equations. In this approximation, the molecular wave function system is simplified as the product of wave functions of electrons and nucleus, where the electron wave function describes the movement of electrons and the nuclear wave function describes the nuclear state. The basic principle of this method is to calculate separately for electrons and the nucleus. First, the electron wave function is resolved, and then it is used as the standard value to solve the wave function of the nucleus.

The Born–Oppenheimer approximation was proposed by Max Born and Robert Oppenheimer in the late 1920s. This method can achieve high accuracy in most cases, especially in the study of molecular structure, electron information, and so on. As an important basis and method in the field of quantum chemistry and molecular dynamics, the Born–Oppenheimer approximation helps researchers in solving many problems [4, 5].

2.1.2.3 Hartree–Fock Approximation

The Hartree–Fock approximation represents wave function as a determinant of the single-electron wave function, and then solves the single-electron wave function by the variational method. In this process, Hartree and Fock almost ignored the interaction between electrons, and only considered the Coulomb interaction between electrons and nucleus.

The Hartree–Fock approximation is another important approximation method in quantum chemistry and quantum molecular dynamics, which was proposed by Douglas Hartree and Wolfgang Pauli in the late 1920s. Compared with the Born–Oppenheimer approximation, this approximation is simple to calculate, so that the calculation results can be obtained faster in many cases. The Hartree–Fock approximation is still a common method in the field of quantum chemistry. It provides better calculation results for many organometallic and transition-metal compounds, especially in dealing with molecular structure and energy issues, and can achieve high accuracy [6, 7].

2.1.3 Design of Novel Solid Electrolyte by Material Genome Method

A summary of methods as an integrated first-principles material genome approach evolved from recent endeavors, is presented in Figure 2.1 [8]. The tasks for theoretical simulation/modeling are classified into two categories for the predictions of

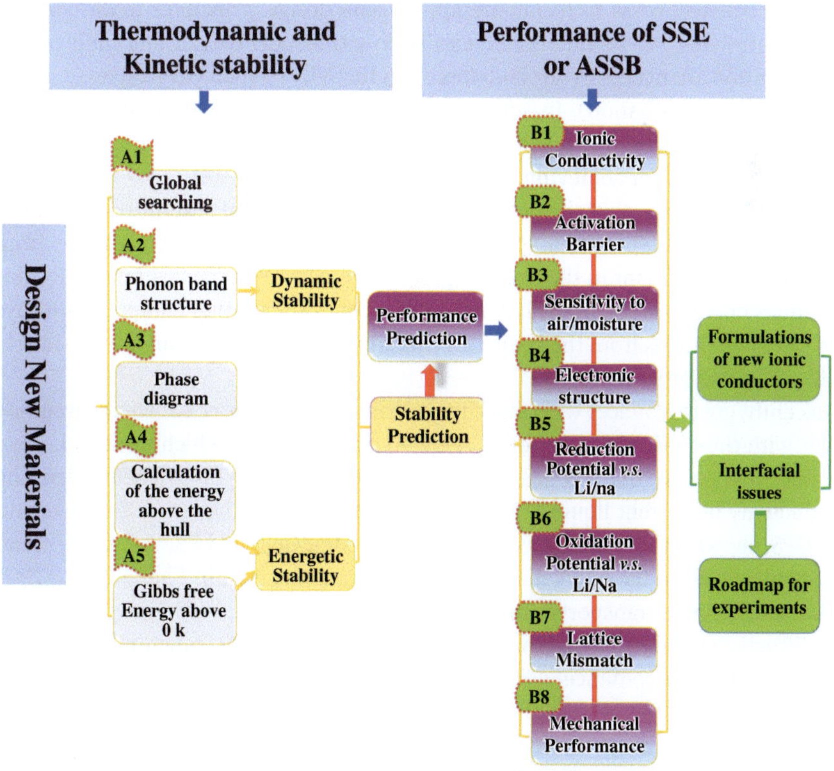

Figure 2.1 A flowchart of the material genome approach. *Source*: Reprinted with permission from [8] / with permission of John Wiley & Sons.

(i) thermodynamic and dynamic stability and (ii) performance of SSE or ASSB. Task (i) is to identify the stable structure for a designated composition, so that each selected phase can be energetically stable with respect to alternative structures of the same composition as well as the constituent phases. Dynamic stability is examined to make sure the phase is also mechanically stable without any vibration modes associated with imaginary phonon frequency. This task is essentially on phase equilibria in the framework of the DFT, without reliance on experimental data or fitting parameters. (From a mathematical point of view, a functional is a function of variable functions.) The subtasks are organized as follows:

1. Global search of low-energy configurations by methods based on the DFT, e.g., universal structure predictor (USPEX) [9, 10].
2. Phonon band structure to check dynamic stability using the supercell approach with frozen-phonon approximation [11, 12].
3. Identify constituent phases in view of phase diagram principles and available data, e.g., consulting handbooks or the Materials Project website (https://www.materialsproject.org) [13, 14].

In order to develop high-performance solid electrolytes for solid-state lithium batteries, it is necessary to deeply understand the fast ion migration mechanism in solid electrolytes, so as to formulate appropriate design principles to design new solid electrolytes. As we mentioned earlier, low transport barrier and high carrier concentration are necessary for high ionic conductivity of solid electrolytes (e.g., the Nernst–Einstein equation). In addition, pre-factors related to anion polarizability, lattice flexibility, carrier concentration, and sub-lattice ordering of migrating ions are also important for achieving high ionic conductivity [15–18].

It has been shown that the activation energies from AIMD (ab-initio molecular dynamics) and CI-NEB often agree well, as long as the identified pathways from the latter are dominant in the statistic process typically experienced in a molecular dynamics process. It has been shown, however, that some rather complex pathways and diffusion modes from the static CI-NEB method may not play an important role in the dynamic process [19–21].

Recently, great advances were made in realizing superb ionic conductivity in sulfide SSEs, with the delivery of $Li_{10}GeP_2S_{12}$ (LGPS) phase [22, 23], which exhibits Li^+-ion conductivity greater than 10 mS cm^{-1} at room temperature [22–24], better than the conductivity of current liquid organic electrolytes for Li-ion batteries. The structure of LGPS, shown in Figure 2.2a, consists of tetrahedral GeS_4 and PS_4 units. The phase diagrams related to the energetic stability of LGPS are shown in Figure 2.2b. AIMD revealed that ionic transport is dictated by a one-dimensional efficient pathway for Li^+ diffusion (Figure 2.2c), so that diffusion along the *c*-axis is much more efficient (Figure 2.2d). In accordance with experimental findings, the alkali ion conductivity was derived from the Nernst–Einstein equation from diffusion coefficient, leading to a high bulk conductivity of 9 mS cm^{-1} at room temperature with a moderate activation energy barrier around 0.25 eV [22, 23]. From a material design point of view, AIMD was shown to be a powerful method in predicting transport properties

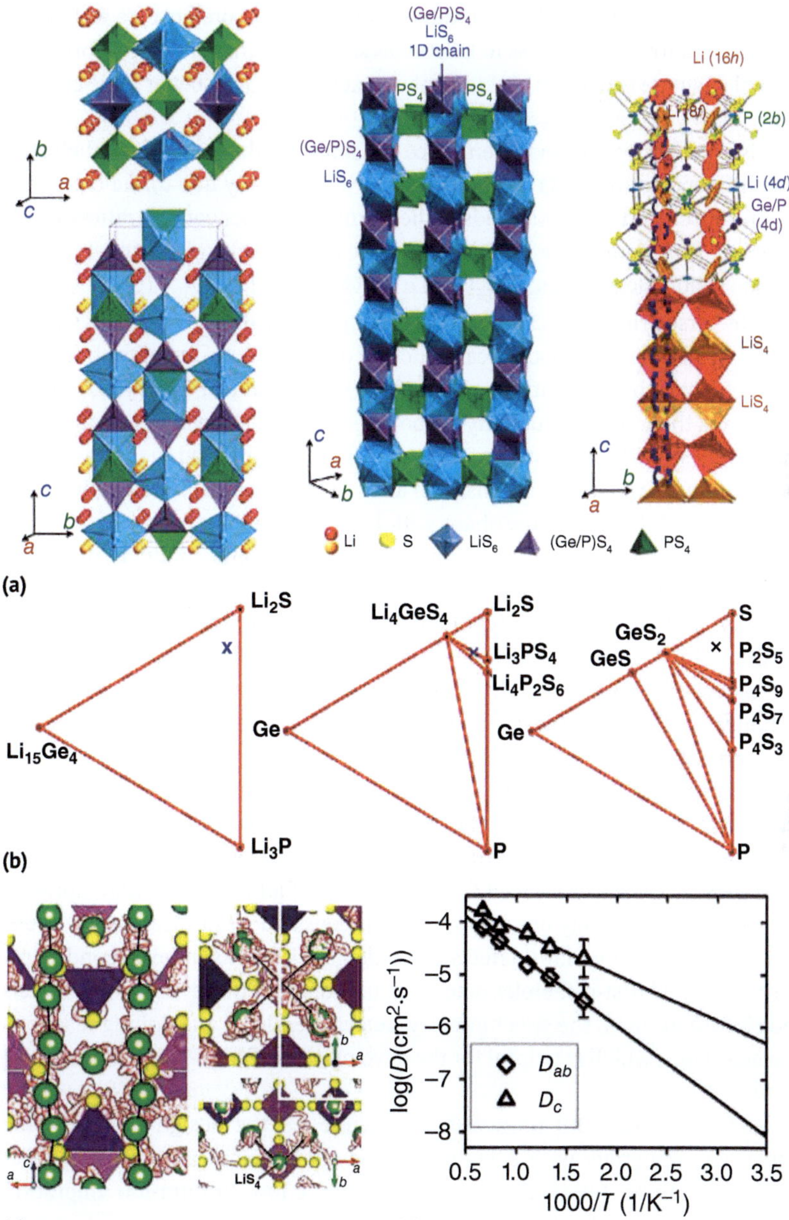

Figure 2.2 (a) Crystal structure of LGPS. *Source*: Reprinted with permission from [22] / with permission of Springer Nature. (b) Phase evolution of the Li–Ge–P–S system as a function of Li chemical potential μ_{Li} at 0, 1.8, and 2.5 eV. $\mu_{Li} = 0$ eV corresponds to metallic Li. Labeled dots denote stable phases at each μ_{Li}. *Source*: [24] / with permission of the American Chemical Society. (c) Trajectories (white) of Li atoms (green) in AIMD simulations at 900 K. The 1D diffusion pathway is along the *c*-direction and diffusion is in the *ab*-plane. *Source*: [24] / with permission of the American Chemical Society. (d) Arrhenius plots of diffusion coefficients in the *c*-direction (D_c) and in the *ab*-plane (D_{ab}). *Source*: [24] / with permission of the American Chemical Society.

via extrapolating diffusion data from elevated temperatures to predict data well below the room temperature. The success in LGPS was so exciting that growing efforts were made in replacing the expensive Ge element and achieving even better ionic conductivities [25, 26].

Doping is an effective method to enhance the ionic conductivity of solid electrolytes. The influence of dopants on the conductivity is mainly due to the change of lithium-ion concentration, which leads to the change of carrier concentration. Ceder and coworkers [27] studied the stability and ionic conductivity of Rb and Ta-doped garnet-type solid electrolytes $Li_{7+2x-y}(La_{3-x}Rb_x)(Zr_{2-y}Ta_y)O_{12}$ ($0 \leq x \leq 0.375$, $0 \leq y \leq 1$) by first-principles calculations. Theoretical calculations showed that doping does not change the topological structure of the lithium-ion migration path, but mainly changes the concentration of lithium ions. Al-doped $LiLaZr_2O_{12}$ significantly improves the ionic conductivity at room temperature by stabilizing the high-temperature phase with high ionic conductivity. This is attributed to the formation of Li vacancies due to charge compensation after Al replaces Li. This stoichiometric change helps to stabilize the cubic phase with high ionic conductivity [28].

Determining the chemical composition with the best ionic conductivity is still a challenge, which requires understanding lattice ion substitution and lithium stoichiometry, and how these component parameters together affect the transport of lithium ions. In order to achieve quantitative control of lithium stoichiometry, it is necessary to understand the natural defect chemistry of materials and how it changes with the synthesis conditions, as well as the response of defects to external doping. Squires et al. [29] used first-principles calculations to analyze the defect chemistry in LLZO. The autonomous defect concentration as a component chemical potential function reflects the influence of synthesis conditions and dopant concentration on the defect chemistry (such as O vacancies, Li vacancies, and Li interstitial sites). The results showed that the method of adjusting the stoichiometry of Li in LLZO by doping may not be directly compensated by Li vacancies, because the main compensation variant defects can change with the synthesis conditions. This work combined first-principles defect calculations and thermodynamic models to provide information on the stoichiometry, defect chemistry, and doping reactions of solid electrolytes, which is helpful for the development of high-performance solid electrolytes.

Ceder and coworkers [30] proposed a principle for the design of superionic conductors, i.e., materials with a body-centered-cubic anion lattice with a face-sharing lithium–anion tetrahedral structure have a low lithium-ion migration barrier (Figure 2.3a and b). Using this design principle and ab-initio calculation method, Ceder and coworkers [31] predicted the new solid electrolytes $Li_{1+2x}Zn_{1-x}PS_4$ ($0 \leq x < 0.5$). The ionic conductivity of the materials is estimated to be about 10^{-2} S cm^{-1}. These materials were synthesized experimentally, and the ionic conductivity is measured to be about 10^{-4}–10^{-3} S cm^{-1} at room temperature [32]. The difference in ionic conductivity is mainly due to the fact that the experiments did not reach the target lithium defect concentration (kinetic and thermodynamic limitations), and the amorphous phase and the impurity phase with poor ionic conductivity were produced during the synthesis process.

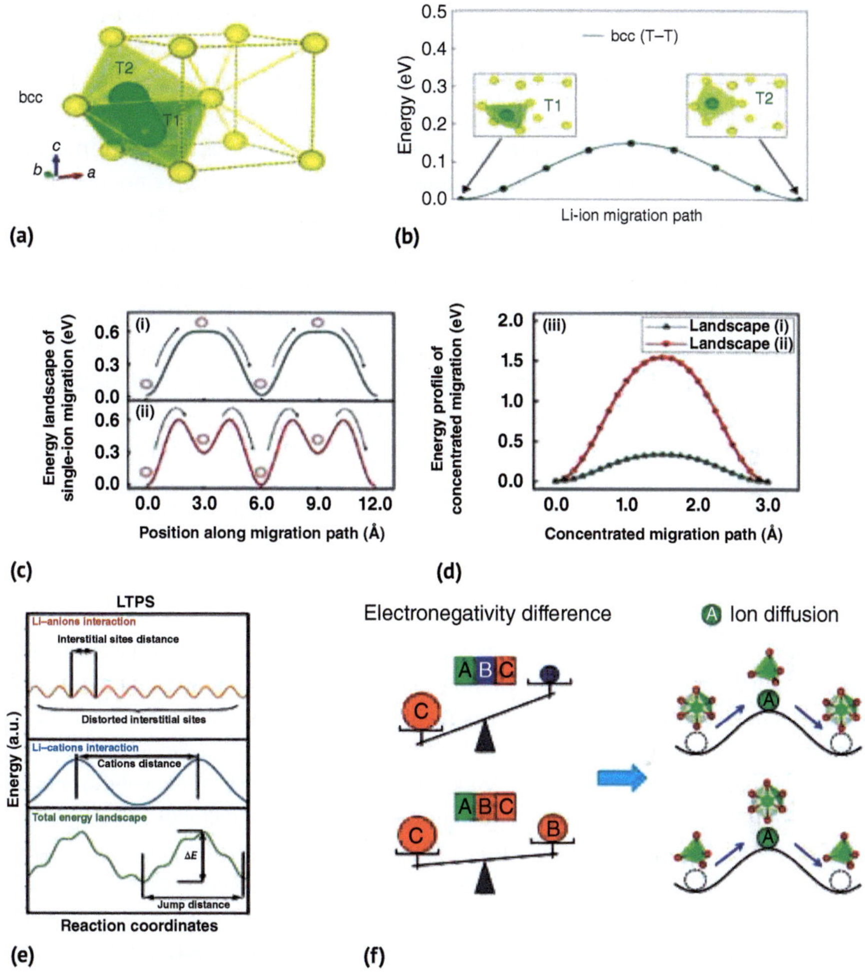

Figure 2.3 Design strategy of new fast ionic conductors: (a) migration path and (b) barrier of lithium ion in BBC (body-centered-cubic) sulfur lattice. *Source*: [30] / with permission of Springer Nature. (c) Atomic energy landscape with low (above) and high (below) migration barriers and (d) barrier. (e) Energy landscape of lithium in one-dimensional model. *Source*: [35] / with permission of Elsevier. (f) A schematic diagram of the effect of the difference in electronegativity between the non-migrating cation B and the anion C on the migration of A ions. *Source*: [38] / Springer Nature / CC BY 4.0.

Mo and coworkers [33] proposed another design strategy, which inserts migrating ions into high-energy sites to activate synergistic ion transport with lower migration barriers (Figure 2.3c and d). Lithium ions were inserted by replacing high-energy sites in $LiTaSiO_5$ and $LiAlSiO_4$ with low-valent non-lithium cations. AIMD simulation results showed that there is a synergistic transport of lithium ions in the doped materials. The ionic conductivity at room temperature is 1–4 mS cm^{-1}, and the migration barrier is significantly reduced to 0.23–0.28 eV.

According to the Meyer–Neldel empirical rule, high pre-exponential factors are usually related to high energy barriers [34]. In alkali metal fast ionic conductors, the coexistence of high pre-exponential factor and low migration barrier is rare. Hautier and coworkers [35] proposed that the $LiTi_2(PS_4)_3$ crystal structure has no regular tetrahedral or octahedral sites for Li occupation, which creates a smooth potential energy surface for lithium-ion–anion interaction, which is conducive not only to the formation of low energy barrier, but also to obtaining higher pre-exponential factor. The energy landscape of lithium in one-dimensional model as shown in Figure 2.3e [35]. Among them, the higher pre-exponential factor is related to the longer jump distance and the higher entropy of the transition state. This work opened up a channel for searching these special crystal frameworks through crystal structure analysis, and inspired us to search for fast ionic conductors with similar structures in the database through high-throughput theoretical calculations.

In alkali metal compounds with ternary, quaternary, or even more elements, anions are usually affected by the electronegativity of non-alkali metal elements [36, 37]. The atomic radius and valence electron configuration of non-alkali metal elements determine their coordination environment and crystal volume. It is expected that the low barrier of alkali metal ion migration can be achieved by adjusting the non-alkali metal elements without changing the crystal structure. Zhu and coworkers [38] proposed a new rule for designing a new ternary ABC fast ionic conductor with a face-centered-cubic anionic sub-lattice structure, as shown in Figure 2.3f. For the structure with stable ion octahedral sites, the large electronegativity difference between the anion C and the nonmobile ion B is beneficial to the rapid migration of A ions. And element B preferentially selects elements with small electronegativity located at the lower left of the periodic table. For the structure with stable tetrahedral sites of A ions, B element should give priority to the elements with large electronegativity at the upper right of the periodic table. The electronegativity of B element is close to that of C element, but less than that of latter.

From the discussion of the above works, we found that there are many factors affecting the ionic conductivity of the solid electrolytes. However, due to the limitations of existing research methods, it is difficult to determine the main factors. If there is a method or formula that can quantify these factors, it will be of great significance for the accurate design of high-ionic-conductivity solid electrolytes.

2.1.4 High-throughput Computational Screening of New Solid Electrolytes

In addition to doping modification of existing solid electrolytes, exploring new crystal structures is also of great significance for the discovery of new solid electrolytes. However, the development of new materials by trial-and-error method in the experiments is time-consuming, labor-intensive, and low in accuracy. It has become an advanced research and development method in the field of materials to use theoretical calculations to screen possible materials before the experimental synthesis of materials. High-throughput computing refers to the use of supercomputing platforms and multiscale integrated, high-throughput concurrent material calculation methods and software to achieve simulation and rapid calculation of large-scale materials. At present, the high-throughput calculation of materials is mainly based on the high-throughput calculation of DFT. By designing a series of calculation processes, the large-scale automatic calculation of the atomic-scale intrinsic properties of materials is realized. High-throughput computing has the advantages of high efficiency, parallelism, and scalability, which can effectively improve the screening efficiency and design and development level of new materials. In the field of solid-state lithium batteries, high-throughput calculations are mainly used to screen, guide, and accelerate the development of inorganic materials such as inorganic solid electrolytes and cathodes. High-throughput computational tasks such as structure selection, input file and parameter preparation, property simulation, output, and data analysis can be performed automatically with less human intervention [39].

Some research teams at home and abroad have developed automatic calculation process software for high-throughput simulations, and also provided databases generated by these high-throughput simulations, e.g., pymatgen and FireWorks [40, 41], Materials Project [13], Automatic-Flow for Materials Discovery (AFLOW) [42], Open Quantum Materials Database (OQMD) [43], Automated Interactive Infrastructure and Database for Computational Science (AiiDA) [44], Novel Materials Discovery (NOMAD) Laboratory [45], Computational Electronic Structure Database (CompES-X), and integrated high-throughput computing.

High-throughput computing combines data mining and machine learning to help understand the big data obtained by it (Figure 2.4), providing a very good

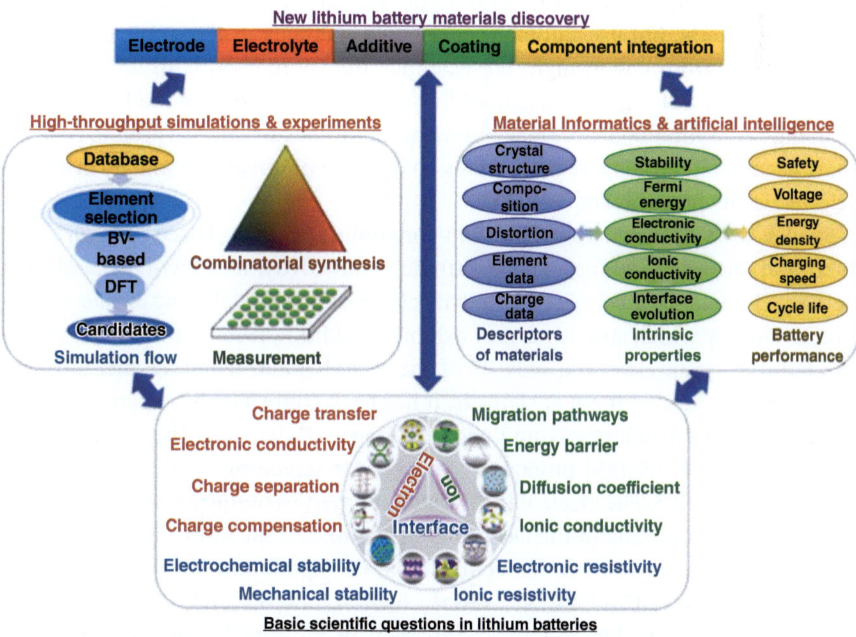

Figure 2.4 Schematic diagram of lithium-ion battery materials developed by high-throughput technology and data science. *Source*: [46] / IOP Publishing / CC BY 3.0.

opportunity to further explore the structure–performance relationship of battery materials and discover new materials. In addition, the comparison between the theoretical calculation results and the experimental data also helps researchers to establish a better theoretical model or carry out experimental measurements [46].

The most important step of high-throughput computing is to determine the prediction target of the model. The screening criteria of solid electrolytes, such as electronic conductivity, ion transport properties, thermal stability, electrochemical window, and mechanical properties, can be achieved by various calculation methods with different bases and accuracies, such as thermodynamics, bond valence (BV), first-principles calculations, molecular dynamics, and Monte Carlo simulations, among others.

High ionic conductivity is a necessary condition for screening solid electrolytes. DFT calculations can provide ion migration paths and energy barriers through AIMD and NEB. However, due to the high computational cost, it is not feasible to use DFT calculations to deal with the high-throughput screening of ion transport characteristics of tens of thousands of compounds. Researchers have proposed two different ideas to solve this problem [47]. One is to establish an overall screening method. First, more than 90% of the candidate materials with high thermal and chemical stability, low electronic conductivity, and good mechanical properties are screened out. Then, DFT method is used to simulate the ion transport characteristics of the materials that meet the above conditions. This screening process uses more constraints to reduce the number of candidate materials and avoid direct simulation of transport properties, but may screen out many metastable structures that may exist in reality. The other idea insists on the transmission characteristics as the screening criteria, but uses different calculation methods to screen the calculation accuracy step from low to high. For example, the BV method with low computational accuracy but high speed is used for prescreening, and then the DFT method with high computational accuracy but higher time consumption is used to perform more accurate ion transport simulations on only the promising candidate materials selected in the previous step [48–51].

Sendek et al. screened 12,831 lithium-containing materials based on high structural and chemical stability, low electronic conductivity, and low cost, and then developed a data-driven ionic conductivity classification model using logistic regression to identify which structures may exhibit rapid lithium-ion transport, as shown in Figure 2.5 [52]. However, only 21 of the 12,831 initial materials met all these requirements. Among them is Li_3InCl_6. It has been confirmed by experiments that it has high ionic conductivity and good interface stability with the oxide cathode [53, 54]. Kahle et al. [55] proposed a process for screening solid electrolytes, as shown in Figure 2.5. The electronic insulation properties of the material were determined by the simulation of the electronic structure, and the diffusion coefficient of lithium ions was predicted by molecular dynamics simulation. They found five fast ionic conductors: Li_5Cl_3O, Li_2CsI_3, $LiGaI_3$, $LiGaBr_3$, and Li_7TaO_6. A detailed study was carried out, and it has an ionic conductivity of 5.7×10^{-4} S cm^{-1} at 300 K.

Gao et al. calculated the lithium-ion transport paths of 1,380 candidate solid electrolyte materials using the BV method [48]. The BV theory simulates the transport

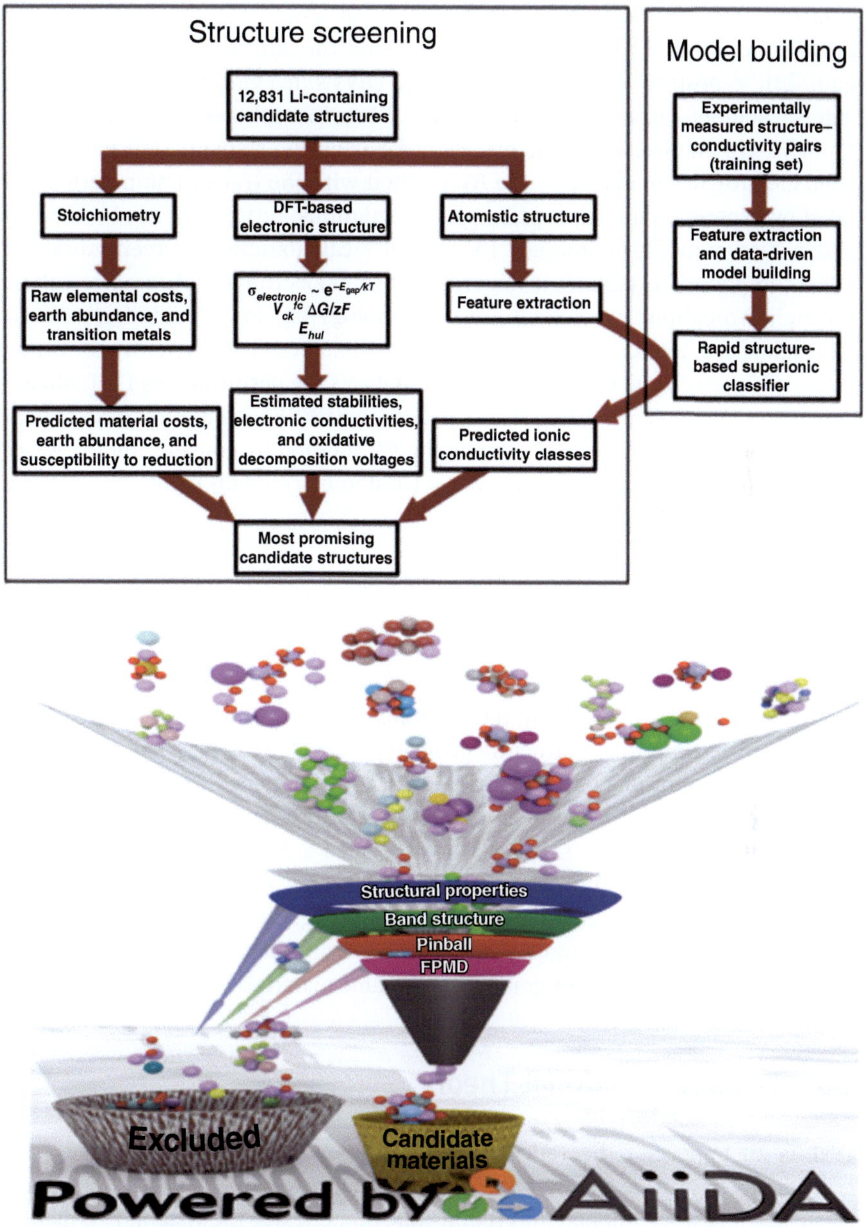

Figure 2.5 Flowchart of solid electrolytes screening. *Source*: [52] / with permission of the Royal Society of Chemistry and [55] / with permission of the Royal Society of Chemistry.

path based on four principles [56]: maximum symmetry, electrical neutrality, local charge neutrality, and equivalence principle. The BV sum can be fitted according to experience:

$$\sum_{X} S_{A-X} = V_A, \quad S_{A-X} = \exp[(r_0 - r_{A-X})/b] \tag{2.2}$$

In the above formula, A is a cation, X is an anion, S_{A-X} represents the BV between anion and cation, and V_A represents the sum of BVs. If A is at the equilibrium position relative to X, then V_A (BVs sum) will be close to the ideal valence state of cation A. r_{A-X} is the actual bond length between A and X, and r_0 and b are empirical parameters. By comparing the transmission path of the solid electrolyte calculated by the BV method with the transmission path calculated by experiments or first-principles calculations, rationality has been recognized. Xiao et al. [57] combined BV and DFT calculations to screen more than 1,000 compounds as candidate materials for solid electrolytes. As a powerful tool for high-throughput prescreening, BV technology has low calculation accuracy but high speed, and can predict migration path and energy barrier trend. On the basis of BV screening, more accurate calculations are performed by DFT. Shi and coworkers [50] developed the band valence site (BVSE) calculation based on the BV method. The BVSE model extends the BV method by considering the Morse potential of cation–cation pairs and the Coulomb repulsion before moving ions and positively charged ions:

$$\mathrm{BVSE}(M) = \frac{D_0}{2}\sum_i \left\{\exp\left[a(R_{\min} - R)\right] - 1\right\} + \sum_{i-1}^{N} E_{\mathrm{Coulomb}}(M - M_i) \tag{2.3}$$

In the above formula, D_0, a, and $R_{\min}$ are empirical parameters of the Morse potential; $E_{\mathrm{Coulomb}}(M - M_i)$ represents the Coulomb repulsion between two different cations and anions. Then, the ion transport network was constructed based on Voronoi decomposition structure geometry (CAVD) and BVSE calculation, and the energy distribution on the migration path is was obtained to analyze the transport properties of ions. The screening process and results are shown in Figure 2.6 Through these calculation methods, they provided a database of crystal structure information, ion migration channel information, and 3D channel maps of about 29,000 inorganic compounds containing Li, Na, K, Ag, Cu, Mg, Zn, Ca, Al, O and F, which accumulates descriptors for machine learning and lays a foundation for large-scale research on ion migration in inorganic materials.

2.2 Density Functional Theory

DFT is an important theoretical method to study multi-electron systems. The core idea is to replace the traditional multi-electron wave function with electron concentration, so as to obtain the ground-state properties of the system. This method greatly simplifies the problem of solving multi-electron systems both conceptually and computationally, because the electron density is a scalar function that depends only on three spatial variables, while the multi-electron wave function contains $3N$ variables (where N is the number of electrons) [58].

The theoretical basis of DFT is the first and second theorems of Hohenberg and Kohn. On this basis, Kohn and Sham established a set of efficient equations (Kohn–Sham equations), which can transform the many-body problem into a series

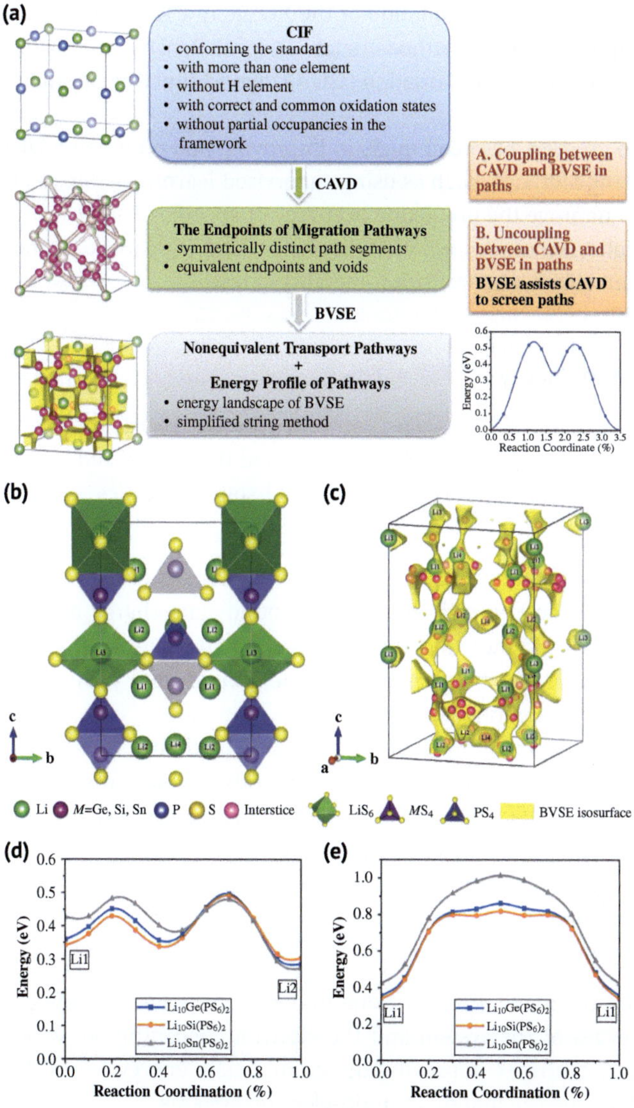

Figure 2.6 Flowchart and results charts of screening solid electrolytes by BVSE method. *Source*: [61] / with permission of the Royal Society of Chemistry.

of single-electron equations. In the Kohn–Sham DFT framework, the interaction between electrons is simplified as an effective potential field, which includes the influence of the external potential field and the Coulomb interaction between electrons, such as exchange correlation (XC).

DFT has important applications in computational chemistry, materials, physics, and many other disciplines. This method can be used to calculate the electronic structure, reaction energy, interactions, and so on. However, the accurate calculation of DFT depends on the accurate approximation of the XC energy. At present, there are many functional approximation methods, such as local density approximation (LDA), generalized gradient approximation (GGA), hybrid functionals, and so on [59].

In recent years, some progress has been made in improving and optimizing DFT through machine learning methods, such as using supervised learning strategies to optimize functionals to improve the prediction accuracy of specific systems. These new methods show higher accuracy and reliability in dealing with weak interactions, non-covalent interactions, and transition-metal systems.

2.2.1 Hohenberg–Kohn Theory

Hohenberg–Kohn theory is the basis of DFT proposed by the Physicists Walter Kohn and Pierre Hohenberg in 1964. The research results of this theory laid a theoretical foundation for us to deeply understand the ground-state properties of multi-electron systems:

1. Hohenberg–Kohn first theorem: according to this theorem, a given external potential field $V_{ext}(r)$ is determined only by the ground-state charge number $n_0(r)$ of the system. This shows that the ground-state characteristics of the system mainly depend on the charge number of the ground state, rather than the multi-electron wave function.
2. Hohenberg–Kohn second theorem: this theorem proposed that there exists a universal energy-density functional $E(n)$. For any given external potential field $V_{ext}(r)$, the ground-state energy of the system corresponds to the energy (global) minimum of this functional, and the charge density corresponding to the minimum energy is the ground-state charge density $n_0(r)$. This theorem provides a theoretical basis for solving the ground-state energy by using the variational principle.

The Hohenberg–Kohn theorems revealed a one-to-one correspondence between the ground-state properties of the system and the electron concentration, which provides the possibility for further improving the ground-state energy level. This result will greatly simplify the calculation of multi-electron systems.

Based on the Hohenberg–Kohn theorems, Kohn and Sham have conducted in-depth research on DFT, established the Kohn–Sham equations, and transformed the many-body interaction problem into a series of single-electron equations. Therefore, DFT is currently the most widely used method in quantum chemistry research. The Kohn–Sham equations described the interaction between electrons by introducing an effective potential field, including the XC energy, which allows for the calculation of the electronic structure. Although the Hohenberg–Kohn theorems have laid a good theoretical foundation for DFT studies, they still need to be approximated in practice because there is no accurate XC energy functional. In recent years, a series of approximation algorithms have been developed, including LDA, GGA, hybrid functionals, and so on.

2.2.2 Thomas–Fermi Theory

Thomas–Fermi (TF) theory first appeared during the early development of quantum mechanics, which is used to simulate multi-electron systems. It was independently proposed by Llewellyn Thomas and Enrico Fermi in 1927 to explain the distribution of electrons in atoms and the volume of atoms.

The core idea of TF theory is to consider the electrons in the atom as a whole electron cloud, rather than considering the behavior of a single electron. It assumes that the electron cloud is uniform, and the repulsive force between electrons can be approximated by an average field. In this model, the kinetic energy of the electron and the repulsion energy between two electrons are approximated by simple mathematical expressions.

Although the TF model is conceptually groundbreaking, it is a rough approximation because it ignores the quantization of electrons and the Pauli incompatibility principle. In addition, it also overestimates the atomic volume and the repulsion energy between electrons. TF theory has an important influence on the development of DFT. The Hohenberg–Kohn theorems in DFT show that the ground-state properties of the system can be uniquely determined by electron density rather than wave function, which echoes the concept of electron cloud in TF theory. However, compared with the TF model, DFT uses more accurate mathematical form and calculation method to provide more accurate predictions.

The TF model still has its application value in some cases, especially in dealing with very large systems or where rapid estimation of the electronic structure is required. However, for accurate quantum chemical calculations, more modern and accurate theories, such as the Hartree–Fock method and the post-Hartree–Fock method, as well as various DFT approximations based on these methods, are more commonly used.

2.2.3 Kohn–Sham Equations

The Kohn–Sham equations are the core equations in DFT, which were proposed by Walter Kohn and Lu Sham in 1965. DFT is a method for calculating the electronic structure of quantum mechanical systems, especially useful in dealing with multi-electron systems, such as molecules and solid materials.

The Kohn–Sham equations were proposed to achieve the following goals. (i) Transforming the multi-body problem into a single-electron problem: Kohn–Sham equations transform complex problems in multi-electron systems into a series of single-electron equations. Although the existing single-electron model (Kohn–Sham equations) can simulate the actual electron density, it needs to be solved by the equivalent potential. (ii) Introducing the effective potential energy: the Kohn–Sham equations adopt the equivalent potential including the external potential energy (such as the Coulomb potential energy of the nucleus) and the XC potential energy. The XC potential energy is the average potential energy that takes into account the exchange and correlation effects between electrons. It cannot be accurately solved and usually requires approximation. (iii) Variational principle: The ground-state charge density of the system is obtained to minimize its energy. The variational method is used to solve the Kohn–Sham equations, for example, by changing the wave function of a single electron to minimize the energy of the system. (iv) Calculation of electron density: the solution of Kohn–Sham equations provides a set of single-electron wave functions, through which the electron density can be calculated, which is the key quantity to describe the properties of the system in DFT [60]. (v) The use of approximate functional: because we do not know the exact expression of the XC potential energy, we must use an approximate functional for its approximation. The systems and properties corresponding to various approximate functionals are different. The selection of appropriate functionals is the key to ensure the accuracy of DFT calculations.

The Kohn–Sham equations are a milestone in the development of DFT, which make it possible to calculate the electronic structure of complex quantum systems and have a profound impact on the fields of chemistry, physics, and materials science. However, DFT and Kohn–Sham equations also have their limitations. For example, they may not be accurate enough in dealing with strong correlation systems, van der Waals effects, etc., and more advanced theoretical methods need to be further developed.

2.3 Molecular Dynamics Simulations

AIMD is a computational method based on the principles of quantum mechanics. It does not require empirical parameters and simulates the dynamic behavior of molecular systems directly from the basic physical laws. This method can provide very accurate results, especially for the study of complex chemical processes and the properties of materials, such as chemical reactions, phase transitions, and electronic structure changes, among others.

AIMD is a "microscope" that observes chemical reactions at the atomic and molecular scales. It calculates the force of the nucleus through electronic structure analysis, and then studies the theory of nuclear motion. This method regards the nucleus as a classical particle under the Born–Oppenheimer approximation, while the electron is described by quantum mechanical behavior.

AIMD is a very important physical phenomenon, which has important applications in condensed matter physics, materials science, polymer chemistry, molecular biology, and many other disciplines. It can be used to simulate the structure of liquid metals at high temperature and pressure, the kinetic characteristics of metal halide perovskites, and the atomic structure evolution of amorphous alloy melts during rapid cooling. The molecular dynamics method based on the first-principles lays the foundation for an in-depth study of the microstructure and properties of molecular systems. With the improvement of computing power and algorithms, AIMD will be even more widely used in scientific research.

The properties of the solid electrolyte have a great influence on the performance of the solid-state battery. In addition, the influence of the properties of the interface between the electrode and the solid electrolyte on the performance of the battery cannot be ignored. The performance of the battery is not only related to the performance of each component, but also strongly affected by the interaction between the components. In addition to the inherent properties of the solid electrolyte, the generation and evolution of the electrode–electrolyte interface phase play a key role in stabilizing the cycle performance and extending the cycle life. Xu et al. [61], by constructing a sandwich structure between a sulfide solid electrolyte and a metal Li anode, used molecular dynamics to calculate the interface stability between Li_6PS_5Cl and Li. In addition, Xu et al. [61], predicted the electrochemical window of li6PS5Cl by using ATAT calculation method, and found that the seven stable electrochemical window range was in the range of 1.7–2.1V (Figure 2.7a–c).

In order to overcome the problem that Li_6PS_5Cl has a very narrow electrochemical window for stable existence (1.7–2.19 V), two functional buffer layers of Li_6PO_5Cl and Li_6PO_4SCl have been identified [61]. As shown in Figure 2.8a, through ATAT scan between Li_6PO_5Cl and Li_6PS_5Cl for stable structural configurations, one can see that Li_6PO_4SCl maintains the same argyrodite structure of the hali-oxide Li_6PO_5Cl. It follows that once oxygen is introduced to substitute 1/5 of the chalcogen site [63], the material can be in direct equilibrium with lithium metal (Figure 2.8b). This makes it an ideal buffer layer to isolate the superb electrolyte based on either $Li_{6.25}PS_{5.25}Cl_{0.75}$ or $Li_{6.25}PTe_{0.125}S_{5.125}Cl_{0.75}$. The high electrochemical potential for Li_6PO_5Cl, on the other hand, makes it an ideal choice as a coating for oxide cathode such as $Li_{0.25}MnO_2$ (LMO) which has an electrochemical potential up to 3.65 V while withholding the oxides framework essential for adequate lithium transport (Figure 2.8c) [64]. The fact that SSE and buffer layers are of the same crystal structure also safeguards small lattice mismatches at the interfaces, with lattice mismatches for $Li_6PO_4SCl|Li_6PS_5Cl|Li_6PO_5Cl$ being 1.5% and 3.1% at the two interfaces, respectively.

Although the ionic conductivity of the solid electrolyte has been significantly improved, the overall impedance of the solid-state battery is difficult to reduce due to the high impedance of the electrode–electrolyte interface, thereby increasing the overpotential, resulting in a decrease in battery energy density and capacity attenuation. The high impedance of the electrode–electrolyte interface comes from the following aspects:

1. Solid–solid physical contact between electrode and electrolyte: although high-pressure or high-temperature sintering can be applied in the manufacturing process of solid-state batteries, there are still some voids and holes in the solid–solid contact, and the mechanical stress due to the volume expansion and contraction caused by the deintercalation of lithium in the battery cycle further deteriorates the physical contact of the interface.
2. Lithium chemical potential difference between electrode and electrolyte: the chemical potential of lithium in sulfide solid electrolyte is often higher than that of oxide cathode, so lithium ions move from electrolyte to electrode and form a space–charge layer (SCL) at the interface, which leads to the enrichment of lithium ions on the cathode side and the depletion of lithium ions on the electrolyte side, hindering the transport of lithium ions at the interface.

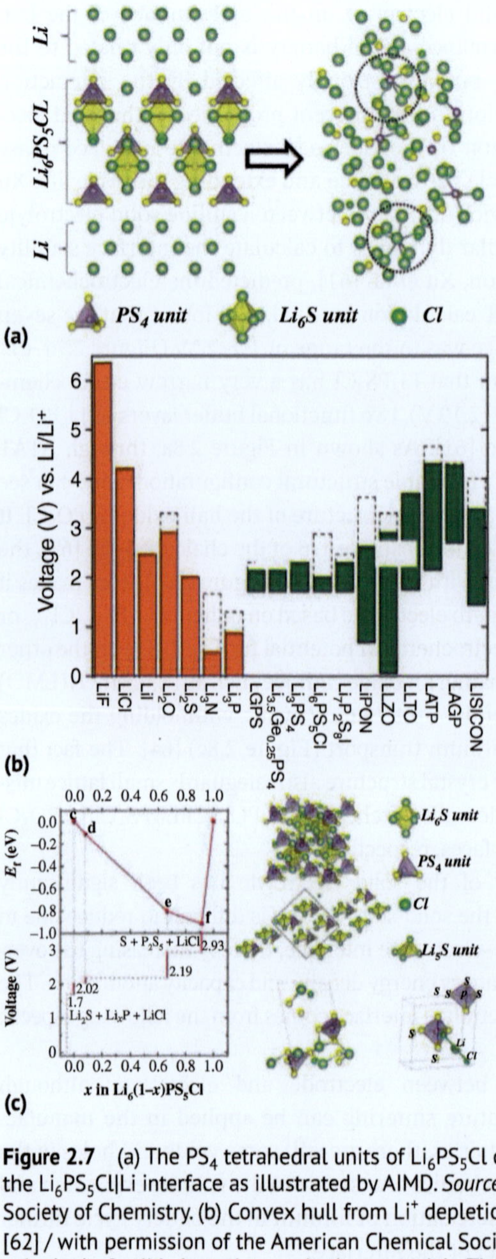

Figure 2.7 (a) The PS_4 tetrahedral units of Li_6PS_5Cl decomposed into Li_xP and Li_2S at the $Li_6PS_5Cl|Li$ interface as illustrated by AIMD. *Source*: [61] / with permission of the Royal Society of Chemistry. (b) Convex hull from Li^+ depletion in $Li_{6(1-x)}PS_5Cl(0 \leq x \leq 1)$. *Source*: [62] / with permission of the American Chemical Society. (c) Electrochemical window (solid color bar) of solid electrolyte and other materials. The oxidation potential to fully delithiate the material is marked by the dashed line. *Source*: [61] / with permission of the Royal Society of Chemistry.

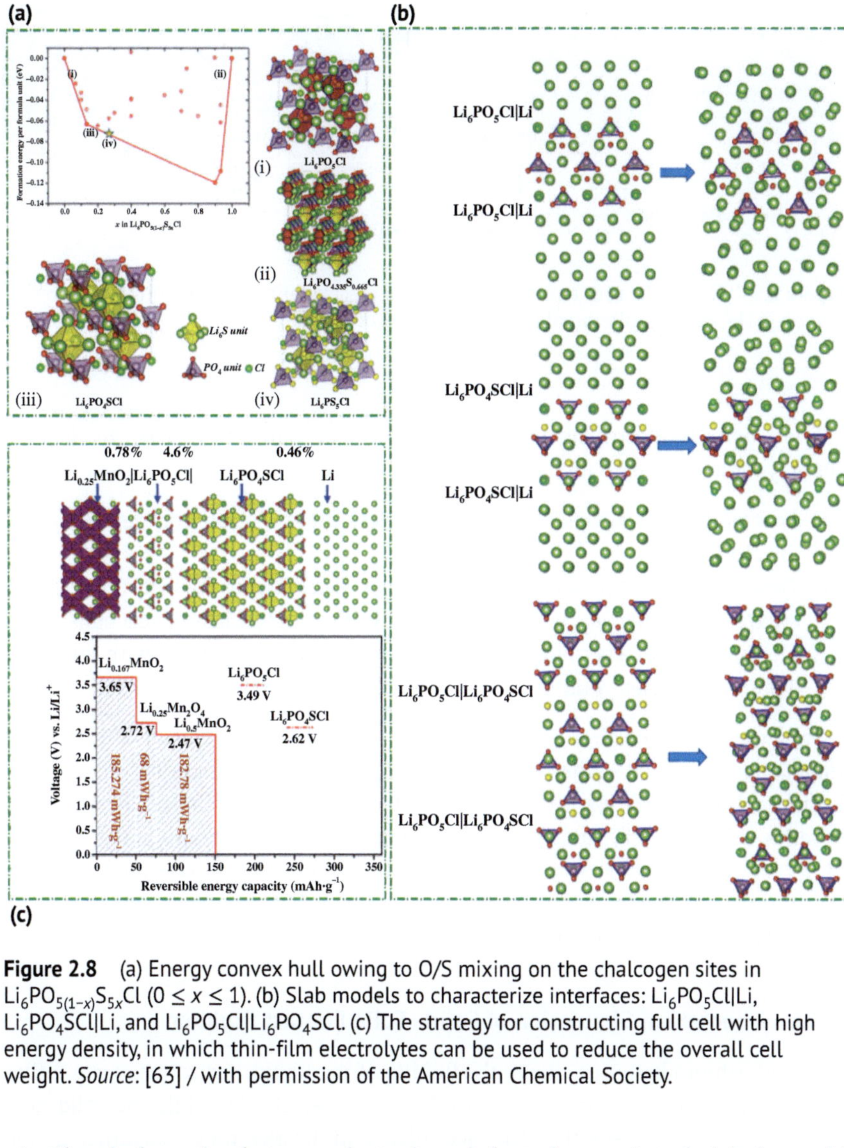

Figure 2.8 (a) Energy convex hull owing to O/S mixing on the chalcogen sites in $Li_6PO_{5(1-x)}S_{5x}Cl$ ($0 \leq x \leq 1$). (b) Slab models to characterize interfaces: $Li_6PO_5Cl|Li$, $Li_6PO_4SCl|Li$, and $Li_6PO_5Cl|Li_6PO_4SCl$. (c) The strategy for constructing full cell with high energy density, in which thin-film electrolytes can be used to reduce the overall cell weight. *Source*: [63] / with permission of the American Chemical Society.

3. Chemical reaction between electrode and electrolyte: an interfacial phase with low ionic conductivity was formed due to the chemical reaction.
4. Electrochemical reaction between electrode and electrolyte: due to the narrow electrochemical stability window of the electrolyte, it leads to decomposition during charging and discharging, or electrochemical side reactions with the electrode, which further deteriorates the interface and hinders the transmission of lithium ions.

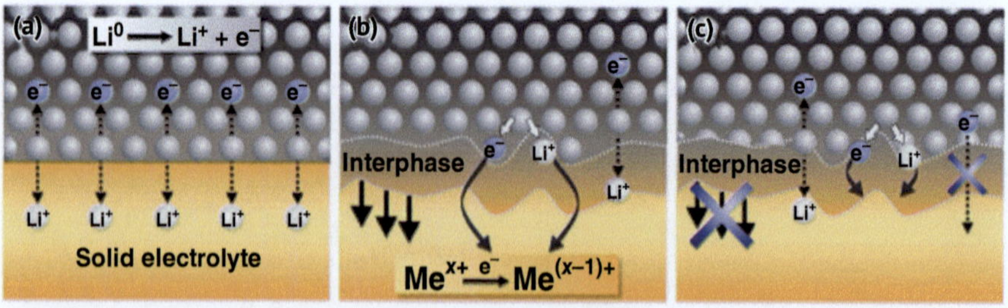

Figure 2.9 Schematic diagrams of three types of electrode–electrolyte interfaces. *Source*: [61] / with permission of the Royal Society of Chemistry.

The electrode–electrolyte interface in solid-state batteries can be divided into three types according to its stability [61], see Figure 2.9:

1. Thermodynamically stable interface, where the electrode does not react with the electrolyte.
2. The interface reacts to form an interface layer with ionic conductivity but negligible electronic conductivity, which can prevent further interfacial side reactions.
3. The interface reacts to form a mixed conductor interface layer with ionic conductivity and electronic conductivity, which cannot prevent the reaction.

We expect that the first two interfaces in the battery are crucial to the overall impedance formation in the battery. Due to the difficulty in accurately detecting the position of the interface and the complex interaction between thermodynamics and kinetics, it is still challenging to experimentally detect the reactions at the interface. However, theoretical calculation has great advantages. It has successfully predicted the interface products, explained the trend of interface reactions, and guided the design and optimization of the interface.

2.3.1 Theoretical Simulation of Solid–Solid Contact Problem

In liquid lithium batteries, due to the fluidity of the liquid electrolyte, it can diffuse and infiltrate the electrode surface in the porous electrode, while the solid–solid physical contact between the solid electrolyte and the electrode is insufficient, and the interface is usually a mixture of point contacts and voids. Hertz contact theory simply describes the contact of two elastic spheres with radii R_1 and R_2 under the action of extrusion force F [65]:

$$r_0 = \left(\frac{R_1 R_2}{R_1 + R_2}\right)^{1/3} \frac{3}{4} F\left(\frac{1 - V_1^2}{E_1} + \frac{1 - V_2^2}{E_2}\right) \tag{2.4}$$

In the above formula, r_0 is the radius of the circular contact area; E_1 and E_2 are the Young's moduli of the two spheres, respectively; and V_1 and V_2 are Poisson's ratios.

In this case, the contact area between the solid electrolyte and the electrode particles is very small, forming a large impedance. Moreover, when the interface is formed, stress and strain will be generated due to the mismatch of crystal parameters on both sides of the material. Strain can change the properties of local materials (such as phase stability, carrier mobility, and potential). Most electrode materials undergo phase transition and lattice expansion/contraction due to lithium insertion and extraction during charge and discharge, and repeated volume changes will further deteriorate the electrode–electrolyte interface contact [66–68]. Based on the 1D Nerman battery model and Poisson contact mechanics theory, Qi and coworker [69] simulated the change of electrode–electrolyte contact area during the cycle. It was found that the loss of discharge voltage and specific capacity is very fast due to the loss of contact area. The contact area of $Li/Li_3PO_4/LiCoO_2$ and $Li/LGPS/TiS_2$ solid-state batteries was calculated by measuring the surface roughness and elastic properties of the electrolyte and the electrode. The variation of the contact area with the applied pressure proves that the applied external pressure helps to restore the contact area and specific capacity loss caused by the cycle.

The electrode material undergoes volume expansion and contraction during the cycle, which will cause mechanical stress on the solid electrolyte. This requires the solid electrolyte to have a low Young's modulus to buffer the stress. Otherwise, cracks or gaps will be formed at the interface to reduce interface contact. Bucci et al. [70] proposed a one-dimensional radial symmetry analysis model based on fracture cohesion theory to analyze the critical conditions of electrode and electrolyte delamination and the conditions of crack propagation. When the electrode undergoes a volume change of about 7.5% during the deintercalation of lithium, it will cause electrode and electrolyte delamination, and the softer solid electrolyte ($E < 25$ GPa) can adapt to up to 25% of the particle volume change and delay the occurrence of delamination. The model provides guidance for improving the contact properties of the electrode–electrolyte interface by controlling parameters such as particle size, material mechanical properties, and adhesion strength.

For the hard oxide solid electrolyte, co-sintering with the cathode material can improve the physical contact between the two, but it will promote element interdiffusion and chemical reaction at the interface, and the volume change of the electrode material during the cycle will further aggravate the contact loss and form cracks and voids at the interface. The sulfide electrolyte is in good contact with the cathode material due to its softness, so only external pressure is required when the solid-state battery is assembled with sulfide electrolyte. However, the choice of pressure needs to be careful. Low pressure is not enough to maintain good contact, and excessive pressure will cause the electrode and electrolyte particles to rupture [71, 72].

According to DFT calculations, the thermodynamic reaction between LLZO and metal Li can be neglected, indicating that the interface between the two is very stable. However, the Li/LLZO interface still has a large impedance, which is related to the poor physical contact of the Li/LLZO interface. It has been reported in the literature [73–75] that pores are formed at the interface between lithium metal and solid electrolyte, especially under high-current-density conditions. Experimental and theoretical calculations have shown that the impurities generated by the reaction of LLZO with air will significantly affect the interface impedance of Li/LLZO.

Sharafi et al. [76] calculated the thermodynamic driving force for the reaction of LLZO with air to form Li_2CO_3 and LiOH by DFT, and found that LLZO is easy to react in humid air. The most favorable reaction pathway involves the protonation of LLZO and the formation of Li_2CO_3. Gao et al. [77] proved that the high impedance of the Li/LLZO interface in the actual battery is derived from the impurity Li_2CO_3 on the surface of LLZO by theoretical calculations. Zheng et al. [78] have different views on this. They believed that the oxide layer on the surface of lithium metal is more likely to affect the properties of the interface than the impurities on the surface of LLZO. They compared the properties of $Li_7La_3Zr_2O_{12}$/Li, $Li_{6.4}La_3Zr_{1.4}Ta_{0.6}O_{12}$/Li, Li_2CO_3/Li, and Li_2O/Li interfaces using DFT calculations, and found that Li_2O on the surface of the metal lithium has a greater effect on the lithium affinity of LLZO than Li_2CO_3 on the surface of LLZO. After removing the impurities on the surface of lithium metal, LLZO exhibits good lithium affinity. Therefore, in addition to the intrinsic properties of the material, the surface impurities of the electrolyte material and the electrode material are also important factors affecting the interface contact.

2.3.2 Theoretical Simulation of SCL

SCL originates from the lithium chemical potential difference between the electrode and the electrolyte. Theoretically, there is an SCL at all electrode–electrolyte interfaces. However, there are only few studies on the SCL between the electrolyte and the lithium metal anode, mainly between the solid electrolyte and the positive electrode. Compared with sulfide electrolytes, the space-charge effect between oxide electrolytes and positive electrodes may be less serious. When the sulfide electrolyte is in contact with the oxide cathode, due to the high chemical potential of lithium in the sulfide electrolyte, lithium ions will migrate to the cathode material, while electrons will remain on the electrolyte side, thus forming a space electric field from the cathode to the electrolyte. When the oxide cathode material has both ionic and electronic conductivity, the positive charge of Li ions accumulated on the cathode side will be compensated by electrons, which leads to the continuous movement of lithium ions from the sulfide electrolyte to the oxide cathode until the interface space electric field strength is sufficient to prevent the migration of lithium ions. When the equilibrium is reached, the lithium ions on the cathode side are enriched and the lithium ions on the electrolyte side are depleted, which seriously hinders the transport of lithium ions at the interface and causes a very large interface impedance [79, 80].

Tateyama and coworkers [79] calculated the interface between $LiCoO_2$ and Li_3PS_4. As shown in Figure 2.10, $LiNbO_2$, as a cathode coating material inserted between $LiCoO_2$ and Li_3PS_4, inhibits the formation of an SCL and the redistribution of lithium ions due to its electronic insulation, thereby reducing the interface impedance.

Due to the multiple effects on the electrode–electrolyte interface, the experimental study of the SCL is still challenging [81]. In terms of theoretical calculations, researchers have tried to predict the SCL with heterogeneous interfaces through lattice models, atomic models, and thermodynamic models [82–86]. However, it is more challenging to predict the SCL at the interface between the solid electrolyte and the electrode, because the redistribution of lithium ions and the reaction

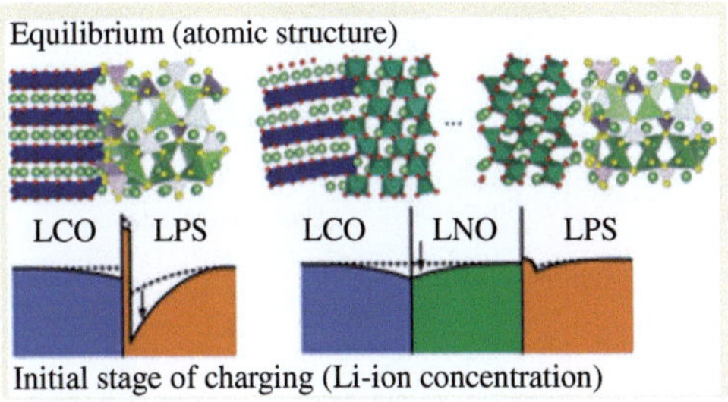

Figure 2.10 The atomic structure and lithium concentration distribution at the LCO/LPS interface at the initial stage of charging. *Source*: [79] / with permission of the American Chemical Society.

between the electrolyte and the electrode material change the arrangement of energy bands at the interface [87]. Researchers simulated the distribution of lithium concentration and potential at the interface. The theoretical results showed that the thickness of the SCL is at nanometer level, and the lithium concentration at the interface will change greatly compared with the bulk phase. However, these models ignore the Coulomb interaction between charged lithium defects caused by the migration of lithium ions [88, 89].

Wagemaker and coworker [85] assumed that only ions are mobile and the electrode–electrolyte interface is chemically stable. By using the solid solution model, the chemical potential distribution as a function of the interface distance was determined from the ion concentration. The calculation results showed that the resistance of the SCL with a thickness of nanometer scale at the interface is less than $1\ \Omega\ cm^2$, and its influence on the performance of the solid-state battery is negligible. The interface impedance is expected to increase significantly only when the lithium ions in the SCL on the solid electrolyte side are completely depleted. The model considers the Coulomb interaction between defects, but does not consider the transfer of electrons (holes) through band bending at the cathode–solid electrolyte interface, which may have a great influence on the formation of space charges, as Qi and coworker [90] predicted for the LiPON–Li_xCoO_2 interface. Qi and coworker established an ab-initio framework to calculate the interfacial thermodynamic driving force and the resulting interfacial net potential drop in an all-solid-state battery model under open-circuit equilibrium conditions. The SCL in the model comes from Li^+ transfer (predicted by defect formation energy) and electron transfer (due to interface band bending). The potential drop and the SCL jointly determine the interface lithium transport barrier.

2.4 Overview of Machine Learning

In recent years, with the accumulation of experimental, theoretical, and computational data and the rapid development of efficient and accurate artificial intelligence technology, materials science research has entered the fourth scientific paradigm,

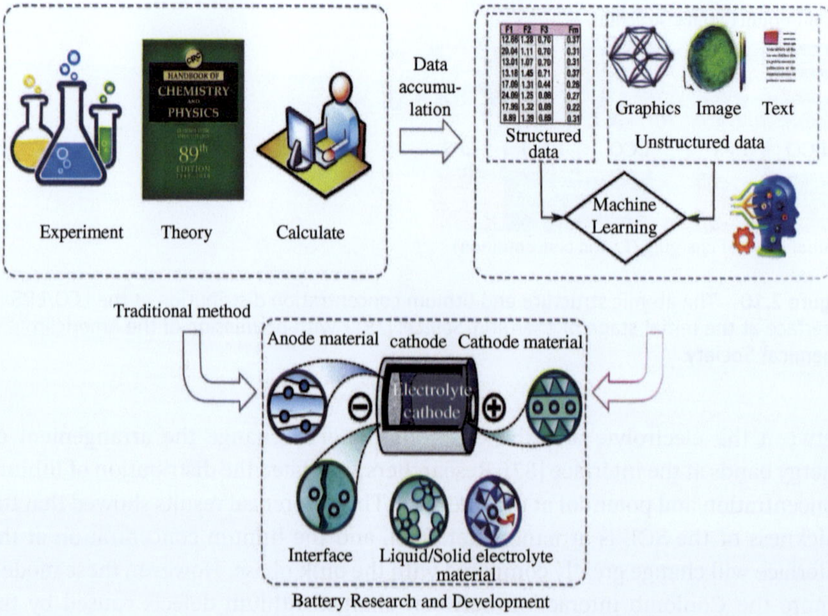

Figure 2.11 Data-driven battery development paradigm diagram [93].

i.e., data-driven materials science research. As shown in Figure 2.11 [86], data-driven materials science uses a large amount of data accumulated by traditional experiments, theories, and computational simulation methods to model and analyze the performance-driven mechanism of electrochemical energy storage materials by means of data-driven artificial intelligence methods to accelerate the design and development of new high-performance electrochemical energy storage materials. At present, as one of the typical representatives of data-driven artificial intelligence methods, machine learning has been widely used in material performance prediction and new material discovery. The application of machine learning in the field of batteries can be traced back to 1999 when Lee et al. [91] used fuzzy logic methods to determine the charging state and health status of the battery. Subsequently, Ceder and coworkers [92] used machine learning technology to predict the crystal structure of materials and used it to discover lithium-based materials for automotive batteries. In 2011, the US government proposed the "Materials Genome Project." One of its goals is to quickly develop materials related to clean energy systems by combining "experiments," "calculations," and "data" through machine learning methods. Since then, data-driven machine learning methods have helped the development of electrochemical energy storage materials (as shown in Figure 2.11).

2.4.1 Machine Learning

The selection of machine learning models plays a key role in the prediction results of artificial intelligence model training. A suitable machine learning model can not

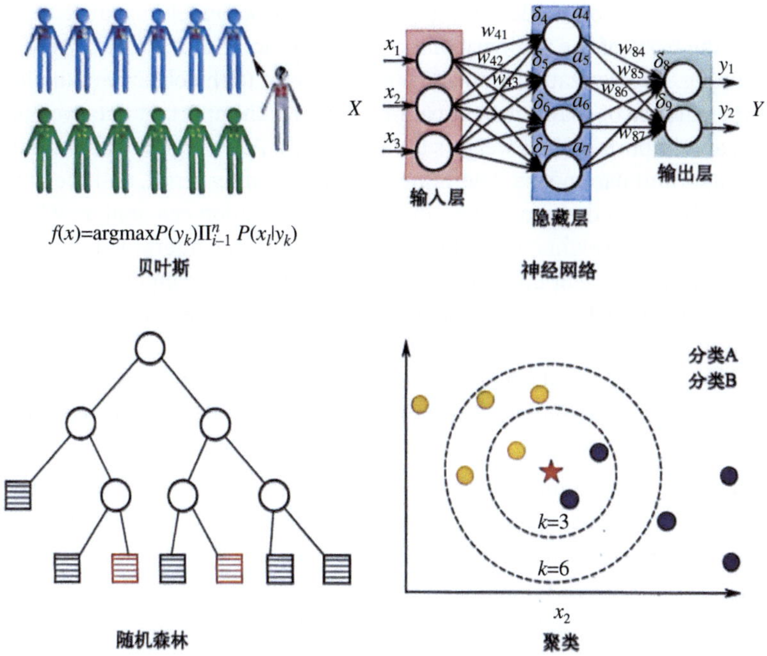

Figure 2.12 Typical machine learning algorithm model [93].

only ensure the reliability of the final output results, but also significantly reduce the time required for training.

Machine learning can be divided into two categories: supervised learning and unsupervised learning (as shown in Figure 2.12). Supervised learning is the most widely used in battery design research. It requires data including input data and corresponding target output data. Supervised learning can establish a learning model from the training data, and then infer new samples based on the model. The purpose is to establish a mapping from the input data to the target output data through machine learning. According to the different target output data, supervised learning can be divided into regression and classification. Unsupervised learning is to automatically classify the input data, and there is no target output data, i.e., it does not require supporting input and output. The advantage of unsupervised learning is that it can interpret the data independently and it is possible to mine the hidden information in the data. In the following, we will introduce the typical machine learning models one by one.

Linear model is one of the simplest models in supervised learning. The purpose is to establish a linear mapping between the input data x and the output data y. The linear model is simple with low computational cost and strong interpretability. It is widely used in mechanism interpretation, battery performance influencing factors, and other research. However, the linear model cannot solve the problem of high complexity. Among them, the lasso regression model and the elastic network model are the linear models that are widely used in the battery field.

Both the lasso regression model and the elastic network model introduce regularization on the basis of the ordinary linear model, in order to reduce the risk of model overfitting. Regularization is to add a penalty term to the objective function to achieve the purpose of sparse weight parameters (i.e., unimportant data weights are reduced, and important data weights are increased). After regularization, only a few more important parameters determine the final learning result, and the anti-noise ability is strong. At the same time, since the regularization can analyze which parameters have more weights, these two models have better interpretability and can explain the importance of different dimensions of data in the input vector.

$LiNiO_2$ has high specific energy and is an excellent choice for high-voltage cathode materials. However, its poor stability seriously limits the cycle life of the battery. For this reason, Yoshida et al. [94] used the lasso regression model to study the effect of doping of different elements on the stability of $LiNiO_2$. They used the interlayer spacing function $\left(\Delta d_{\text{ave(calc.)}} = \sum i(d_{id} - d_{ic})/3\right)$ of $LiNiO_2$ as the cathode stability parameter. The atomic number, covalent radius, electronegativity, ionization energy, atomic mass, thermal expansion coefficient, substitution position of Co (Co is another doping element of $LiNiO_2$), and density were used as the descriptors of the regression model. First, the equation of the lasso regression model was obtained by using the existing training data. Then, the regression model was tested by using the calculated data obtained by DFT calculations. It was found that the root-mean-square error between $\Delta d_{\text{ave(calc.)}}$ predicted by the lasso regression model and the $\Delta d_{\text{ave(calc.)}}$ calculated by DFT was only 8.3×10^{-3} Å (Figure 2.13). Based on this

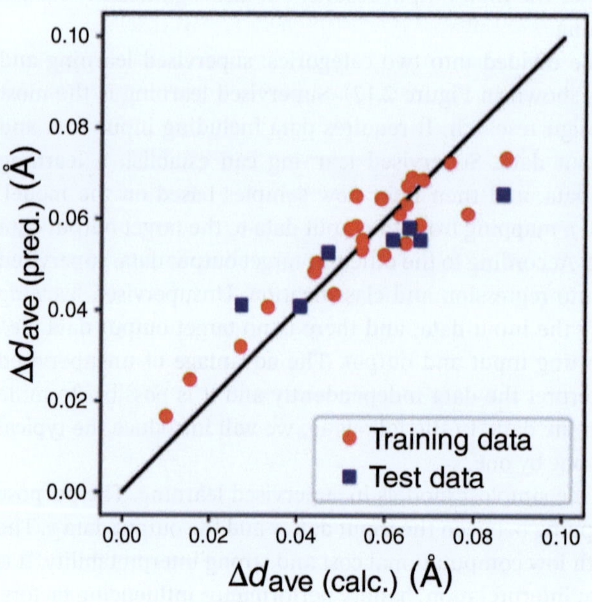

Figure 2.13 Layer spacing of $LiNiO_2$ predicted by lasso regression model. *Source*: [94] / with permission of the American Chemical Society.

lasso regression model, they found that the electrical density between the cathode material layers is a key factor affecting crystal shrinkage and cathode stability.

Compared with the lasso regression model, the biggest advantage of the elastic network model is that it can deal with the problem of multifeature correlation more effectively. Severson et al. [95] successfully predicted the cycle life of commercial lithium iron phosphate/graphite batteries using the elastic network model. First, they used 72 fast charging conditions to obtain 124 battery pack data with different cycle lives as a dataset. Then, the difference between the capacities in the discharge voltage curves of the 100th and 10th cycles of the battery was regarded as the function $Q_{100-10}(V)$ of voltage, the discharge capacity of the second cycle of the battery and the difference between the discharge capacity of the second cycle and the maximum discharge capacity were used as descriptors, and the "elastic network" model was generated by linear and nonlinear transformations. Finally, based on the discharge voltage curve of the early cycle, the model was used to realize the regression prediction of the battery life using the cycle data of the first 100 cycles of the battery and the classification prediction of the battery life using the cycle data of the first five cycles.

2.4.1.1 Bayesian Learning

Bayesian learning is a process of learning and reasoning through probability rules. Taking Naive Bayes as an example, its basic principle is to learn the joint probability from input data to target output data through training dataset. For the input data to be classified, the probability distribution of each category is calculated by the learned joint probability formula combined with the conditional probability formula. The ultimate goal of Bayesian learning is to calculate the probability of a category under a specific input condition, and finally, the category with the highest probability under this condition will be identified as the final output. Bayesian learning performs well in small sample data analysis, so it is widely used in battery health status assessment.

Jiang et al. [96] developed a prediction model that can evaluate battery life under different charging protocols using the hierarchical Bayesian model (HBM) method. As shown in Figure 2.14, the goal of HBM is to model the hierarchical structure of the dataset and use the Bayesian method for reasoning. In this model, the training data is the battery life under different charging protocols (represented by y^i, i is the ith cycle protocol). What the researchers want to predict is the period distribution of the battery life under the nth new charging protocol (i.e., θ_n). In order to perform this predictive inference, HBM needs to use training data to learn at two levels. As shown in Figure 2.14, the first level is about the life cycle distribution of each charging protocol, i.e., θ_n. The second level is abstract knowledge, which is used to describe the life cycle variability between different charging protocols and the overall life cycle distribution of these protocols, which can be represented by two parameters α and β. Among them, α is the degree to which the battery life tends to be uniform in each charging protocol, and β represents the average battery life distribution in all charging protocols.

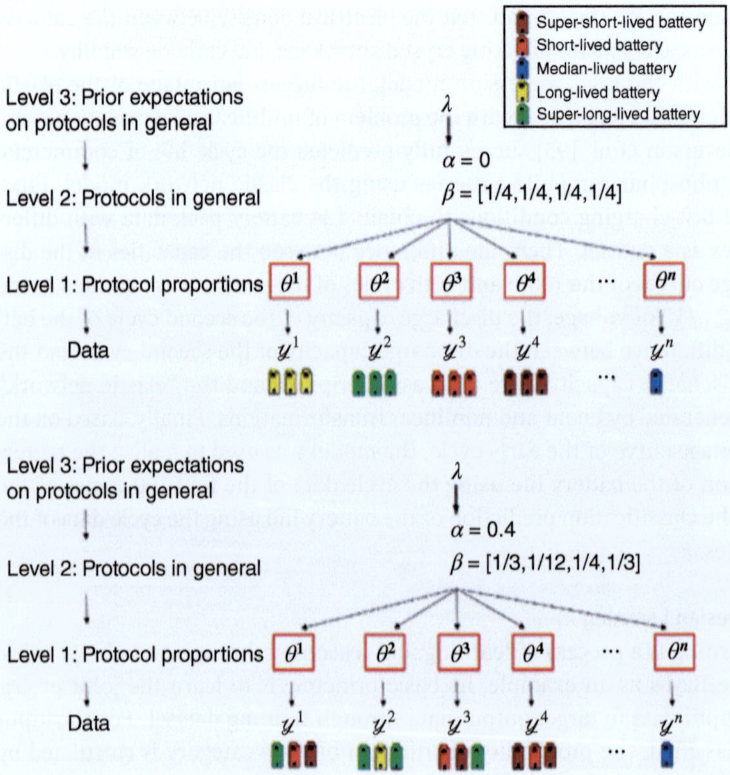

Figure 2.14 HBM of battery life based on different charging protocols. *Source*: [96] / with permission of Elsevier.

2.4.1.2 Neural Networks

As a very popular nonlinear model, neural network algorithm can be competent for different machine learning methods such as supervised learning and unsupervised learning. The basic principle of the neural network is to learn the weight of each neuron in each network layer and combine with the activation function to obtain the output of the layer, and finally obtain the mapping from the input layer to the output layer. The neural network model has strong adaptive ability and can deal with high uncertainty problems. It is suitable for developing end-to-end prediction models. It has high robustness and fault tolerance, strong ability to solve complex problems, and can fully approximate the potential mapping of complex problems. Due to the excellent function approximation ability of neural network, it has been widely used in battery health assessment and new materials design. However, neural network is a typical black box model, which is poorly interpreted and cannot explain the material mechanism, feature importance, etc. The representative neural network algorithms in the field of battery design include artificial neural network, convolutional neural network, and graph neural network.

Artificial neural network is to learn the weight coefficient between any two neurons between layers by optimizing the loss function, and to approximate the target mapping from input to output by combining with the activation function. The parameters between each neuron are independent and not shared when the artificial neural network is trained with a fully connected structure, which is suitable for solving the problem of mutual independence between training data. He et al. [97] established a battery SOC model based on artificial neural network and unscented Kalman filter. In this model, the working current, voltage, and temperature of the battery are used as the input of the neural network, while the SOC of the battery is used as the output of the neural network. At the same time, the unscented Kalman filter is used to filter out the outliers in the neural network estimation to reduce the estimation error of the neural network. Based on this, they used the open-source battery data of the Advanced Battery Alliance to control the prediction error of the battery SOC within 3.5%.

Convolutional neural network derives its name from the convolution process during training. Compared with artificial neural network, convolutional neural network has more grid search (convolution) and pooling process. The data of input layer is generally a kind of data with grid topology. Through the search of convolution kernel, the input of each hidden layer is the feature extraction data of the previous layer, i.e., the data of the hidden layer is a new grid data containing the characteristics of the previous layer data. After that, the features are further extracted through the pooling layer data, and the input data of the next layer is obtained. After the training of the hidden layer, the convolutional neural network will vectorize the final grid data to obtain a feature data with a higher dimension. This feature data will be used as the input data of the neural network structure of the fully connected layer to learn the final target mapping function. Convolutional neural networks are good at processing data with grid-like topologies (such as images, diffraction structures, or crystal structures), and can achieve pixel-level classification of large datasets. In the field of batteries, convolutional neural networks have obvious application advantages in feature segmentation of high-resolution transmission electron microscope images, defect analysis based on X-ray diffraction images, and quantitative analysis of powder X-ray diffraction.

Lithium metal has the highest theoretical capacity among all anode materials, but because it forms lithium branches, it will not only seriously affect the cycle life of the battery, but also cause safety hazards. However, due to the low atomic number of lithium element, the interaction between lithium and experimental probes is weak, which makes it difficult for conventional characterization techniques to perform high-resolution imaging of lithium deposition morphology. To this end, Dixit et al. [98] used synchrotron radiation X-ray tomography combined with deep convolutional neural network technology to quantitatively track the morphological changes of the interface between the lithium metal electrode and the LLZO solid electrolyte during lithium deposition. As shown in Figure 2.15, compared with the traditional binarization process, the deep convolutional neural network processes a single cross-sectional image of lithium metal, and the resulting segmented image

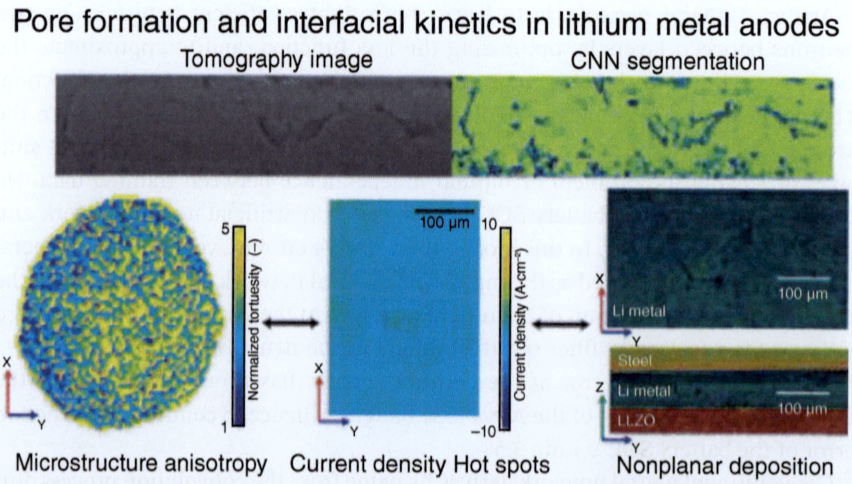

Figure 2.15 The X-ray tomography images of the Li/LLZO interface and the control images after CNN convolution processing. *Source*: [98] / with permission of the American Chemical Society.

has higher confidence, indicating that it can more effectively segment lithium metal and pores in low-contrast images of X-ray tomography. Researchers used 800 lithium metal anode images generated in a single electrochemical cycle to segment and train the neural network, and verified the confidence on another 200 additional images from the same electrode. It was found that it takes about 0.3 seconds to segment a single slice using this method. The segmentation time can be shortened by an order of magnitude compared with the time required to manually mark these images, and the confidence level is more than 80%, which is comparable to the segmentation confidence level of the most advanced neural network. As shown in Figure 2.15, based on this study, they also found that the formation of hot spots on the surface of lithium metal anodes is related to the anisotropic microstructure in solid electrolytes.

Compared with artificial neural network and convolutional neural network, graph neural network is a machine learning algorithm based on graph structure. For general neural network algorithms, the data must be structured, i.e., serialized or gridded data, including text, images, etc. The graph neural network is good at solving unstructured input data, including social networks, knowledge graphs, etc.

Xie and Grossman [99] proposed a crystal graph convolutional neural network. Each crystal is represented by a crystal graph, and satisfies the invariance of atomic index replacement and unit cell selection. The crystal graph is used as the input data of the graph neural network to accurately predict the formation energy, bandgap, Fermi energy, and elastic properties of the crystal structure. Aiming at the shortcomings of time consumption and low accuracy of molecular dynamics simulation of polymer electrolytes, Xie et al. [100] improved the molecular dynamics simulation of ion transport properties of polymer electrolytes by using multitask graph neural network and based on a large number of noisy, non-convergent short molecular dynamics data and a small amount of convergent long molecular dynamics data.

2.4.1.3 Gaussian Process Regression

Gaussian process regression is a data-driven nonparametric machine learning model with hyperparameter-adaptive characteristics, which is easy to implement and does not involve parameter presets. The data-driven properties and non-parametricity of Gaussian regression enable it to effectively adapt to complex input conditions. Unlike linear models and nonlinear models, Gaussian process regression is a non-presupposed machine learning algorithm, i.e., Gaussian regression does not presuppose whether the problem is linear or nonlinear before training. Under the condition that the kernel function form is not limited, Gaussian regression can theoretically achieve arbitrary approximation of any different function curve.

Ha et al. [101] combined sparse Gaussian process regression with DFT to study the effect of aluminum doping on the capacity of high-energy-density $Li_{1.22}Ru_{0.61}Ni_{0.11}Al_{0.06}O_2$ (Al-LRNO) cathode material. Since the sparse Gaussian process regression involves related factors such as the local chemical environment of the compound, the intermolecular interactions can be considered. At the same time, since the local chemical environments of Al-LRNO and LRNO are very similar, energy prediction can be completed in only a few steps, which makes the calculation speed of the model hundreds of times faster than conventional DFT calculations. The authors used this model to study the structure of Al-doped cathode materials and the X-ray absorption near-edge structure spectrum and the diffusion of lithium ions during charge and discharge. They found that in the positive electrode containing 4d-orbital elements, the doping of aluminum can effectively suppress the voltage/capacity attenuation of the battery.

2.4.1.4 Ensemble Learning

Ensemble learning method is the integration of multiple machine learning algorithms, i.e., through the combination of multiple weak base learners, to complete the final learning task. The types of ensemble learning methods also depend on the selection of different base learners, which can be divided into three different types of self-help methods, lifting methods, and stacking methods. Ensemble learning first constructs a base learner in a parallel or sequential manner, and then combines the base learners through some combination algorithm (such as majority voting and weight average, among others). By combining different models, ensemble learning has higher prediction performance and is easier to implement, and the base learner is generally simple, so it can effectively avoid overfitting problems. However, since ensemble learning is a multi-model hybrid structure, the prediction results of ensemble learning models are often difficult to understand, i.e., the interpretability is relatively low. Among them, random forest and extreme gradient boosting tree are two ensemble learning algorithms widely used in the field of lithium batteries.

Random forest is a self-help ensemble learning method based on decision tree model. The model prediction steps of random forest are as follows: first, the data of the dataset is put back and randomly selected, and some features are randomly selected as the input of training to obtain N sub-datasets. Then, the final N base learners are obtained by training based on the sub-datasets. When predicting, each base learner will output its own prediction to generate an N-dimensional vector; finally, the average value of the data in the vector or vote is calculated to produce the final prediction result. Due to the random division of datasets, random forest is highly adaptable to class-imbalanced data and can handle a large number of data inputs. At present, the random forest algorithm is mainly applied for battery capacity prediction, battery health status (SOH) evaluation, and so on.

The extreme gradient boosting tree has made many improvements on the basis of the gradient boosting tree. For example, the regularization term is added to the loss function, and the column sampling is used when training the base learner, which can effectively prevent the overfitting of the learner and supplement the missing data. For missing data, the limit gradient boosting tree will assign it to different subtrees and calculate the information gain. Finally, the subtree with the highest information gain is used as the classification of the missing value. For overfitting, the extreme gradient boosting tree uses methods such as learning rate shrinkage, early stop optimization, and constituency tree construction to avoid overfitting of the model, and has higher prediction accuracy than other traditional models. Aiming at the problem of interface resistance between metal/semiconductor heterojunctions, Chen et al. [102] established a complete set of heterojunctions with a sample size of 1,092 based on the periodic table of elements, and predicted the Schottky barrier between these heterojunctions by combining the first-principles calculation and the limit gradient boosting tree model. Finally, they screened out six kinds of two-dimensional van der Waals metal/semiconductor heterojunctions with ohmic contact and high tunneling probability. This study inspired us with the idea that the limit gradient boosting tree should be very suitable for studying the interface problems of solid-state batteries.

2.4.1.5 Clustering (Group Analysis)

Clustering, also known as group analysis, refers to the process of gathering samples without categories into different groups according to the principle of "clustering by category." Such sets of data objects are called clusters, and each such cluster is described. Its purpose is to divide the data into several groups, so that the similarity between data in each group is as large as possible, and the similarity between different groups is as small as possible. As a typical unsupervised learning algorithm, it can automatically group data without knowing the label of data in advance. It is usually used for exploratory analysis of data and discovery of hidden patterns in data. Common clustering model algorithms include K-means, hierarchical clustering, density clustering, and other algorithms.

The development of inorganic fast ionic conductors with high ion transport performance is an important means to promote the rapid development of all-solid-state batteries. At present, inorganic fast ionic conductors with high ion transport properties are mainly concentrated in several types of inorganic compounds such as sulfides, oxides, and halides. Most inorganic compounds lack conductivity data. However, the conventional supervised learning model requires a large amount of

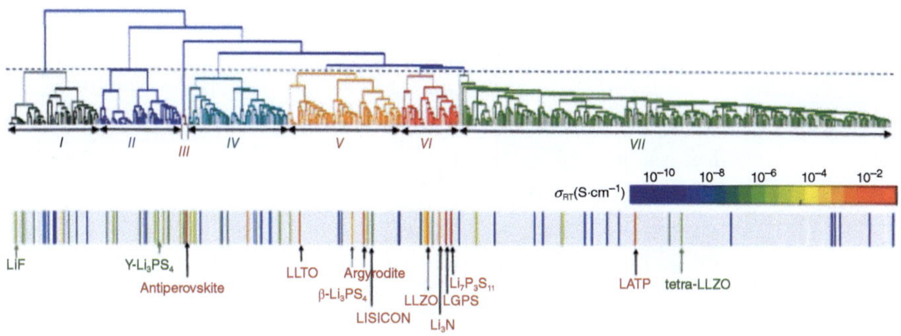

Figure 2.16 The lithium-containing compounds are grouped by clustering method, and the schematic diagram of new fast ionic conductor materials is screened out. *Source*: [103] / Springer Nature / CC BY 4.0.

data to train the model in order to achieve better performance prediction. As a typical unsupervised learning method, clustering does not need to use a large amount of conductivity data, which can reduce the challenge of data scarcity. Based on this, researchers from Toyota Research Institute in North America and the University of Maryland [103] used the clustering method, employing the XRD pattern in the ICSD inorganic crystal structure database as the characteristic parameter, and the hierarchical clustering method to screen and classify lithium-containing materials with similar mXRD representation for screening new inorganic fast ionic conductors with high ion transport performance, as shown in Figure 2.16. This clustering method can be used to successfully group lithium-containing materials with high lithium ionic conductivity and low lithium ionic conductivity, and select new material systems with high ion transport performance from the groups.

2.4.2 Databases

The establishment of the battery materials database can lay a solid foundation for the application of artificial intelligence technologies such as machine learning and data mining in batteries. At present, the data sources of the materials database include experimental data and computational data. In addition to the large amount of data that can be collected from the literature and existing data, high-throughput experiments and calculations also provide increasingly considerable data. The collected data usually needs to be classified according to its acquisition method, accuracy, and related physical properties, and the corresponding labels are matched for the data to realize the establishment of the database (Figure 2.17). After that, the eligible materials can be directly screened according to the application requirements, and the machine learning algorithm can be used to mine the relationship between the macroscopic properties and microstructure of the materials.

The data of battery materials mainly come from two aspects: experiments and calculations. Among the many properties of battery materials, lithium deintercalation/intercalation potential, thermodynamic stability, and chemical

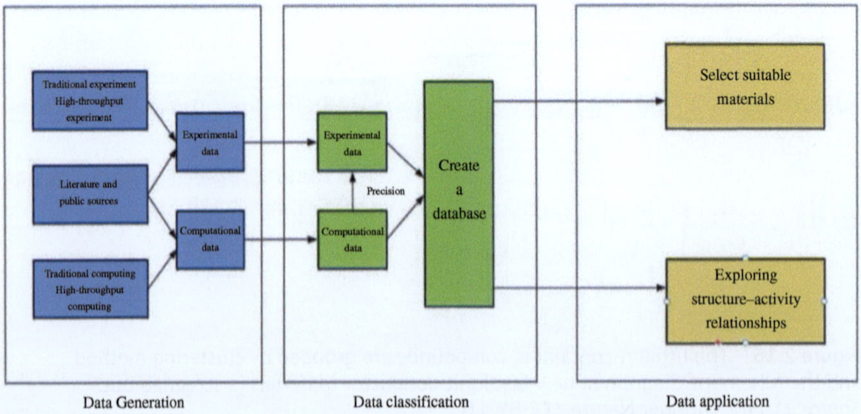

Figure 2.17 Schematic diagram of material data generation, classification, and application process [93].

stability can be obtained from energy, electronic structure, and other information calculated by DFT. Therefore, the general materials database containing high-throughput DFT calculation results can provide a large number of datasets for the battery materials database. At present, a materials database has been launched containing theoretical calculation information such as system energy, band structure, mechanical modulus, and thermodynamic phase diagram. The source of the material structure includes both the existing materials in the inorganic crystal database and a large number of virtual structures derived from the existing materials.

In addition to obtaining the battery material data from the general materials database, there are also databases for specific properties of the battery materials. Among them, the lithium-ion transport kinetics database calculated by geometric and semiempirical methods is the main one, including the battery material ion transport database launched by the Institute of Physics of the Chinese Academy of Sciences and the ion transport characteristic database of Shanghai University.

The collection and collation of experimental data mainly come from various published works in the literature. In the early stages of the development of the battery materials database, Ghadbeigi et al. [104] manually collected a large amount of battery material data from the scientific literature and constructed a database. With the development of computer technology, Huang and Cole [105] established a scheme for automatically collecting battery data from the literature by using ChemDataExtractor written by themselves, and constructed a database with a total of more than 290,000 entries, including capacity, internal resistance, Coulomb efficiency, energy density, and voltage.

In the process of data acquisition, it is necessary to pay attention to the conditions of data generation and the error of data. For experimental data, the measurement environment (including temperature, pressure, etc.) and measurement methods often affect the size of the value, so the subsequent data mining needs to classify the data, and the data measured under the same conditions can be compared more scientifically. For theoretical simulation data, it is easier to set the same simulation parameters. For example, in high-throughput calculations based on DFT, data accuracy can be controlled in the same range by setting the same correlation function, integral density, and convergence conditions. The comprehensive and accurate data type is the basis for large-scale analysis and mining of battery material data. Therefore, the database combining experimental data and computational data is an important direction for future database development.

References

1 Lü, C.-Y., Chen, Y.-W., Xie, M.-T., Li, X.-Y., Yu, H.-Y., Zhong, Y. and Xiang, H.-J. (2023) First-principles calculation method for periodic system under external electromagnetic field. *Acta Physica Sinica*, 72(23), 237102.

2 Wu, M., Wen, Z., Liu, Y., Wang, X. and Huang, L. (2011) Electrochemical behaviors of a Li_3N modified Li metal electrode in secondary lithium batteries. *Journal of Power Sources*, 196(19), 8091–8097.

3 Feng, W.-L., Wang, F., Zhou, X., Ji, X., Han, F.-D. and Wang, C.-S. (2020) Stability of interphase between solid state electrolyte and electrode. *Acta Physica Sinica*, 69(22), 228206.

4 Cheng, Z., Liu, T., Zhao, B., Shen, F., Jin, H. and Han, X. (2021) Recent advances in organic-inorganic composite solid electrolytes for all-solid-state lithium batteries. *Energy Storage Materials*, 34, 388–416.

5 Zhang, D., Xu, X., Qin, Y., Ji, S., Huo, Y., Wang, Z., Liu, Z., Shen, J. and Liu, J. (2019) Recent progress in organic–inorganic composite solid electrolytes for all-solid-state lithium batteries. *Chemistry—A European Journal*, 26(8), 1720–1736.

6 Levi, E., Levi, M. D., Salitra, G., Aurbach, D., Oesten, R., Heider, U. and Heider, L. (1999) Electrochemical and in-situ XRD characterization of $LiNiO_2$ and $LiCo_{0.2}Ni_{0.8}O_2$ electrodes for rechargeable lithium cells. *Solid State Ionics*, 126, 97–108.

7 Zhu, K., Wu, T., Zhu, Y., Li, X., Li, M., Lu, R., Wang, J., Zhu, X. and Yang, W. (2017) Layered Fe-substituted $LiNiO_2$ electrocatalysts for high-efficiency oxygen evolution reaction. *ACS Energy Letters*, 2(7), 1654–1660.

8 Xu, H., Yu, Y., Wang, Z. and Shao, G. (2019) First principle material genome approach for all solid-state batteries. *Energy & Environmental Materials*, 2(4), 234–250.

9 Oganov, A. R. and Glass, C. W. (2006) Crystal structure prediction using ab initio evolutionary techniques: principles and applications. *The Journal of Chemical Physics*, 124(24), 244704.

10 Glass, C. W., Oganov, A. R. and Hansen, N. (2006). USPEX—evolutionary crystal structure prediction. *Computer Physics Communications*, 175 (11–12), 713–720.

11 Parlinski, K., Li, Z. Q. and Kawazoe, Y. (1997) First-principles determination of the soft mode in cubic ZrO_2. *Physical Review Letters*, 78(21), 4063–4066.

12 Togo, A., Oba, F. and Tanaka, I. (2008) First-principles calculations of the ferroelastic transition between rutile-type and $CaCl_2$-type SiO_2 at high pressures. *Physical Review B*, 78(13), 134106.

13 Jain, A., Ong, S. P., Hautier, G., Chen, W., Richards, W. D., Dacek, S., Cholia, S., Gunter, D., Skinner, D., Ceder, G. and Persson, K. A. (2013) Commentary: The Materials Project: a materials genome approach to accelerating materials innovation. *APL Materials*, 1(1), 011002.

14 Jain, A., Persson, K. A. and Ceder, G. (2016) Research update: the materials genome initiative: data sharing and the impact of collaborative *ab initio* databases. *APL Materials*, 4(5), 053102.

15 Kraft, M. A., Culver, S. P., Calderon, M., Böcher, F., Krauskopf, T., Senyshyn, A., Dietrich, C., Zevalkink, A., Janek, J. and Zeier, W. G. (2017) Influence of lattice polarizability on the ionic conductivity in the lithium superionic argyrodites Li_6PS_5X (X = Cl, Br, I). *Journal of the American Chemical Society*, 139(31), 10909–10918.

16 Culver, S. P., Koerver, R., Krauskopf, T. and Zeier, W. G. (2018) Designing ionic conductors: the interplay between structural phenomena and interfaces in thiophosphate-based solid-state batteries. *Chemistry of Materials*, 30(13), 4179–4192.

17 Krauskopf, T., Pompe, C., Kraft, M. A. and Zeier, W. G. (2017) Influence of lattice dynamics on Na^+ transport in the solid electrolyte $Na_3PS_{4-x}Se_x$. *Chemistry of Materials*, 29(20), 8859–8869.

18 Muy, S., Bachman, J. C., Giordano, L., Chang, H.-H., Abernathy, D. L., Bansal, D., Delaire, O., Hori, S., Kanno, R., Maglia, F., Lupart, S., Lamp, P. and Yang, S.-H. (2018) Tuning mobility and stability of lithium ion conductors based on lattice dynamics. *Energy & Environmental Science*, 11(4), 850–859.

19 Wang, Z., Xu, H., Xuan, M. and Shao, G. (2018) From anti-perovskite to double anti-perovskite: tuning lattice chemistry to achieve super-fast Li^+ transport in cubic solid lithium halogen–chalcogenides. *Journal of Materials Chemistry A*, 6(1), 73–83.

20 Wang, Z., Deng, M., Xia, X., Gao, Y. and Shao, G. (2018) Fundamental basis for distinctive sensing of H_2 in humid environment. *Energy & Environmental Materials*, 1(3), 174–178.

21 Wang, Z. and Shao, G. (2018) High-capacity cathodes for magnesium lithium chlorine tri-ion batteries through chloride intercalation in layered MoS_2: a computational study. *Journal of Materials Chemistry A*, 6(16), 6830–6839.

22 Noriaki, K., Kenji, H., Yuichiro, Y., Masaaki, H., Ryoji, K., Masao, Y., Takashi, K., Yuki, K., Shigenori, H., Koji, K. and Akio, M. (2011) A lithium superionic conductor. *Nature Materials*, 10(9), 682–686.

23 Kuhn, A., Duppel, V. and Lotsch, B. V. (2013) Tetragonal $Li_{10}GeP_2S_{12}$ and Li_7GePS_8—exploring the Li ion dynamics in LGPS Li electrolytes. *Energy & Environmental Science*, 6(12), 3548–3552.

24 Mo, Y., Ong, S. P. and Ceder, G. (2011) First principles study of the $Li_{10}GeP_2S_{12}$ lithium super ionic conductor material. *Chemistry of Materials*, 24(1), 15–17.

25 Kato, Y., Hori, S., Saito, T., Suzuki, K., Hirayama, M., Mitsui, A., Yonemura, M., Iba, H. and Kanno, R. (2016) High-power all-solid-state batteries using sulfide superionic conductors. *Nature Energy*, 1(4), 16030.

26 Hu, Y.-S. (2016) Batteries: Getting solid. *Nature Energy*, 1(4), 16042.

27 Miara, L. J., Ong, S. P., Mo, Y., Richards, W. D., Park, Y., Lee, J.-M., Lee, H. S. and Ceder, G. (2013) Effect of Rb and Ta doping on the ionic conductivity and stability of the garnet $Li_{7+2x-y}(La_{3-x}Rb_x)(Zr_{2-y}Ta_y)O_{12}$ ($0 \leq x \leq 0.375$, $0 \leq y \leq 1$) superionic conductor: a first principles investigation. *Chemistry of Materials*, 25(15), 3048–3055.

28 Rettenwander, D., Blaha, P., Laskowski, R., Schwarz, K., Bottke, P., Wilkening, M., Geiger, C. A. and Amthauer, G. (2014) DFT study of the role of Al^{3+} in the fast ion-conductor $Li_{7-3x}Al^{3+}{}_xLa_3Zr_2O_{12}$ garnet. *Chemistry of Materials*, 26(8), 2617–2623.

29 Squires, A. G., Scanlon, D. O. and Morgan, B. J. (2019) Native defects and their doping response in the lithium solid electrolyte $Li_7La_3Zr_2O_{12}$. *Chemistry of Materials*, 32(5), 1876–1886.

30 Wang, Y., Richards, W. D., Ong, S. P., Miara, L. J., Kim, J. C., Mo, Y. and Ceder, G. (2015) Design principles for solid-state lithium superionic conductors. *Nature Materials*, 14(10), 1026–1031.

31 Richards, W. D., Wang, Y., Miara, L. J., Kim, J. C. and Ceder, G. (2016) Design of $Li_{1+2x}Zn_{1-x}PS_4$, a new lithium ion conductor. *Energy & Environmental Science*, 9(10), 3272–3278.

32 Kaup, K., Lalère, F., Huq, A., Shyamsunder, A., Adermann, T., Hartmann, P. and Nazar, L. F. (2018) Correlation of structure and fast ion conductivity in the solid solution series $Li_{1+2x}Zn_{1-x}PS_4$. *Chemistry of Materials*, 30(3), 592–596.

33 Di Stefano, D., Miglio, A., Robeyns, K., Filinchuk, Y., Lechartier, M., Senyshyn, A., Ishida, H., Spannenberger, S., Prutsch, D., Lunghammer, S., Rettenwander, D., Wilkening, M., Roling, B., Kato, Y. and Hautier, G. (2019) Superionic diffusion through frustrated energy landscape. *Chem*, 5(9), 2450–2460.

34 Xu, Z., Chen, X., Chen, R., Li, X. and Zhu, H. (2020) Anion charge and lattice volume dependent lithium ion migration in compounds with fcc anion sublattices. *npj Computational Materials*, 6(1), 47.

35 He, X., Zhu, Y. and Mo, Y. (2017) Origin of fast ion diffusion in super-ionic conductors. *Nature Communications*, 8(1), 15893.

36 Ngai, K. L. (1998) Meyer–Neldel rule and anti Meyer–Neldel rule of ionic conductivity: conclusions from the coupling model. *Solid State Ionics*, 105, 231–235.

37 Krauskopf, T., Culver, S. P. and Zeier, W. G. (2018) Bottleneck of diffusion and inductive effects in $Li_{10}Ge_{1-x}Sn_xP_2S_{12}$. *Chemistry of Materials*, 30(5), 1791–1798.

38 Xu, Z.-M., Bo, S.-H. and Zhu, H. (2018) $LiCrS_2$ and $LiMnS_2$ cathodes with extraordinary mixed electron–ion conductivities and favorable interfacial compatibilities with sulfide electrolyte. *ACS Applied Materials & Interfaces*, 10(43), 36941–36953.

39 Morgan, D. and Ceder, G. (2005) High-throughput and data mining with ab initio methods. *Measurement Science & Technology*, 16(2), 296–301.

40 Ong, S. P., Richards, W. D., Jain, A., Hautier, G., Kocher, M., Cholia, S., Gunter, D., Chevrier, V. L., Persson, K. A. and Ceder, G. (2013) Python materials genomics (pymatgen): a robust, open-source python library for materials analysis. *Computational Materials Science*, 68, 314–319.

41 Jain, A., Ong, S. P., Chen, W., Medasani, B., Qu, X., Kocher, M., Brafman, M., Petretto, G., Rignanese, G. M., Hautier, G., Gunter, D. and Persson, K. A. (2015) FireWorks: a dynamic workflow system designed for high-throughput applications. *Concurrency and Computation: Practice and Experience*, 27(17), 5037–5059.

42 Curtarolo, S., Setyawan, W., Hart, G. L. W., Jahnatek, M., Chepulskii, R. V., Taylor, R. H., Wang, S., Xue, J., Yang, K., Levy, O., Mehl, M. J., Stokes, H. T., Demchenko, D. O. and Morgan, D. (2012) AFLOW: an automatic framework for high-throughput materials discovery. *Computational Materials Science*, 58, 218–226.

43 Kirklin, S., Saal, J. E., Meredig, B., Thompson, A., Doak, J. W., Aykol, M., Rühl, S. and Wolverton, C. (2015) The Open Quantum Materials Database (OQMD): assessing the accuracy of DFT formation energies. *npj Computational Materials*, 1(1), 15010.

44 Pizzi, G., Cepellotti, A., Sabatini, R., Marzari, N. and Kozinsky, B. (2016) AiiDA: automated interactive infrastructure and database for computational science. *Computational Materials Science*, 111, 218–230.

45 Ghiringhelli, L. M., Carbogno, C., Levchenko, S., Mohamed, F., Huhs, G., Lüders, M., Oliveira, M. and Scheffler, M. (2017) Towards efficient data exchange and sharing for big-data driven materials science: metadata and data formats. *npj Computational Materials*, 3(1), 46.

46 Alberi, K., Nardelli, M. B., Zakutayev, A., Mitas, L., Curtarolo, S., Jain, A., Fornari, M., Marzari, N., Takeuchi, I., Green, M. L., Kanatzidis, M., Toney, M. F., Butenko, S., Meredig, B., Lany, S., Kattner, U., Davydov, A., Toberer, E. S., Stevanovic, V., Walsh, A., Park, N.-G., Aspuru-Guzik, A., Tabor, D. P., Nelson, J., Murphy, J., Setlur, A., Gregoire, J., Li, H., Xiao, R., Ludwig, A., Martin, L. W., Rappe, A. M., Wei, S.-H. and Perkins, J. (2019) The 2019 materials by design roadmap. *Journal of Physics D: Applied Physics*, 52(1), 013001.

47 Wang, X., Xiao, R., Li, H. and Chen, L. (2018) Discovery and design of lithium battery materials via high-throughput modeling. *Chinese Physics B*, 27(12), 128801.

48 Gao, J., Chu, G., He, M., Zhang, S., Xiao, R., Li, H. and Chen, L. (2014) Screening possible solid electrolytes by calculating the conduction pathways using bond valence method. *Science China Physics, Mechanics & Astronomy*, 57(8), 1526–1536.

49 He, B., Chi, S., Ye, A., Mi, P., Zhang, L., Pu, B., Zou, Z., Ran, Y., Zhao, Q., Wang, D., Zhang, W., Zhao, J., Adams, S., Avdeev, M. and Shi, S. (2020) High-throughput screening platform for solid electrolytes combining hierarchical ion-transport prediction algorithms. *Scientific Data*, 7(1), 151.

50 Zhang, L., He, B., Zhao, Q., Zou, Z., Chi, S., Mi, P., Ye, A., Li, Y., Wang, D., Avdeev, M., Adams, S. and Shi, S. (2020) A database of ionic transport characteristics for over 29 000 inorganic compounds. *Advanced Functional Materials*, 30(35), 2003087.

51 Katcho, N. A., Carrete, J., Reynaud, M., Rousse, G., Casas-Cabanas, M., Mingo, N., Rodriguez-Carvajal, J. and Carrasco, J. (2019) An investigation of the structural properties of Li and Na fast ion conductors using high-throughput bond-valence calculations and machine learning. *Journal of Applied Crystallography*, 52(1), 148–157.

52 Sendek, A. D., Yang, Q., Cubuk, E. D., Duerloo, K.-A. N., Cui, Y. and Reed, E. J. (2017) Holistic computational structure screening of more than 12 000 candidates for solid lithium-ion conductor materials. *Energy & Environmental Science*, 10(1), 306–320.

53 Li, X., Liang, J., Chen, N., Luo, J., Adair, K. R., Wang, C., Banis, M. N., Sham, T. K., Zhang, L., Zhao, S., Lu, S., Huang, H., Li, R. and Sun, X. (2019) Water-mediated synthesis of a superionic halide solid electrolyte. *Angewandte Chemie International Edition*, 58(46), 16427–16432.

54 Li, X., Liang, J., Luo, J., Norouzi Banis, M., Wang, C., Li, W., Deng, S., Yu, C., Zhao, F., Hu, Y., Sham, T.-K., Zhang, L., Zhao, S., Lu, S., Huang, H., Li, R., Adair, K. R. and Sun, X. (2019) Air-stable Li_3InCl_6 electrolyte with high voltage compatibility for all-solid-state batteries. *Energy & Environmental Science*, 12(9), 2665–2671.

55 Kahle, L., Marcolongo, A. and Marzari, N. (2020) High-throughput computational screening for solid-state Li-ion conductors. *Energy & Environmental Science*, 13(3), 928–948.

56 Brown, I. D. (2009) Recent developments in the methods and applications of the bond valence model. *Chemical Reviews*, 109(12), 6858–6919.

57 Xiao, R., Li, H. and Chen, L. (2015) High-throughput design and optimization of fast lithium ion conductors by the combination of bond-valence method and density functional theory. *Scientific Reports*, 5(1), 14227.

58 Zhang, Z. and Lei, T. M. (2011) First-principles study of N-doped 4H-SiC. Master's thesis. Xidian University.

59 Du, Y.-T., Gan, L.-Y. and Yang, F. (2019) Theoretical exploration of MXene/graphene heterostructures as potential electrodes for ion batteries. Master's thesis. Southwest Jiaotong University.

60 Hu, X. Y. and Xia, C. R. (2021) Study on cathode of solid oxide fuel cell with carbon dioxide tolerance. Master's thesis. University of Science and Technology of China.

61 Xu, H., Yu, Y., Wang, Z. and Shao, G. (2019) A theoretical approach to address interfacial problems in all-solid-state lithium ion batteries: tuning materials chemistry for electrolyte and buffer coatings based on Li_6PA_5Cl hali-chalcogenides. *Journal of Materials Chemistry A*, 7(10), 5239–5247.

62 Zhu, Y., He, X. and Mo, Y. (2015) Origin of outstanding stability in the lithium solid electrolyte materials: insights from thermodynamic analyses based on first-principles calculations. *ACS Applied Materials & Interfaces*, 7(42), 23685–23693.

63 Xu, H., Xuan, M., Xiao, W., Shen, Y., Li, Z., Wang, Z., Hu, J. and Shao, G. (2019) Lithium ion conductivity in double antiperovskite $Li_{6.5}OS_{1.5}I_{1.5}$: alloying and boundary effects. *ACS Applied Energy Materials*, 2(9), 6288–6294.

64 Wenzel, S., Leichtweiss, T., Krüger, D., Sann, J. and Janek, J. (2015) Interphase formation on lithium solid electrolytes—an in situ approach to study interfacial reactions by photoelectron spectroscopy. *Solid State Ionics*, 278, 98–105.

65 Persson, B. N. J. (2006) Contact mechanics for randomly rough surfaces. *Surface Science Reports*, 61(4), 201–227.

66 Koerver, R., Aygün, I., Leichtweiß, T., Dietrich, C., Zhang, W., Binder, J. O., Hartmann, P., Zeier, W. G. and Janek, J. (2017) Capacity fade in solid-state batteries: interphase formation and chemomechanical processes in nickel-rich layered oxide cathodes and lithium thiophosphate solid electrolytes. *Chemistry of Materials*, 29(13), 5574–5582.

67 Koerver, R., Zhang, W., de Biasi, L., Schweidler, S., Kondrakov, A. O., Kolling, S., Brezesinski, T., Hartmann, P., Zeier, W. G. and Janek, J. (2018) Chemo-mechanical expansion of lithium electrode materials—on the route to mechanically optimized all-solid-state batteries. *Energy & Environmental Science*, 11(8), 2142–2158.

68 Shi, T., Zhang, Y.-Q., Tu, Q., Wang, Y., Scott, M. C. and Ceder, G. (2020) Characterization of mechanical degradation in an all-solid-state battery cathode. *Journal of Materials Chemistry A*, 8(34), 17399–17404.

69 Tian, H.-K. and Qi, Y. (2017) Simulation of the effect of contact area loss in all-solid-state Li-ion batteries. *Journal of The Electrochemical Society*, 164(11), E3512–E3521.

70 Bucci, G., Talamini, B., Balakrishna, A. R., Chiang, Y.-M. and Carter, W. C. (2018) Mechanical instability of electrode-electrolyte interfaces in solid-state batteries. *Physical Review Materials*, 2(10), 105407.

71 Yamamoto, M., Takahashi, M., Terauchi, Y., Kobayashi, Y., Ikeda, S. and Sakuda, A. (2017) Fabrication of composite positive electrode sheet with high active material content and effect of fabrication pressure for all-solid-state battery. *Journal of the Ceramic Society of Japan*, 125(5), 391–395.

72 Choi, S., Jeon, M., Ahn, J., Jung, W. D., Choi, S. M., Kim, J.-S., Lim, J., Jang, Y.-J., Jung, H.-G., Lee, J.-H., Sang, B.-I. and Kim, H. (2018) Quantitative analysis of microstructures and reaction interfaces on composite cathodes in all-solid-state batteries using a three-dimensional reconstruction technique. *ACS Applied Materials & Interfaces*, 10(28), 23740–23747.

73 Wang, M. J., Choudhury, R. and Sakamoto, J. (2019) Characterizing the Li-solid-electrolyte interface dynamics as a function of stack pressure and current density. *Joule*, 3(9), 2165–2178.

74 Krauskopf, T., Mogwitz, B., Rosenbach, C., Zeier, W. G. and Janek, J. (2019) Diffusion limitation of lithium metal and Li–Mg alloy anodes on LLZO type solid electrolytes as a function of temperature and pressure. *Advanced Energy Materials*, 9(44), 1902568.

75 Lewis, J. A., Cortes, F. J. Q., Liu, Y., Miers, J. C., Verma, A., Vishnugopi, B. S., Tippens, J., Prakash, D., Marchese, T. S., Han, S. Y., Lee, C., Shetty, P. P., Lee, H.-W., Shevchenko, P., De Carlo, F., Saldana, C., Mukherjee, P. P. and McDowell, M. T. (2021) Linking void and interphase evolution to electrochemistry in solid-state batteries using operando X-ray tomography. *Nature Materials*, 20(4), 503–510.

76 Sharafi, A., Yu, S., Naguib, M., Lee, M., Ma, C., Meyer, H. M., Nanda, J., Chi, M., Siegel, D. J. and Sakamoto, J. (2017) Impact of air exposure and surface chemistry on Li–$Li_7La_3Zr_2O_{12}$ interfacial resistance. *Journal of Materials Chemistry A*, 5(26), 13475–13487.

77 Gao, J., Guo, X., Li, Y., Ma, Z., Guo, X., Li, H., Zhu, Y. and Zhou, W. (2019) The ab initio calculations on the areal specific resistance of Li-metal/$Li_7La_3Zr_2O_{12}$ interphase. *Advanced Theory and Simulations*, 2(6), 1900028.

78 Zheng, H., Wu, S., Tian, R., Xu, Z., Zhu, H., Duan, H. and Liu, H. (2019) Intrinsic lithiophilicity of Li–garnet electrolytes enabling high-rate lithium cycling. *Advanced Functional Materials*, 30(6), 1906189.

79 Haruyama, J., Sodeyama, K., Han, L., Takada, K. and Tateyama, Y. (2014) Space–charge layer effect at interface between oxide cathode and sulfide electrolyte in all-solid-state lithium-ion battery. *Chemistry of Materials*, 26(14), 4248–4255.

80 Gao, B., Jalem, R., Ma, Y. and Tateyama, Y. (2019) Li^+ transport mechanism at the heterogeneous cathode/solid electrolyte interface in an all-solid-state battery via the first-principles structure prediction scheme. *Chemistry of Materials*, 32(1), 85–96.

81 Wang, L., Xie, R., Chen, B., Yu, X., Ma, J., Li, C., Hu, Z., Sun, X., Xu, C., Dong, S., Chan, T.-S., Luo, J., Cui, G. and Chen, L. (2020) In-situ visualization of the space-charge-layer effect on interfacial lithium-ion transport in all-solid-state batteries. *Nature Communications*, 11(1), 5889.

82 Morgan, B. J. and Madden, P. A. (2011) Effects of lattice polarity on interfacial space charges and defect disorder in ionically conducting AgI heterostructures. *Physical Review Letters*, 107(20), 206102.

83 Stegmaier, S., Voss, J., Reuter, K. and Luntz, A. C. (2017) Li^+ defects in a solid-state Li ion battery: theoretical insights with a Li_3OCl electrolyte. *Chemistry of Materials*, 29(10), 4330–4340.

84 Fu, L., Chen, C.-C., Samuelis, D. and Maier, J. (2014) Thermodynamics of lithium storage at abrupt junctions: modeling and experimental evidence. *Physical Review Letters*, 112(20), 208301.

85 de Klerk, N. J. J. and Wagemaker, M. (2018) Space-charge layers in all-solid-state batteries; important or negligible? *ACS Applied Energy Materials*, 1(10), 5609–5618.

86 Zhang, Q., Pan, J., Lu, P., Liu, Z., Verbrugge, M. W., Sheldon, B. W., Cheng, Y.-T., Qi, Y. and Xiao, X. (2016) Synergetic effects of inorganic components in solid electrolyte interphase on high cycle efficiency of lithium ion batteries. *Nano Letters*, 16(3), 2011–2016.

87 Kasamatsu, S., Tada, T. and Watanabe, S. (2012) Parallel-sheets model analysis of space charge layer formation at metal/ionic conductor interfaces. *Solid State Ionics*, 226, 62–70.

88 Landstorfer, M., Funken, S. and Jacob, T. (2011) An advanced model framework for solid electrolyte intercalation batteries. *Physical Chemistry Chemical Physics*, 13, 12817–12825.

89 Braun, S., Yada, C. and Latz, A. (2015) Thermodynamically consistent model for space-charge-layer formation in a solid electrolyte. *The Journal of Physical Chemistry C*, 119, 22281–22288.

90 Swift, M. W. and Qi, Y. (2019) First-principles prediction of potentials and space-charge layers in all-solid-state batteries. *Physical Review Letters*, 122(16), 167701.

91 Im, C., Park, D., Kim, H. and Lee, J. (2018) Al-incorporation into $Li_7La_3Zr_2O_{12}$ solid electrolyte keeping stabilized cubic phase for all-solid-state Li batteries. *Journal of Energy Chemistry*, 27(5), 1501–1508.

92 Curtarolo, S., Morgan, D., Persson, K., Rodgers, J. and Ceder, G. (2003) Predicting crystal structures with data mining of quantum calculations. *Physical Review Letters*, 91(13), 135503.

93 Cui, G. (2024) *High Energy Density Solid-State Lithium Batteries*. Beijing: Chemical Industry Press.

94 Yoshida, T., Hongo, K. and Maezono, R. (2019) First-principles study of structural transitions in $LiNiO_2$ and high-throughput screening for long life battery. *The Journal of Physical Chemistry C*, 123(23), 14126–14131.

95 Severson, K. A., Attia, P. M., Jin, N., Perkins, N., Jiang, B., Yang, Z., Chen, M. H., Aykol, M., Herring, P. K., Fraggedakis, D., Bazant, M. Z., Harris, S. J., Chueh, W. C. and Braatz, R. D. (2019) Data-driven prediction of battery cycle life before capacity degradation. *Nature Energy*, 4(5), 383–391.

96 Jiang, B., Gent, W. E., Mohr, F., Das, S., Berliner, M. D., Forsuelo, M., Zhao, H., Attia, P. M., Grover, A., Herring, P. K., Bazant, M. Z., Harris, S. J., Ermon, S., Chueh, W. C. and Braatz, R. D. (2021) Bayesian learning for rapid prediction of lithium-ion battery-cycling protocols. *Joule*, 5(12), 3187–3203.

97 He, W., Williard, N., Chen, C. and Pecht, M. (2014) State of charge estimation for Li-ion batteries using neural network modeling and unscented Kalman filter-based error cancellation. *International Journal of Electrical Power & Energy Systems*, 62, 783–791.

98 Dixit, M. B., Verma, A., Zaman, W., Zhong, X., Kenesei, P., Park, J. S., Almer, J., Mukherjee, P. P. and Hatzell, K. B. (2020) Synchrotron imaging of pore formation in Li metal solid-state batteries aided by machine learning. *ACS Applied Energy Materials*, 3(10), 9534–9542.

99 Xie, T. and Grossman, J. C. (2018) Crystal graph convolutional neural networks for an accurate and interpretable prediction of material properties. *Physical Review Letters*, 120(14), 145301.

100 Xie, T., France-Lanord, A., Wang, Y., Lopez, J., Stolberg, M. A., Hill, M., Leverick, G. M., Gomez-Bombarelli, R., Johnson, J. A., Yang, S.-H. and Grossman, J. C. (2022) Accelerating amorphous polymer electrolyte screening by learning to reduce errors in molecular dynamics simulated properties. *Nature Communications*, 13(1), 3415.

101 Deringer, V. L., Bartók, A. P., Bernstein, N., Wilkins, D. M., Ceriotti, M. and Csányi, G. (2021) Gaussian process regression for materials and molecules. *Chemical Reviews*, 121(16), 10073–10141.

102 Chen, A., Wang, Z., Zhang, X., Chen, L., Hu, X., Han, Y., Cai, J., Zhou, Z. and Li, J. (2022) Accelerated mining of 2D van der Waals heterojunctions by integrating supervised and unsupervised learning. *Chemistry of Materials*, 34(12), 5571–5583.

103 Zhang, Y., He, X., Chen, Z., Bai, Q., Nolan, A. M., Roberts, C. A., Banerjee, D., Matsunaga, T., Mo, Y. and Ling, C. (2019) Unsupervised discovery of solid-state lithium ion conductors. *Nature Communications*, 10, 5260.

104 Ghadbeigi, L., Sparks, T. D., Harada, J. K. and Lettiere, B. R. (2015). Data-mining approach for battery materials. *Proceedings of the 2015 IEEE Conference on Technologies for Sustainability (SusTech)*, 239–244. New York: Curran Associates.

105 Huang, S. and Cole, J. M. (2020) A database of battery materials auto-generated using ChemDataExtractor. *Scientific Data*, 7(1), 260.

3

Designer Materials and Structure for High-performance All-solid-state Battery

3.1 Research on Cathode Materials

The development of all-solid-state lithium batteries (ASSLBs) represents a pivotal shift in energy storage technology, promising enhanced safety and energy density compared to conventional lithium-ion batteries (LIBs). However, the transition from liquid to solid electrolytes introduces significant challenges, particularly at the cathode interface, where sluggish ion transport, poor interfacial contact, and undesirable side reactions impede electrochemical performance. The cathode in ASSLBs is a composite structure typically comprising cathode active materials (CAMs), SEs, and conductive additives. Unlike liquid electrolytes that permeate porous electrodes and ensure intimate contact, solid electrolytes rely on point contacts with CAMs, leading to high interfacial resistance and limited active material utilization. This section comprehensively reviews recent advances in cathode material design, interfacial engineering, and electrode architectural strategies aimed at overcoming these limitations and achieving high-rate capability, long cycle life, and high energy density in ASSLBs.

3.1.1 Crystallographic and Morphological Engineering of CAMs

The intrinsic properties of CAMs, including crystallographic orientation, particle morphology, and grain boundary density, profoundly influence Li^+ transport kinetics and mechanical stability in ASSLBs. Layered oxide materials such as $LiCoO_2$ (LCO), $LiNi_xCo_yMn_zO_2$ (NCM), and $LiNi_xCo_yAl_zO_2$ (NCA) are widely studied due to their high capacity and operating voltage. However, their polycrystalline (PC) nature often leads to microcracking during cycling due to anisotropic volume changes, resulting in contact loss with the SE and rapid capacity fade. To address this, single-crystalline (SC) CAMs have emerged as a promising alternative. SC materials lack grain boundaries, which are typical sites for crack initiation and Li^+ transport obstruction. For instance, Wang et al. demonstrated that SC-NCM532 exhibited Li^+ diffusion coefficients 6–14 times higher than its PC counterpart in ASSLBs using $Li_{10}GeP_2S_{12}$ (LGPS) as the SEs [1]. The absence of grain boundaries in SC materials facilitates continuous Li^+ pathways, enhancing rate capability and cycling stability. However, SC CAMs often exhibit lower specific surface area, which may limit interfacial reaction sites. Thus, optimizing particle size and morphology is critical to balancing Li^+ diffusion length and active surface area. Furthermore, concentration-gradient designs, such as radially aligned rod-shaped grains in FCG75 NCM, have shown improved mechanical integrity and Li^+ transport compared to randomly oriented grains [2]. These designs mitigate strain-induced cracking by accommodating volume changes more effectively, thereby maintaining interfacial contact during cycling.

3.1.2 The Role of Conducting Additives and Electronic Percolation

Electronic conductivity within the cathode composite is essential for achieving high-rate performance. In conventional LIBs, carbon additives such as carbon black (CB) and vapor-grown carbon fiber (VGCF) are used to form percolating networks that facilitate electron transport. However, in ASSLBs, carbon materials can catalyze the decomposition of sulfide-based SEs, leading to the formation of insulating by-products (e.g., Li_2S, Li_3P) and increased interfacial resistance. Zhang et al. systematically investigated the interfacial stability between LGPS and various carbon additives and found that decomposition was not directly correlated with specific surface area but rather with the intrinsic electronic properties and microstructure of carbon [3]. For instance, one-dimensional (1D) VGCF with a lower surface area (24 $m^2\ g^{-1}$) showed less severe decomposition compared to 0D CB (80 $m^2\ g^{-1}$), attributed to reduced contact area with the SEs.

To mitigate these side reactions, surface coating of carbon additives has been explored. Deng et al. applied a poly(3,4-ethylenedioxythiophene) (PEDOT) coating on carbon nanotubes via molecular layer deposition, which significantly suppressed LGPS decomposition while maintaining electronic percolation [4]. Similarly, Al_2O_3-coated VGCF via atomic layer deposition (ALD) improved cyclability and capacity retention in NCM-based ASSLBs [5]. These coatings act as electronic insulators while allowing Li^+ transport, thereby decoupling electronic and ionic pathways and reducing parasitic reactions. Moreover, the morphology of conductive additives influences the uniformity of the electrode microstructure. Randau et al. demonstrated that 1D fibers form more effective percolating networks than 0D particles, reducing the risk of isolated CAMs and improving rate capability. The optimal content of conductive additives is typically around 3 wt.% for VGCF, balancing electronic conductivity and minimized side reactions.

3.1.3 Interfacial Engineering and Li^+ Percolation Pathways

Achieving intimate and stable interfacial contact between CAMs and SEs is paramount for efficient Li^+ transport. Conventional composite cathodes prepared by simple mixing and cold pressing often exhibit high porosity (20–30%) and poor contact, leading to high charge-transfer resistance. To address this, various coating strategies have been developed to form core–shell structures where CAM particles are uniformly coated with a thin SE layer. For example, Park et al. used a liquid-phase synthesis to coat LCO with 0.4LiI–0.6Li_3SnS_4, achieving full coverage and significantly improved rate performance compared to physically mixed electrodes [6]. The coated electrodes exhibited larger contact areas and more continuous Li^+ pathways, reducing interfacial resistance.

Dry coating methods, such as mechanical milling, offer a solvent-free alternative for SE coating. Nakamura et al. prepared Li_3PS_4 (LPS)-coated NCM particles (NCM@LPS) through high-energy ball milling, resulting in a conformal SE layer that enhanced Li^+ percolation and enabled high-rate operation up to 5C without carbon additives [7]. This approach not only improves interfacial contact but also allows for reduced SE content in the composite, increasing energy density.

Beyond material-level modifications, electrode architecture plays a critical role in Li^+ percolation. The use of three-dimensional (3D) current collectors or patterned electrodes can enlarge the interfacial area and reduce local current density. For instance, Xu et al. demonstrated that 3D-micropatterned SE layers improved Li^+ flux and reduced void formation at the interface, enhancing rate capability and cycle life. These structural innovations highlight the importance of designing electrodes with optimized tortuosity and porosity to facilitate rapid ion transport [8].

Significant progress has been made in the design and engineering of cathode materials for ASSLBs, yet challenges remain in achieving commercial-level performance. The ideal cathode should combine SC or facet-controlled CAMs to minimize cracking, coated conductive additives to ensure electronic percolation without side reactions, and intimate SE/CAM interfaces optimized through advanced coating or infiltration techniques. Electrode fabrication must evolve toward scalable, solvent-free processes that maintain interfacial integrity and high ionic conductivity. Future research should focus on understanding the fundamental electrochemomechanical interactions at interfaces, developing multifunctional binders, and integrating computational tools like machine learning to optimize electrode architectures. With continued innovation in material synthesis and electrode engineering, ASSLBs are poised to meet the demands of high-energy and high-power applications, including electric vehicles (EVs) and grid storage.

3.2 Types of Common Cathode Materials

3.2.1 Structural Fundamentals, Modification Strategies of LCO Cathodes, and Applications in Solid-state Batteries

Rechargeable LIBs have reshaped modern energy storage systems, powering consumer electronics, electric mobility, and grid energy storage since their commercialization in 1991 [9]. Among the various cathode materials, layered lithium cobalt oxide (LCO) stands as the most successful prototype, first proposed by Goodenough and coworkers in 1980 for its reversible Li^+ intercalation and deintercalation behavior, which enabled battery working voltages to exceed 4 V (vs. Li^+/Li) and eliminated the reliance on unsafe Li metal anodes [10]. Sony's adoption of LCO in the first commercial LIBs marked a critical milestone, driving the rapid development of LIB technologies and materials over the past decades [9]. Despite the emergence of alternatives such as $LiFePO_4$ (LFP), $LiMn_2O_4$ (LMO), Ni-rich ternary oxides (NCA/NCM), and Li- and Mn-rich layered oxides (LMRO), LCO remains extensively applied in major commercial LIBs, particularly in the consumer electronics market. Its distinct advantages include high working voltage, superior volumetric energy density (e.g., 2,812 Wh L^{-1} at a cutoff voltage of 4.4 V, compared to ≈2,600 Wh L^{-1} for NCM811 and ≈2,700 Wh L^{-1} for NCA), excellent conductivity, ease of preparation, and mature manufacturing processes [11]. Moreover, LCO has become a model material for exploring advanced battery technologies like solid-state electrolytes (SSEs) and fast charging, highlighting its enduring significance in energy storage research [12].

LCO possesses an α-$NaFeO_2$-type layered structure with the *R*3*m* space group, featuring unit-cell parameters of $a = 2.816$ Å and $c = 14.08$ Å [13]. In this structure, oxygen atoms form a cubic closed-packed (ccp) arrangement at the 6c site, while Li^+ and Co^{3+} are orderly located at the 3b and 3a sites in the (111) plane, respectively, coordinating with oxygen to form edge-shared LiO_6 and CoO_6 octahedra [14]. The theoretical density of LCO, calculated from its crystal structure, is ≈5.05 g cm^{-3}, and its practical compact density can reach ≈4.2 g cm^{-3} due to high thermal stability (up to 1,000 °C) and low formation energy, which facilitates the synthesis of large crystallites (≈ 10 μm) [15]. Electronically, Co^{3+} in LCO exhibits a low-spin electronic configuration ($3d^6$: $t_{2g}^6e_g^0$) with no unpaired electrons, contributing to LCO's nonmagnetic property [16]. The strong covalent interaction between Co^{3+}-3d and O^{2-}-2p states induces short Co—O bond lengths (≈1.92 Å), enhancing structural rigidity and promoting fast 2D Li^+ diffusion via a tetrahedral-site hop mechanism [17]. However, LCO's application in high-energy-density batteries is hindered by intrinsic high-voltage instability. As operating voltage increases to boost capacity and energy density (e.g., 220 mAh g^{-1} at 4.6 V vs. 140 mAh g^{-1} at 4.2 V), LCO undergoes a series of irreversible bulk phase transitions (H1 → H2 → M1 → H3 → H1-3 → O1) accompanied by significant anisotropic lattice strain—for instance, the *c*-axis length shortens by ≈5.3% from H3 to H1-3 and a further ≈5.9% upon transition to O1 [18]. Additionally, twin boundaries (even in pristine LCO particles) act as initiation points for cracks and irreversible phase transitions, with coherent twin boundaries exhibiting an intersecting angle that expands to ≈121° at high voltages (>4.4 V) and inducing void formation [19]. CoO_2 layer gliding, driven by dislocation dissociation, further exacerbates structural instability, leading to irreversible phase transitions to spinel or rock-salt phases, especially at voltages above 4.7 V [20]. Surface instability is another critical issue: interfacial side reactions between LCO and electrolytes form a cathode electrolyte interface (CEI) layer composed of inorganic (e.g., LiF, Li_2CO_3) and organic components, which thickens with cycles and decomposes above 4.5 V [21, 22]. High-valent Co^{4+} on the LCO surface is reduced to Co^{2+}, causing Co dissolution that not only degrades the cathode but also damages the solid electrolyte interface (SEI) on the anode [23]. Oxygen redox, triggered by deep delithiation ($x \leq 0.5$ in Li_xCoO_2), leads to lattice oxygen oxidation

($O_2^- \rightarrow O_2^{n-}$) and potential O release, further compromising structural stability and safety [24, 25].

To address these challenges, extensive research has focused on structural modification strategies for LCO. Elemental doping is effective for stabilizing the bulk structure: Ni doping expands the LCO lattice, forms Li/Ni anti-sites that act as pillars to inhibit CoO layer gliding, and suppresses unfavorable H1-3 and O1 phase transitions [26]; Ti doping, despite its low solubility (theoretical maximum of 0.15%), segregates at grain boundaries to relieve intragranular stress and stabilize the surface by restraining oxygen activity [27]; and Mg doping, either at Co or Li sites, enhances Li^+ diffusivity and electronic conductivity, with Mg at Li sites (due to comparable ionic radius, ≈0.72 Å) forming a "pillaring" effect to prevent CoO_2 layer gliding [28]. Multielement co-doping, such as Ti–Mg–Al co-doping (TMA–LCO), integrates the advantages of individual dopants: Mg and Al stabilize the bulk by inhibiting adverse phase transitions, while Ti segregates at grain boundaries to modify microstructure and suppress interfacial side reactions. Surface modification, another key strategy, includes both electrochemically inert and active coatings. Inert coatings like $LiAlO_2$ form a dual-layer structure ($LiAlO_2/LiCo_{1-x}Al_xO_2$) upon calcination, where $LiAlO_2$ acts as a physical barrier against electrolytes and $LiCo_{1-x}Al_xO_2$ promotes Li^+ diffusion [29]. Active coatings, such as $LiCoPO_4$, epitaxially grow on LCO surfaces, maintaining structural integrity even after long-term cycling at 4.6 V and inhibiting nanovoid formation and irreversible spinel transitions [30]. Additionally, Al^{3+}, F^-, and PO_4^{3-} co-modification can form a spinel-like coherent coating, further facilitating Li^+ diffusion kinetics and surface stabilization [31].

The development of ASSLBs has opened new avenues for LCO applications, as SSEs address the safety issues of liquid electrolytes and complement LCO's high energy density. LCO exhibits good compatibility with various SSEs, such as Y- and Ta-doped $Li_7La_3Zr_2O_{12}$ ($Li_{6.75}La_{2.84}Y_{0.16}Zr_{1.75}Ta_{0.25}O_{12}$). In a microscale all-solid-state cell with $Li_7La_3Zr_2O_{12}$ (LLZO)–SSEs and Au anode, LCO maintained structural stability even after charging to 5.1 V (vs. Li^+/Li), with suppressed Co dissolution and interfacial side reactions [32]. Interface engineering is crucial for optimizing LCO-ASSLBs performance: surface coatings like $LiCoPO_4$ improve interfacial compatibility between LCO and SSEs, reducing interfacial impedance and enhancing cycling stability [30]. Moreover, LCO's high volumetric energy density translates to superior performance in ASSLBs, making it a promising candidate for high-energy-density applications such as miniaturized consumer electronics. Despite significant progress, challenges remain, including the need for a deeper understanding of LCO–SSEs interface dynamics, the design of more effective modification strategies to enhance high-voltage stability, and the development of scalable synthesis methods for modified LCO. Future research should leverage advanced in-situ characterization techniques (e.g., continuous rotation electron diffraction [cRED] and in-situ STEM) to probe LCO's structure during electrochemical cycling [33], combined with theoretical calculations based on high-throughput machine learning to guide modification design. By addressing these gaps, LCO is poised to play a pivotal role in the next generation of high-performance solid-state batteries.

3.2.2 From Material Development to Device Application: LFP Cathodes in High-voltage All-solid-state Batteries

Olivine-type lithium iron phosphate (LFP) has emerged as a pivotal cathode material for LIBs since its first report by Padhi et al. in 1997, owing to its exceptional thermal stability, low toxicity, abundant raw material reserves (Fe and P), and ultra-long cyclic life [34]. With a theoretical specific capacity of 170 mAh g^{-1} and a flat operating voltage of ~3.45 V vs. Li^+/Li (well within the stability window of most electrolytes), LFP addresses critical safety concerns of layered cathodes (e.g., LCO, Ni-rich NCM) while offering cost-effectiveness. However, intrinsic limitations—including low electronic conductivity (~10^{-10} S cm^{-1}) and 1D Li^+ diffusion channels (along the [010] direction in the *Pnma* space group) that are prone to blockage by defects—have historically restricted its rate capability [35]. Over the past two decades, targeted modifications (e.g., conductive coating, nanocrystallization, and heterogeneous doping) have drastically advanced LFP performance, enabling its integration into high-power applications such as EVs and stationary energy storage [36].

The crystal structure of LFP is fundamental to its electrochemical behavior: oxygen atoms form a distorted hexagonal close-packed framework, with Fe^{3+}/Fe^{2+} occupying octahedral 4a sites, Li^+ residing in octahedral 4c sites, and P^{5+} in tetrahedral 4b sites [37]. This ordered structure ensures minimal volume change (~6.8%) during Li^+ insertion/extraction, mitigating particle cracking and capacity fade [38]. However, Li^+ diffusion is confined to 1D channels along the *b*-axis, and the insulating PO_4^{3-} polyanions limit electronic transport. Early studies by Andersson and Thomas confirmed that unmodified LFP exhibited negligible capacity at high rates (e.g., <50 mAh g^{-1} at 1C), attributed to sluggish Li^+ diffusion (10^{-14}–10^{-13} cm^2 s^{-1}) and poor electron transfer [39]. To overcome these bottlenecks, carbon coating has become the most widely adopted strategy: Ravet et al. demonstrated that in-situ carbon deposition using sucrose as a precursor enhanced LFP's electronic conductivity by seven orders of magnitude, enabling a discharge capacity of ~160 mAh g^{-1} at 0.1C [36]. Subsequent advances in carbon coating—such as core–shell nanostructures (e.g., LFP@C nanoparticles with a uniform 2–5 nm carbon layer) reported by Wang et al.—further improved rate capability, with capacities of 90 mAh g^{-1} retained at 60C [40].

Nanocrystallization has also proven effective in shortening Li^+ diffusion paths and increasing electrolyte–cathode contact area. Gibot et al. showed that LFP particles with a critical size < 45 nm could transition from a two-phase (LFP/$FePO_4$) reaction to a single-phase solid-solution behavior at room temperature, eliminating polarization caused by phase boundary migration [41]. This structural transition, validated by in-situ XRD, yielded a sloped charge–discharge profile and a rate capability of 122 mAh g^{-1} at 5C. Heterogeneous doping, though controversial, has been explored to modulate LFP's electronic structure: Chung et al. claimed that Nb^{5+}/Ti^{4+} doping increased bulk conductivity to 10^{-2} S cm^{-1} [42]. Regardless, strategic co-doping (e.g., Mg^{2+}/Al^{3+}) has been shown to stabilize 1D Li^+ channels by reducing anti-site defects (Fe^{2+} occupying Li^+ sites), further enhancing cyclic stability [43].

The rise of ASSLBs—enabled by nonflammable SSEs and improved energy density—has opened new avenues for LFP, as its low operating voltage and high stability align with the compatibility requirements of most SSEs. ASSLBs address the flammability issue of liquid electrolytes while enabling high-voltage operation, but their performance with LFP cathodes is governed by cathode–SSE interface dynamics. Polymer electrolytes (e.g., polyethylene oxide [PEO]) are widely studied for ASSLBs due to their flexibility, but pristine PEO suffers from low oxidation stability (< 4.2 V) and poor ionic conductivity at room temperature [44]. Yang et al. resolved this by functionalizing PEO with $-OCH_3$ groups (replacing labile $-OH$) to expand its stability window to 4.3 V, and when paired with carbon-coated LFP, the ASSLB achieved 93.2% capacity retention after 50 cycles [45]. Auvergniot et al. revealed that Li_6PS_5Cl oxidizes to LiCl, P_2S_5, and elemental S at the LFP interface, increasing interfacial resistance [46]; this was mitigated by Sun and coworkers via an in-situ $Li_3P_{1+x}O_4S$ gradient coating on LFP, which suppressed space–charge layer (SCL) formation and improved capacity retention to 80% after 250 cycles [47].

Oxide SSEs (e.g., $Li_{1.4}Al_{0.4}Ti_{1.6}(PO_4)_3$ [LATP], LLZO) offer high oxidation stability but suffer from poor interfacial contact with LFP due to rigidity. To enhance contact, Yang et al. synthesized LATP-coated LFP via a solution-based method: the 20-nm-thick LATP layer not only blocked direct reaction between LFP and PEO but also provided a fast Li^+ transport pathway, leading to an areal capacity of 1.2 mAh cm^{-2} in ASSLBs [30]. Additionally, low-temperature sintering (e.g., 675 °C via FAST/SPS) had been used to fabricate dense LFP/LLZO composite cathodes, reducing element diffusion (e.g., La/Zr into LFP) and side reactions that degrade cyclic performance [48].

Despite significant progress, challenges remain for LFP-based ASSLBs: (i) the low tap density of LFP (~1.5 g cm^{-3} vs. 2.2 g cm^{-3} for LCO) limits volumetric energy density, which can be addressed by designing spherical LFP particles or 3D porous structures [49, 50]; (ii) low-temperature performance (e.g., <50% capacity retention at 0 °C) requires electrolyte optimization (e.g., fluorinated solvents) to enhance Li^+ conductivity; and (iii) interfacial impedance between LFP and SSEs demands advanced interface engineering (e.g., composite coatings of $LiNbO_3$ and conductive polymers) [51]. Future directions will focus on integrating LFP with composite SSEs (e.g., polymer-in-salt electrolytes) to balance flexibility and conductivity, as well as exploring "zero-strain" LFP composites to mitigate mechanical stress during cycling.

In conclusion, LFP has evolved from a low-rate cathode to a high-performance material via structural modifications, and its compatibility with diverse SSEs makes it a promising candidate for ASSLBs. Further advances in interface engineering and electrolyte design will unlock LFP's full potential in next-generation safe, low-cost, and high-energy solid-state batteries.

3.2.3 Advancing High-Ni NCM Cathodes: From Single-crystal/ Polycrystal Design to Solid-state Battery Applications

Layered NCM materials were developed as a cost-effective alternative to LCO—the first commercial cathode—to address LCO's high Co cost, limited capacity (~140 mAh g^{-1} at 4.2 V), and safety risks. Since their commercialization in the 2000s, NCM has undergone iterative Ni content optimization: early low-Ni variants (e.g., NCM333, $x = 0.33$) balanced stability and processability but lacked energy density, while the shift to medium-Ni (NCM622, $x = 0.6$) and high-Ni (NCM811, $x = 0.8$) designs was driven by the demand for EV batteries with longer range. Structurally, NCM adopts an α-$NaFeO_2$-type layered framework (space group $R\bar{3}m$) with ordered Li^+/transition-metal (TM) sites, which underpins its reversible Li^+ intercalation—but also introduces anisotropic volume change that becomes a critical limitation as Ni content rises, laying the groundwork for subsequent microstructural modifications.

Layered NCM materials have emerged as core cathode materials for high-energy-density LIBs and ASSLBs, with their performance tightly linked to microstructural design and Ni content optimization. High-Ni NCM ($x > 0.8$, e.g., NCM811, NCM90) delivers superior specific capacity (up to 220 mAh g^{-1}) via $Ni^{2+}/Ni^{3+}/Ni^{4+}$ redox reactions, but it suffers from intrinsic degradation: during Li^+ insertion/extraction, the H2 → H3 phase transition induces severe anisotropic lattice strain (*c*-axis shrinkage reaching 7.6% for $LiNiO_2$) that generates intergranular microcracks in PC particles [52]. These cracks act as channels for electrolyte infiltration, accelerating surface reactions (e.g., formation of insulating NiO-like phases) and capacity fade—evidenced by PC-NCM811 retaining only 63% capacity after 1,000 cycles in liquid electrolytes, while its SC counterpart maintains structural integrity [53, 54]. Early modification strategies like atomic doping (Al^{3+}, Zr^{4+}) and surface coating (AlF_3, Li_3PO_4) delayed degradation by suppressing phase transitions and blocking side reactions, but they failed to address the root cause of microcracks from random primary particle orientation in polycrystals [55, 56].

Recent advances in microstructure engineering, highlighted in both references, have revolutionized NCM performance. Concentration gradient NCM (CG-NCM), featuring an Ni-rich core (high capacity) and an Mn-rich shell (high stability) with radially aligned primary particles, dissipates tensile stress during cycling: Kim et al. reported a CG-NCM90 with 88% capacity retention after 1,000 cycles at 1C, as the shell's columnar grains inhibited (H2 → H3)-induced cracking [57]. SC-NCM, without grain boundaries, further eliminates intergranular damage and reduces interface impedance. Doerrer et al. demonstrated SC-NCM811 paired with Li_6PS_5Cl sulfide SSEs achieving 210 mAh g^{-1} at 30 °C and 85% first-cycle coulombic efficiency under 0.2 MPa stack pressure, outperforming PC-NCM in ASSLBs [58]. Heterostructured NCM, integrating layered and spinel/rock-salt phases, also enhances stability: W-doped NCM (NCW90) with a surface rock-salt layer retained 92% capacity after 1,000 cycles in pouch cells, a marked improvement over conventional NCA89 (63% retention) [54].

In ASSLBs, NCM faces unique interface challenges stemming from solid–solid contact issues and SSE compatibility. Rigid oxide SSEs (e.g., LLZO) form point-to-point contact with NCM, while sulfide SSEs (e.g., Li_6PS_5Cl) develop SCL due to Li^+ migration from SSEs to NCM (driven by chemical potential differences), increasing impedance by two to three orders of magnitude [59]. Liu et al. focused on $LiNi_{0.8}Co_{0.15}Al_{0.05}O_2$ full-gradient cathodes (GCs) with excess LLZTO at the cathode/electrolyte interface. This design effectively alleviates the expansion and shrinkage of the NCA lattice *c*-axis. More importantly, via HRTEM/EELS characterization, the team first confirmed the existence of surface local cation migration, which in turn hinders the formation of spinel distortion and rock-salt phase. Additionally, residual Li—generated from enhanced Li^+ diffusion—further blocks lattice distortion. Eventually, the cathode achieves a 50-cycle capacity retention of 89%, while the retention rates of reverse gradient cathode (R-GC) and uniformly doped cathode (UDC) cell are 45% and 58.6%, respectively [60]. Additionally, NCM's anisotropic volume change (4–9% for NCM90) causes particle cracking and void formation at the NCM/SSEs interface during cycling; Han et al. observed severe interfacial separation in PC-NCM/sulfide ASSLBs after 50 cycles, whereas SC-NCM maintained continuous contact with SSEs [53]. To address these issues, interface engineering strategies from both references have proven effective: Liu et al. applied an $LiNbO_3$ buffer layer between NCM811 and Li_6PS_5Cl, reducing interfacial impedance by 60% and enabling 64.5% capacity retention after 100 cycles [61], while Fan et al. fabricated LLZO-coated SC-NCM622 that avoided SSEs/NCM side reactions and retained 84.4% capacity after 200 cycles in ASSLBs [62]. For polymer ASSLBs, Yi et al. coated SC-NCM83 with LATP to improve compatibility with PEO electrolytes, achieving 90.5% capacity retention after 100 cycles—nearly 30% higher than bare SC-NCM83 [63].

The synergy between microstructural design and interface engineering, as outlined in the two studies, defines the future of NCM in ASSLBs. SC-NCM and CG-NCM address the fundamental limitations of PC materials by eliminating grain boundaries and mitigating strain, while tailored coatings and buffer layers resolve SSB-specific interface issues. Further optimization—such as integrating doping with SC fabrication or developing gradient SC-NCM—could unlock higher energy density and longer cycle life, solidifying NCM's role in next-generation ASSLBs. As demonstrated by the referenced works, these strategies not only overcome NCM's intrinsic challenges but also leverage ASSLBs' safety advantages, paving the way for their application in EVs and grid storage.

3.2.4 Lithium Manganate Cathodes and Their Applications in Solid-state Batteries

Spinel lithium manganate (LMO) has long been regarded as a promising cathode material for LIBs due to its inherent advantages, including earth-abundant and non-toxic manganese sources, good thermal stability, and a unique 3D framework that enables fast Li^+ diffusion [64]. The electrochemical performance of LMO is characterized by two typical voltage plateaus: the 3–4.3 V range (vs. Li^+/Li) corresponding to the $Mn^{4+}/Mn^{3.5+}$ redox couple, delivering a theoretical specific capacity of 148.2 mAh g^{-1}, and the ~3 V range associated with $Mn^{3.5+}/Mn^{3+}$ redox and formation of $Li_2Mn_2O_4$, which offers a higher theoretical capacity of 285.5 mAh g^{-1} [65]. LMO has attracted sustained attention for low-cost and environmentally friendly energy storage systems. However, its commercialization is hindered by severe capacity decay during cycling, mainly caused by three intrinsic issues: Jahn–Teller (JT) distortion of Mn^{3+}-centered MnO_6 octahedra, Mn(III) disproportionation leading to Mn dissolution, and oxygen vacancy formation inducing Mn migration [66].

To address these challenges, extensive modifications have been developed over the past decades, driving the evolution of LMO cathodes. Bulk element doping is one of the most effective strategies: for instance, Al-doped LMO ($LiAl_{0.1}Mn_{1.9}O_4$) exhibited 88.7% capacity retention after 1,000 cycles at 1C, attributed to enhanced lattice stability and suppressed JT distortion [67]; rare-earth Eu^{3+} doping ($LiMn_{1.98}Eu_{0.02}O_4$) tuned the electronic environment of MnO_6 octahedra, achieving 87.3% retention at 10C for 600 cycles even at 50 °C [68]. Recently, cationic disorder (e.g., Li/Mn exchange, Li disorder) has emerged as a novel approach to intrinsically disrupt JT distortion: Lin et al. realized in-situ Li disorder in spent LMO via an upcycling strategy, enabling 94.4% capacity retention after 60 cycles and accessing stable 3-V-region performance [69]. Surface engineering, especially epitaxial coating, also plays a critical role: Lee et al. prepared LMO with a hierarchical epitaxial layer ($Li_{1.15}Al_{0.32}Mn_{1.53}O_4/Li_{1.15}Mg_{0.2}Mn_{1.65}O_4$), which suppressed Mn dissolution and maintained 90% capacity retention after 800 cycles at 60 °C by avoiding interfacial phase separation [70]. Additionally, exposing stable (111) facets of LMO (with the lowest surface energy) reduced electrolyte contact and Mn dissolution, as demonstrated by hollow fusiform LMO with (111) facets showing 83.3% retention after 1,000 cycles at 1C [71].

The development of ASSLBs provides new opportunities for LMO cathodes, as SSEs can mitigate Mn dissolution and interface side reactions that plague liquid electrolyte systems. LMO exhibits good compatibility with various SEs, such as garnet-type LLZO and NASICON-type LATP. Bi et al. coated LMO with Nb-doped LLZO ($Li_{6.375}La_3Zr_{1.375}Nb_{0.625}O_{12}$), forming a stable CEI that suppressed Mn reactivity and achieved 81.3% capacity retention after 100 cycles [72]. Delluva et al. investigated thin-film LMO/LLZO interfaces, confirming reduced charge-transfer resistance and improved cycling stability compared to liquid electrolyte cells [73]. However, challenges remain, including high CEI impedance between LMO and SEs, and limited Li^+ transport at the interface, which require further optimization via interface engineering (e.g., buffer layers) or SEs modification.

In conclusion, LMO cathodes have evolved significantly through doping, cationic disorder, and surface engineering, addressing intrinsic stability issues. Their application in ASSLBs leverages the advantages of SEs to further enhance performance, making LMO a competitive candidate for next-generation low-cost and high-safety energy storage. Future research should focus on synergistic modifications (e.g., combining epitaxial coating with cationic disorder) and optimizing LMO/SE interfaces to realize practical ASSLB systems based on LMO cathodes.

3.2.5 Li-rich Mn-based Cathodes and Their Applications in Solid-state Batteries

The rapid expansion of renewable energy systems and EVs has driven an urgent demand for high-energy-density, safe, and cost-effective energy storage technologies [74]. Conventional LIBs with liquid electrolytes and cathodes like LCO (theoretical capacity: 274 mAh g^{-1}) or LFP (theoretical capacity: 170 mAh g^{-1}) fail to satisfy the growing need for long EV driving ranges and grid-scale storage. Li-rich Mn-based oxide (LRMO) cathodes, with a general formula of $x\text{Li}_2\text{MnO}_3 \cdot (1-x)\text{LiMO}_2$ (M = Ni, Co, and Mn), have emerged as promising alternatives due to their exceptional specific capacity (>250 mAh g^{-1}) and energy density (≈1,000 Wh kg^{-1})—far exceeding those of Ni-rich NCM/NCA (≈200 mAh g^{-1}, 800 Wh kg^{-1}) [75]. This high capacity originates from the synergistic contribution of cationic redox (Ni^{2+}/Ni^{3+}/Ni^{4+}, Co^{3+}/Co^{4+}) in the $LiMO_2$ phase and anionic oxygen redox (O^{2-}) in the Li_2MnO_3 phase, enabled by the unique "Li—O—Li" superlattice configuration [76]. The low content of expensive Co/Ni further enhances LRMO's cost competitiveness, making it suitable for large-scale applications [77].

Despite these advantages, LRMO cathodes face critical challenges that have hindered their commercialization in liquid LIBs. First, irreversible oxygen escape during cycling (triggered by O^{2-} oxidation at voltages >4.5 V) leads to structural degradation, including TM ion migration (e.g., Mn^{3+} diffusion into Li vacancies) and phase transformation from layered to spinel/rock-salt structures [78]. This results in severe voltage decay (typically 0.2–0.5 V after 100 cycles) and poor cycling stability [79]. Second, the extremely low electronic conductivity (≈10^{-8} S cm^{-1}) of LRMO limits rate performance, while TM dissolution (e.g., Mn^{3+}) in liquid electrolytes exacerbates anode corrosion and capacity fading [80]. Early efforts to mitigate these issues, such as surface coating with Al_2O_3 or Li_3PO_4, showed limited success in liquid systems, prompting the shift to ASSLBs [81].

The development of LRMO cathodes spans three decades, with key milestones shaping their understanding and optimization. In 1991, Rossouw and Thackeray first reported LRMO by introducing excess Li and Mn into layered $LiMO_2$, identifying its two-phase ($LiMO_2$/*R*3*m* and Li_2MnO_3/*C*2/*m*) structure [82]. In 1996, Yamada investigated lattice instability in Li-rich manganese spinels, including TM ion migration behavior [83]. Subsequent studies clarified the role of oxygen redox: Seo et al. used first-principles calculations to confirm that the linear "Li—O—Li" configuration in Li_2MnO_3 enables reversible O^{2-} redox via nonbonding O-2p orbitals [76]. By 2010, researchers began addressing LRMO's drawbacks—for example, Xia and coworkers adjusted TM spin states to reduce voltage decay, achieving an energy efficiency of 93%, but these improvements remained constrained by liquid electrolyte limitations [84].

ASSLBs have emerged as a game changer for LRMO cathodes, leveraging SEs to overcome liquid LIB shortcomings. Unlike organic liquid electrolytes, SEs (e.g., sulfides, halides, and oxides) offer high thermal stability, wide electrochemical windows (2–4.8 V for halides), and effective suppression of TM dissolution and oxygen escape [85]. For instance, LRMO-based ASSLBs with Li_3InCl_6 (halide SE) showed reduced Mn^{3+} dissolution and improved CEI film uniformity compared to liquid systems, leading to 79.7% capacity retention after 100 cycles [86]. These attributes make LRMO-ASSLBs a promising candidate for high-energy-density storage, with theoretical energy densities exceeding 600 Wh kg^{-1} when paired with Li-metal anodes.

Nevertheless, LRMO-ASSLBs face unique interfacial challenges. Mechanically, PC-LRMO particles (≈ 10 μm) form microcracks during cycling due to anisotropic volume changes (expansion along *c*-axis, contraction along *a*/*b*-axes), creating voids (up to 9.5% after 50 cycles) that disrupt Li^+ transport [87]. Chemically, high-voltage operation (>4.5 V) oxidizes sulfide SEs, forming resistive interfacial layers, while the electrochemical potential difference between LRMO and SEs generates SCLs that reduce Li^+ conductivity. To address these issues, several strategies have been developed: (i) single-crystal/nanosized LRMO (500 nm–1 μm) minimizes grain boundaries and crack formation, improving structural integrity [88]; (ii) zero-strain LRMO designs, achieved via cation disorder engineering, eliminate volume changes during cycling [89]; (iii) artificial interface layers (e.g., Li_3PO_4) stabilize lattice oxygen and inhibit SEs oxidation, as demonstrated by Sun and coworkers who achieved a discharge capacity of 230.7 mAh g^{-1} at 0.1C and 431 cycles [90]; and (iv) composite cathodes with carbon additives construct continuous electron transport networks, enhancing redox kinetics [91].

In conclusion, LRMO cathodes represent a viable solution for high-energy-density ASSLBs, but their commercialization requires further advances in material design and interface engineering. Future research should focus on optimizing single-crystal LRMO synthesis, developing SEs compatible with LRMO's high-voltage window, and integrating in-situ characterization to unravel interfacial evolution. With interdisciplinary efforts in materials science and device engineering, LRMO-ASSLBs are poised to overcome the energy density and safety limitations of conventional batteries, driving innovations in EVs, aerospace, and grid storage.

3.3 Preparation and Properties of High-voltage Ternary Cathode Materials

Ternary cathode materials (e.g., NCM, NCA) have become the core of high-energy LIBs due to their tunable specific capacity and moderate cost. To meet the growing demand for energy density in EVs and grid storage (targeting >350 Wh kg^{-1} for LIB packs), developing high-voltage ternary cathodes (operating voltage ≥4.3 V vs. Li/Li^+) has become a critical research focus. However, high-voltage cycling induces severe challenges, including lattice phase transitions (e.g., H2 → H3 for Ni-rich NCM), TM dissolution, interfacial electrolyte oxidation, and mechanical cracking, which deteriorate electrochemical performance. This review systematically summarizes the crystal structure fundamentals of high-voltage ternary cathodes, key preparation methods (e.g., coprecipitation, ALD), performance degradation mechanisms, and advanced modification strategies (doping, coating). Finally, future directions for overcoming scalability and stability issues are proposed, aiming to provide guidelines for the practical application of high-voltage ternary cathodes. LIBs dominate the energy storage market for portable electronics and EVs, owing to their high energy density, long cycle life, and low self-discharge. The cathode material, as a key component determining LIB performance, accounts for ~40% of the total battery cost and directly influences energy density [92]. Among various cathode types, ternary materials (NCM and NCA) have emerged as preferred candidates for high-energy applications, as their Ni content can be tailored to balance capacity and stability: increasing Ni content (e.g., NCM811, NCA80) elevates specific capacity (>200 mAh g^{-1}) but exacerbates structural instability under high voltage.

To surpass the energy density of commercial LIBs (currently ~250 Wh kg^{-1} for EV packs), shifting to high-voltage operation (≥4.3 V) is essential. For example, increasing the cutoff voltage of NCM811 from 4.2 to 4.5 V raises its energy density by ~20%, but this also triggers detrimental processes [93]: (i) anisotropic volume contraction (up to 7.6% for $LiNiO_2$ during H2 → H3 phase transition) leads to micro-cracking and active material isolation; (ii) oxidative lattice oxygen release reacts with organic electrolytes, generating high-impedance by-products (e.g., Li_2CO_3, PO_x); and (iii) TM ions (Ni^{4+}, Co^{3+}) dissolve into the electrolyte and deposit on the anode, accelerating capacity fade. These issues have hindered the commercialization of high-voltage ternary cathodes.

Recent advances in material synthesis and modification have addressed some of these challenges. For instance, coprecipitation with precise pH control enables uniform precursor particle formation, while ALD of $LiNbO_3$ coatings suppresses interfacial side reactions. Additionally, density functional theory (DFT) calculations have guided doping strategies (e.g., Zr, Mg) to stabilize the layered structure. This review focuses on the preparation methods, degradation mechanisms, and modification strategies of high-voltage ternary cathodes, integrating experimental data and theoretical insights from recent literature to provide a comprehensive overview.

3.3.1 Crystal Structure and Degradation Mechanisms of High-voltage Ternary Cathodes

High-voltage ternary cathodes adopt a layered α-$NaFeO_2$ structure ($R\bar{3}m$ space group), consisting of alternating Li^+ layers (3a sites) and TM–O octahedral layers (3b sites) stacked along the *c*-axis. The ideal structure requires no Li^+/TM^{n+} cation mixing, as this would block Li^+ diffusion channels and reduce capacity. For NCM, Mn^{4+} stabilizes the lattice via strong Mn—O bonds, while Co^{3+} enhances electronic conductivity; $Ni^{2+}/Ni^{3+}/Ni^{4+}$ redox couples contribute to high capacity. For NCA, Al^{3+} replaces Mn^{4+} to improve thermal stability but slightly reduces capacity.

Under high voltage (>4.3 V), the layered structure undergoes sequential phase transitions: H1 (hexagonal, 2.7–4.2 V) → H2 (hexagonal, 4.2–4.3 V) → H3 (hexagonal, >4.3 V). The H2 → H3 transition is particularly problematic, as it involves a 4–5% contraction of the *c*-axis and 2–3% expansion of the *a*-axis, generating internal stress that fractures secondary particles. For Ni-rich NCM (Ni ≥ 80%), this phase transition is more severe due to the higher Ni^{2+} content, which weakens TM—O bonds and promotes oxygen release.

3.3.2 Degradation Mechanisms Under High Voltage

Furthermore, the interface stability of solid electrolytes and oxide cathodes under high voltage is crucial for achieving ASSLBs with high safety, high energy/power density, and high cycle stability. The ion and electron transport mechanisms in

ASSLBs resemble those in conventional liquid LIBs, featuring simultaneous directed movement of Li^+ and electrons within the composite electrode during charging and discharging. The composite electrode of solid-state batteries typically comprises active material, solid electrolyte, and a certain amount of conductive additive. The active material and solid electrolyte connect to form an ion conduction network, while the active material and conductive additive form an electron conduction network.

Interface issues between sulfide electrolytes and composite cathodes primarily involve both physical and chemical aspects. First, lithium insertion/extraction causes significant volumetric changes in active material particles, leading to contact degradation. Electrochemical strain in the electrolyte also induces localized stresses, impairing ionic conductivity and causing uneven charge distribution, which accelerates interface failure. Additionally, due to the rigidity of the active material and solid electrolyte, along with surface roughness, contact typically occurs in a point-to-point manner. This loose solid–solid contact impedes Li^+ transport at the interface. Chemical interface stability is equally critical. Internal chemical issues in composite electrodes primarily include self-decomposition of the sulfide electrolyte, the SCL problem between the sulfide electrolyte and CAMs, and electrochemical reaction issues.

3.3.2.1 Physical Interface Issues in Composite Cathodes

In traditional liquid LIBs, the liquid electrolyte exhibits excellent fluidity and wettability, enabling it to penetrate voids within composite electrodes and form continuous interface contact with active materials. Additionally, cathode materials typically undergo structural transformations during cycling, accompanied by volume expansion/contraction. Stress accumulation from repeated charge–discharge cycles can cause severe cracking in cathode particles. The liquid electrolyte mitigates locally generated stresses by distributing pressure uniformly throughout the battery. High-voltage delithiation reduces the electrostatic repulsion between TM–O layers, causing the *c*-axis to contract. For NCM811, the *c*-axis length decreases from ~14.2 Å (fully lithiated) to ~13.8 Å (4.5 V), leading to intergranular cracking. These cracks act as pathways for electrolyte infiltration, accelerating TM dissolution and further structural collapse. Repeated volume changes during cycling induce cumulative stress in ternary particles. For NCA cathodes in polymer electrolytes, intercrystalline cracking occurs after 20 cycles, more severe than in liquid electrolytes, due to the poor mechanical flexibility of solid electrolytes. In ASSLBs, Koerver et al. investigated the interface evolution of NCM811 and β-Li_3PS_4 composite cathodes [94]. Initially, NCM811 particles and β-Li_3PS_4 maintain close contact before cycling. However, during the first cycle, an irreversible interfacial layer forms at the cathode–electrolyte boundary. Subsequently, volume contraction of NCM811 particles during charging causes loss of contact with electrolyte particles. Since no additional solid electrolyte is available to fill the resulting voids, contact loss accumulates with cycling, leading to poor rate performance and capacity retention in the corresponding all-solid-state battery. Janek and coworkers conducted a quantitative study on mechanical stresses induced by volume changes in cathode materials within ASSLBs [95]. First, XRD characterized volume changes at different charge states: LCO expands during charging and contracts during discharging, while NCM811 contracts during charging and expands during discharging. To address volume variation, the authors proposed blending these oppositely contracting cathode materials to create a stress–strain-neutral composite cathode. Beyond this approach, applying external pressure can partially suppress volume changes, thereby improving solid–solid contact between CAMs and the electrolyte during charge–discharge cycles.

Due to the narrow thermodynamic stability window of sulfide electrolytes, sulfide-based electrolytes also face issues of self-decomposition and volume changes during charging and discharging. Electrochemical strain in sulfide electrolytes generates localized stresses, leading to crack formation, particle deformation, and contact failure. Tan et al. investigated the decomposition mechanism of LPSCl at different potentials [96]. They found LPSCl oxidizes to S and P_2S_5 during charging. Upon discharge, these oxidation products are reduced to the LPS intermediate in the first cycle, subsequently forming Li_2S and Li_3P. The redox reaction is reversible and exhibits a specific capacity as high as 965 mAh g^{-1}. Since the oxidation products S and P_2S_5 are insulating, the impedance increase primarily stems from the oxidation of the electrolyte at high voltages. DFT calculations can predict interfacial stability by determining the volume change ΔV associated with heterogeneous reactions through comparing the total volume of products to that of reactants [97]. A negative ΔV value indicates a decrease in total volume post-reaction, potentially creating voids that reduce interfacial contact. A positive ΔV value signifies an increase in total volume post-reaction, which may induce additional strain and mechanical failure at the interface.

Furthermore, to achieve electrochemical performance comparable to conventional liquid LIBs, ASSLBs must ensure sufficient solid–solid contact between active materials, electrolyte, and conductive additives. This requires maximizing the contact area between carbon-based CAMs, electrolyte, and conductive additives while minimizing charge-transfer resistance at the interface. Although sulfide solid electrolytes are more flexible than oxide electrolytes, charge transfer occurs via physically adjacent particles. Under high pressure, sulfide electrolytes still struggle to fill all voids within composite electrodes, thereby hindering efficient Li^+ transport between CAMs and electrolytes. Research indicated that achieving a balance between ionic and electronic conduction within composite electrodes is critical for realizing high-performance ASSLBs [98]. When ionic pathways outnumber electronic pathways, it leads to low active material utilization in the cathode, forming "dead zones" disconnected from the electronic framework. Conversely, when electronic conduction pathways dominate over ionic ones, increased Li^+ diffusion paths cause uneven distribution of electrochemical reactions across the composite electrode, impairing the rate and cycling capabilities of ASSLBs. A balanced ion/electron permeation pathway within a reasonable range is essential for achieving high-energy-density/high-power-density sulfide-based ASSLBs.

3.3.2.2 Composite Cathode Interface Chemistry Issues

In 2006, Takada and coworkers first proposed the existence of an SCL between the oxide cathode and the sulfide solid electrolyte, with this interface layer being the source of high interfacial impedance [99]. They suggested that due to differences in chemical potential, the LCO cathode exerts a stronger force on Li^+ than the β-Li_3PS_4 solid electrolyte. When the electrolyte contacts the LCO cathode, a portion of Li^+ migrates from the β-Li_3PS_4 surface toward the oxide cathode side, forming the SCL. The LCO cathode material acts as a mixed conductor for both ions and electrons. This allows the Li^+ concentration at the LCO interface to be depleted by electron conduction. Consequently, Li^+ from the solid electrolyte surface migrates back toward the LCO, further forming an Li^+-depleted SCL. This leads to interface degradation, high interfacial impedance, and diminished battery performance. However, some researchers contended that the impact of the SCL is overestimated. Ion transport also depends on numerous other factors, such as crystal structure, lithium distribution between different crystal sites, and vacancy concentration. Therefore, integrating theoretical calculations with experimental results is essential to further elucidate the SCL formation mechanism and guide the development of ASSLBs.

Determining the energy difference between the electrolyte's lowest occupied molecular orbital (LUMO; reduction potential) and highest occupied molecular orbital (HOMO; oxidation potential) is crucial. The Fermi levels of both electrode materials must lie within the electrolyte's "electrochemical window" to maintain thermodynamic stability. Otherwise, the electrolyte will oxidize at the cathode interface to form a CEI layer or reduce at the anode interface to form an SEI layer [100]. Due to the relatively limited electrochemical stability window of sulfide electrolytes (approximately 1.6–2.3 V vs. Li^+/Li), sulfide electrolytes with narrow stability windows undergo self-decomposition at the electrode–electrolyte interface to form interfacial layers when paired with high-voltage (~ 4 V) LCO, NCM, and LMO. Sakuda et al. first observed a 10 nm interfacial layer at the LCO/Li_2S–P_2S_5 interface via transmission electron microscopy (TEM), resulting from mutual diffusion of Co, P, and S elements [101]. First-principles calculations revealed negative exchange energies between Co and P, driving mutual diffusion at the sulfide electrolyte/LCO interface. Banerjee et al. discovered that the spontaneous chemical reaction between NCA and Li_6PS_5Cl generates oxidation products of Ni_3S_4, LiCl, and Li_3PO_4. The oxidation products of Li_6PS_5Cl typically include elemental S, sulfites, and highly oxidized P_2S_x compounds [102]. These interfaces with low ionic conductivity significantly impede Li^+ transport at the interface. Based on DFT calculations, binary eutectic phase diagrams for various mixtures can be established. High reaction energies indicate chemical instability when NCA cathodes contact Li_6PS_5Cl, with poor compatibility leading to the generation of a series of by-products.

3.3.3 Preparation Methods of High-voltage Ternary Cathodes

The preparation of high-voltage ternary cathodes involves two key steps: (i) synthesis of TM hydroxide/carbonate precursors; and (ii) calcination with Li sources (e.g., $LiOH \cdot H_2O$, Li_2CO_3) to form the layered structure. The choice of preparation method directly affects precursor morphology, particle size distribution (PSD), and final cathode performance.

Coprecipitation is the most widely used industrial method for preparing ternary precursors, due to its scalability, low cost, and ability to control particle morphology. The process involves dissolving Ni^{2+}, Co^{2+}, and Mn^{2+} (or Al^{3+}) salts in deionized water, then coprecipitating with a chelating agent (e.g., NH_4OH) and precipitant (e.g., NaOH) under inert atmosphere (N_2/Ar). Key parameters include pH (8.5–11), stirring rate (500–1,500 rpm), and temperature (40–80 °C), which dictate precursor PSD and tap density. The sol–gel method enables molecular-level mixing of precursors, leading to uniform TM distribution and fine particle sizes (50–200 nm). A typical process uses citric acid or ethylene glycol as a chelating agent to form a gel with TM salts, followed by drying and calcination. This method is particularly suitable for preparing cathodes with dopants (e.g., Zr, Mg), as dopants can be homogeneously incorporated into the gel network. ALD is a precision coating technique used to deposit ultrathin (1–10 nm), uniform films on cathode surfaces, which protects against electrolyte oxidation and TM dissolution. ALD operates via sequential self-limiting reactions between gaseous precursors and the cathode surface, enabling conformal coating even on complex particle morphologies. Spray pyrolysis is a one-step method for preparing spherical ternary particles with high uniformity and porosity. A precursor solution (TM salts + Li salt) is atomized into droplets, which are pyrolyzed in a high-temperature furnace (600–1,000 °C) to form cathode particles. This method is suitable for large-scale production, as it avoids the washing and drying steps required for coprecipitation.

3.3.4 Performance Modification Strategies for High-voltage Ternary Cathodes

To address the degradation issues of high-voltage ternary cathodes, various modification strategies have been developed, focusing on lattice stabilization, interfacial protection, and mechanical reinforcement.

Doping introduces foreign ions into the ternary lattice to enhance structural stability and suppress phase transition. Doping ions are typically divided into two categories: (i) TM-site dopants (e.g., Zr^{4+}, Mg^{2+}) that strengthen TM—O bonds; and (ii) Li-site dopants (e.g., Na^+, K^+) that reduce Li^+/Ni^{2+} mixing. Zr^{4+} is a widely used TM-site dopant, as its large ionic radius (0.84 Å) and high valence stabilize the layered structure by increasing TM—O bond energy. Mg^{2+} doping is another effective strategy, as Mg^{2+} has a similar ionic radius to Ni^{2+} (0.65 Å vs. 0.69 Å) and forms stable Mg—O bonds.

Surface coating forms a physical barrier between the cathode and electrolyte, inhibiting interfacial reactions and TM dissolution. Ideal coating materials generally possess the following characteristics: (i) good compatibility with both the CAMs and electrolyte, featuring a high decomposition voltage to maintain electrochemical stability during charging and discharging; (ii) excellent Li^+ conductivity and low electron conductivity, where high ionic conductivity ensures efficient Li^+ transport while low electron conductivity suppresses side reactions at the interface; and (iii) exhibit favorable mechanical properties, low cost, and environmental friendliness.

To date, lithium-ion conductive coatings prepared via sol–gel, hydrothermal synthesis, or ALD methods, including $LiNbO_3$ [103], $LiNb_{0.5}Ta_{0.5}O_3$ [104], $Li_4Ti_5O_{12}$ [99], Li_3PO_4 [105], Li_2SiO_3 [106], $BaTiO_3$ [98], Li_2ZrO_3 [107], $Li_{0.35}La_{0.5}Sr_{0.05}TiO_3$ [108], and $LiZr_2(PO_4)$ [109], have been reported to significantly reduce the interfacial resistance between sulfide electrolytes and oxide cathodes. Among these, $LiNbO_3$ exhibits relatively high ionic conductivity (10^{-5} S cm^{-1}) and low electronic conductivity (10^{-11} S cm^{-1}) in its amorphous state, making it the most commonly used coating material for high-nickel cathodes in ASSLBs [103]. Li et al. developed a dual buffer layer strategy to enhance the interfacial stability and electrochemical performance of NCA. Researchers first constructed a core–shell structured NCA material, featuring nickel-rich $LiNi_{0.85}Co_{0.15}O_2$ particles internally and an aluminum-rich $LiNi_{0.6}Co_{0.15}Al_{0.25}O_2$ coating externally [110]. Subsequently, a thin $LiNbO_3$ layer was deposited onto the surface of the core–shell structure to further enhance interfacial stability. This dual buffer layer strategy delivered outstanding electrochemical performance in all-solid-state batteries, achieving an ICE of 90.2% at 0.06C and maintaining a high capacity retention of 89.4% after 400 cycles at a current density of 0.3C and a temperature of 60 °C. Hybrid coatings combine the advantages of inorganic and organic materials, providing both chemical stability and mechanical flexibility. Fan and coworkers developed a "polymer-patched inorganic" coating for NCM811, consisting of LATP nanoparticles and cyclized PAN (cPAN). The LATP provided Li^+ conductivity, while cPAN repaired defects in the LATP layer. The hybrid-coated NCM811 exhibited 72.7% capacity retention after 500 cycles at 0.5 V and 87.3 mAh g^{-1} at 2C [111].

In addition to designing interface coatings, synthesizing solid electrolytes stable under high pressure can also enhance the interface stability of sulfide-based ASSLBs. In recent years, halide electrolytes have seen rapid development. Halide electrolytes possess moderate ionic conductivity and high oxidation stability, enabling oxidation potentials exceeding 4 V (vs. Li/Li^+). They can serve as ionic additives in high-voltage composite cathode materials, where cathode CAMs achieve ultrafast interfacial Li^+ transport without requiring any additional surface coatings.

Han et al. investigated four composite cathodes using SC and PC $Li(Ni_{0.88}Co_{0.11}Al_{0.01})O_2$ (NCA88) as the CAM, with Li_3YCl_6 (LYC) and $Li_6PS_5Cl_{0.5}Br_{0.5}$ (LPSClBr) as the electrolytes. Among all composite cathode combinations, the full cell with SC-NCA88 paired with LYC exhibited the highest first-cycle coulombic efficiency and cycling performance. However, when LYC electrolyte was combined with either SC- or PC-NCA88 active materials, it maintained low specific capacity at high rates. The lower ionic conductivity of LYC compared to LPSClBr may hinder rapid Li^+ transport within the composite cathode. Additionally, LYC's smaller particle size and uneven distribution within the composite cathode disrupt Li^+ diffusion pathways, leading to a mismatch between cycling performance and rate capability. Therefore, comprehensive consideration of charge-transport pathways—including Li^+ migration within the active material, charge transfer at interfaces, and diffusion pathways throughout the entire electrode—is essential. Zhou et al. reported that the $Li_2In_{1/3}Sc_{1/3}Cl_4$ electrolyte matched uncoated LCO and $LiNi_{0.85}Co_{0.1}Mn_{0.05}O_2$ with excellent rate capability and high-voltage cycling stability up to 4.8 V.

ASSLBs exhibit long-term cycling stability with >3,000 cycles and 80% capacity retention at 250 MPa pressure [112].

To further enhance the electrochemical stability of halide electrolytes, Zhang et al. investigated the structural evolution and electrochemical stability of the F-doped halide electrolyte $Li_3InCl_{4.8}F_{1.2}$ [113]. Linear sweep voltammetry (LSV) results indicated significantly improved electrochemical stability of $Li_3InCl_{4.8}F_{1.2}$. First-principles calculations revealed an oxidation limit of 4.42 V, with intermediate phase products such as LiF, $LiInF_4$, and InF_3 forming at high voltages. These products possess a high electrochemical window, effectively passivating the halide electrolyte/CAMs interface and suppressing further decomposition of $Li_3InCl_{4.8}F_{1.2}$. ASSLBs employing $Li_3InCl_{4.8}F_{1.2}$ and LCO cathodes exhibit outstanding electrochemical performance. At a cutoff voltage of 4.47 V, the full cell achieves a first-cycle discharge specific capacity of 160.6 mAh g^{-1} with an ICE as high as 92%, alongside excellent cycling stability.

Despite significant progress, high-voltage ternary cathodes still face key challenges for commercialization: cycle stability at ultrahigh voltage (>4.5 V): current cathodes show rapid capacity fade above 4.5 V, due to severe oxygen release and electrolyte decomposition. Future research should focus on developing oxygen-stabilized structures (e.g., cation-disordered rock salt) or solid electrolytes (e.g., sulfides, oxides) that are stable against high-voltage cathodes. Scalability of modification techniques: ALD and sol–gel methods are expensive for industrial production. Developing low-cost, scalable coating/doping methods (e.g., dry coating, spray pyrolysis) is essential. Cost reduction: high-Ni cathodes (Ni ≥ 80%) are costly due to Ni raw material prices. Exploring low-Ni high-voltage cathodes (e.g., NCM622 with doping) or recycling TM from spent cathodes could reduce costs.

Compatibility with ASSLBs: high-voltage ternary cathodes react with sulfide solid electrolytes, forming SCLs. Developing interface modifiers to improve compatibility is critical for ASSLBs commercialization. Future directions should integrate experimental and theoretical approaches: (i) use machine learning to screen optimal dopants/coatings; (ii) develop in situ characterization techniques to study atomic-scale degradation; and (iii) scale up single-crystal and FCG (full-concentration gradient) cathodes for EV applications. With these advances, high-voltage ternary cathodes are expected to achieve energy densities >400 Wh kg^{-1}, enabling long-range EVs and high-capacity grid storage.

3.4 The Development Prospect of Cathode Materials

Cathode materials are the core components determining the energy density, cycling stability, and safety of LIBs and their derivatives (e.g., ASSLBs), which are pivotal for powering EVs, grid energy storage, and portable electronics. With the growing demand for higher energy density (targeting >600 Wh kg^{-1} for ASSLBs) and longer cycle life (≥1,000 cycles with >80% capacity retention), the development of cathode materials is shifting from simple component optimization to multiscale engineering (atomic, microstructural, and interfacial) [114]. This perspective summarizes the key development prospects of cathode materials, focusing on structural innovation, interfacial regulation, performance expansion, and engineering application, with insights from recent advances in layered nickel-rich oxides, sulfur-based cathodes, and their compatibility with solid electrolytes.

Layered Ni-rich oxides (e.g., NCM, NCA) remain the most promising cathode candidates for high-energy LIBs due to their high specific capacity (>200 mAh g^{-1}) and relatively low cost. However, the trade-off between Ni content (capacity) and stability (cycling/thermal) limits their practical application, especially when Ni content exceeds 80% (e.g., NCM811, NCA89). The root cause lies in the severe anisotropic volume change during the H2 → H3 phase transition (*c*-axis contraction up to 7.6% for $LiNiO_2$) and microcrack generation along grain boundaries, which accelerates electrolyte infiltration and parasitic reactions.

Future development will focus on microstructural tailoring to mitigate internal stress and heterostructure design to enhance surface stability. CG cathodes (e.g., core–shell gradient [CSG], FCG) have shown great potential by placing high-Ni cores (for capacity) and Mn-rich shells (for stability) in a single particle [115]. Computational tools (e.g., DFT) and machine learning will play a key role in optimizing these structures. Such "computation-guided experiment" strategies will accelerate the development of Ni-rich cathodes with both high capacity (>220 mAh g^{-1}) and long cycle life (>3,000 cycles), meeting the 10-year service requirement for EVs. The interface between cathodes and solid electrolytes (e.g., sulfides, oxides) is a critical bottleneck for ASSLBs, as severe chemical/electrochemical reactions and poor solid–solid contact lead to high interfacial resistance and capacity fade. For sulfide solid electrolytes (e.g., Li_6PS_5Cl), the main issues include the formation of SCLs, TM diffusion (e.g., Co diffusion depth up to 50 nm in LCO/Li_2S–P_2S_5 interfaces), and oxidative decomposition of SSEs at high voltages (>2.5 V vs. Li/Li^+). Future interface engineering will focus on passivation layer design, doping modification, and in situ interfacial phase construction. Surface coating is a mature strategy to isolate cathodes from SSEs. $LiNbO_3$, Li_2ZrO_3, and gradient oxyphosphosulfide coatings have been shown to inhibit TM diffusion and SCL formation. In situ construction of conductive and stable interfacial phases is a promising emerging direction. These strategies highlight the need for precise control of interfacial chemistry to achieve low-resistance, long-term stable cathode/SSE contacts. Moreover, practical applications (e.g., EVs, grid storage) require cathodes to achieve high areal capacity (>5 mAh cm^{-2}) and stable performance across a wide temperature range (−40–60 °C). For high-loading cathodes, the key challenge is optimizing ion/electron transport in thick electrodes (≥ 20 μm), which is often limited by poor solid–solid contact and tortuous pathways.

References

1 Wang, C., Yu, R., Hwang, S., Liang, J., Li, X., Zhao, C., Sun, Y., Wang, J., Holmes, N., Li, R., Huang, H., Zhao, S., Zhang, L., Lu, S., Su, D. and Sun, X. (2020) Single crystal cathodes enabling high-performance all-solid-state lithium-ion batteries. *Energy Storage Materials*, 30, 98–103.

2 Jung, S. H., Kim, U.-H., Kim, J.-H., Jun, S., Yoon, C. S., Jung, Y. S. and Sun, Y.-K. (2020) Ni-rich layered cathode materials with electrochemo-mechanically

compliant microstructures for all-solid-state Li batteries. *Advanced Energy Materials*, 10(6), 1903360.

3 Zhang, W., Leichtweiss, T., Culver, S. P., Koerver, R., Das, D., Weber, D. A., Zeier, W. G. and Janek, J. (2017) The detrimental effects of carbon additives in $Li_{10}GeP_2S_{12}$-based solid-state batteries. *ACS Applied Materials & Interfaces*, 9(41), 35888–35896.

4 Deng, S., Sun, Y., Li, X., Ren, Z., Liang, J., Doyle-Davis, K., Liang, J., Li, W., Banis, M. N., Sun, Q., Li, R., Hu, Y., Huang, H., Zhang, L., Lu, S., Luo, J. and Sun, X. (2020) Eliminating the detrimental effects of conductive agents in sulfide-based solid-state batteries. *ACS Energy Letters*, 5(4), 1243–1251.

5 Randau, S., Walther, F., Neumann, A., Schneider, Y., Negi, R. S., Mogwitz, B., Sann, J., Becker-Steinberger, K., Danner, T., Hein, S., Latz, A., Richter, F. H. and Janek, J. (2021) On the additive microstructure in composite cathodes and alumina-coated carbon microwires for improved all-solid-state batteries. *Chemistry of Materials*, 33(4), 1380–1393.

6 Park, K. H., Oh, D. Y., Choi, Y. E., Nam, Y. J., Han, L., Kim, J.-Y., Xin, H., Lin, F., Oh, S. M. and Jung, Y. S. (2016) Solution-processable glass LiI-Li_4SnS_4 superionic conductors for all-solid-state Li-ion batteries. *Advanced Materials*, 28(9), 1874–1883.

7 Nakamura, H., Kawaguchi, T., Masuyama, T., Sakuda, A., Saito, T., Kuratani, K., Ohsaki, S. and Watano, S. (2020) Dry coating of active material particles with sulfide solid electrolytes for an all-solid-state lithium battery. *Journal of Power Sources*, 448, 227579.

8 Xu, R., Liu, F., Ye, Y., Chen, H., Yang, R. R., Ma, Y., Huang, W., Wan, J. and Cui, Y. (2021) A morphologically stable Li/electrolyte interface for all-solid-state batteries enabled by 3D-micropatterned garnet. *Advanced Materials*, 33(49), 2104009.

9 Manthiram, A. and Goodenough, J. B. (2021) Layered lithium cobalt oxide cathodes. *Nature Energy*, 6(3), 323.

10 Mizushima, K., Jones, P. C., Wiseman, P. J. and Goodenough, J. B. (1980) Li_xCoO_2 $(0 < x < -1)$: a new cathode material for batteries of high energy density. *Materials Research Bulletin*, 15(6), 783–789.

11 Yang, Q., Huang, J., Li, Y., Wang, Y., Qiu, J., Zhang, J., Yu, H., Yu, X., Li, H. and Chen, L. (2018) Surface-protected $LiCoO_2$ with ultrathin solid oxide electrolyte film for high-voltage lithium ion batteries and lithium polymer batteries. *Journal of Power Sources*, 388, 65–70.

12 Wang, C.-W., Zhang, S.-J., Lin, C., Xue, S., Deng, Y.-P., Zhang, B., Yang, L., Yao, X., Zeng, L., Li, J.-T., Pan, F. and Yin, Z.-W. (2023) Mechanochemical reactions between polyanionic borate and residue Li_2CO_3 on $LiCoO_2$ to stabilize cathode/electrolyte interface in sulfide-based all-solid-state batteries. *Nano Energy*, 108, 108192.

13 Johnston, W. D., Heikes, R. R. and Sestrich, D. (1958) The preparation, crystallography, and magnetic properties of the $Li_xCo_{(1-x)}O$ system. *Journal of Physics and Chemistry of Solids*, 7(1), 1–13.

14 Jiang, Y., Qin, C., Yan, P. and Sui, M. (2019) Origins of capacity and voltage fading of $LiCoO_2$ upon high voltage cycling. *Journal of Materials Chemistry A*, 7(36), 20824–20831.

15 Wang, X., Wang, X. and Lu, Y. (2019) Realizing high voltage lithium cobalt oxide in lithium-ion batteries. *Industrial & Engineering Chemistry Research*, 58(24), 10119–10139.

16 Yang, J. H., Kim, H. and Ceder, G. (2021) Insights into layered oxide cathodes for rechargeable batteries. *Molecules*, 26(11), 3173.

17 Hertz, J. T., Huang, Q., McQueen, T., Klimczuk, T., Bos, J. W. G., Viciu, L. and Cava, R. J. (2008) Magnetism and structure of Li_xCoO_2 and comparison to Na_xCoO_2. *Physical Review B*, 77(7), 075119.

18 Amatucci, G. G., Tarascon, J. M. and Klein, L. C. (1996) CoO_2, the end member of the Li_xCoO_2 solid solution. *Journal of The Electrochemical Society*, 143(3), 1114.

19 Jiang, Y., Yan, P., Yu, M., Li, J., Jiao, H., Zhou, B. and Sui, M. (2020) Atomistic mechanism of cracking degradation at twin boundary of $LiCoO_2$. *Nano Energy*, 78, 105364.

20 Li, S., Sun, Y., Gao, A., Zhang, Q., Lu, X. and Lu, X. (2022) Sustainable $LiCoO_2$ by collective glide of CoO_6 slabs upon charge/discharge. *Proceedings of the National Academy of Sciences of the United States of America*, 119(20), e2120060119.

21 Zhang, J.-N., Li, Q., Wang, Y., Zheng, J., Yu, X. and Li, H. (2018) Dynamic evolution of cathode electrolyte interphase (CEI) on high voltage $LiCoO_2$ cathode and its interaction with Li anode. *Energy Storage Materials*, 14, 1–7.

22 Seong, W. M., Yoon, K., Lee, M. H., Jung, S.-K. and Kang, K. (2019) Unveiling the intrinsic cycle reversibility of a $LiCoO_2$ electrode at 4.8-V cutoff voltage through subtractive surface modification for lithium-ion batteries. *Nano Letters*, 19(1), 29–37.

23 Amatucci, G. G., Tarascon, J. M. and Klein, L. C. (1996) Cobalt dissolution in $LiCoO_2$-based non-aqueous rechargeable batteries. *Solid State Ionics*, 83(1), 167–173.

24 Kong, W., Zhang, J., Wong, D., Yang, W., Yang, J., Schulz, C. and Liu, X. (2021) Tailoring Co_{3d} and O_{2p} band centers to inhibit oxygen escape for stable 4.6 V $LiCoO_2$ cathodes. *Angewandte Chemie International Edition*, 60(52), 27102–27112.

25 Sun, C., Liao, X., Xia, F., Zhao, Y., Zhang, L., Mu, S., Shi, S., Li, Y., Peng, H., Van Tendeloo, G., Zhao, K. and Wu, J. (2020) High-voltage cycling induced thermal vulnerability in $LiCoO_2$ cathode: cation loss and oxygen release driven by oxygen vacancy migration. *ACS Nano*, 14(5), 6181–6190.

26 Qin, N., Gan, Q., Zhuang, Z., Wang, Y., Li, Y., Li, Z., Hussain, I., Zeng, C., Liu, G., Bai, Y., Zhang, K. and Lu, Z. (2022) Hierarchical doping engineering with active/inert dual elements stabilizes $LiCoO_2$ to 4.6 V. *Advanced Energy Materials*, 12(31), 2201549.

27 Zhang, J.-N., Li, Q., Ouyang, C., Yu, X., Ge, M., Huang, X., Hu, E., Ma, C., Li, S., Xiao, R., Yang, W., Chu, Y., Liu, Y., Yu, H., Yang, X.-Q., Huang, X., Chen, L. and Li, H. (2019) Trace doping of multiple elements enables stable battery cycling of $LiCoO_2$ at 4.6 V. *Nature Energy*, 4(7), 594–603.

28 Huang, Y., Zhu, Y., Fu, H., Ou, M., Hu, C., Yu, S., Hu, Z., Chen, C.-T., Jiang, G., Gu, H., Lin, H., Luo, W. and Huang, Y. (2021) Mg-pillared $LiCoO_2$: towards stable cycling at 4.6 V. *Angewandte Chemie International Edition*, 60(9), 4682–4688.

29 Shao, L., Zhou, L., Yang, L., Jia, C., Wang, C., Hu, S., Zeng, X., Yang, C., Huang, C., Zhou, Y. and Xi, X. (2019) Enhanced 4.5 V/55 °C cycling performance of $LiCoO_2$ cathode via $LiAlO_2$–$LiCo_{1-x}Al_xO_2$ double-layer coatings. *Electrochimica Acta*, 297, 742–748.

30 Yang, X., Wang, C., Yan, P., Jiao, T., Hao, J., Jiang, Y., Ren, F., Zhang, W., Zheng, J., Cheng, Y., Wang, X., Yang, W., Zhu, J., Pan, S., Lin, M., Zeng, L., Gong, Z., Li, J. and Yang, Y. (2022) Pushing lithium cobalt oxides to 4.7 V by lattice-matched interfacial engineering. *Advanced Energy Materials*, 12(23), 2200197.

31 Huang, W., Zhao, Q., Zhang, M., Xu, S., Xue, H., Zhu, C., Fang, J., Zhao, W., Ren, G., Qin, R., Zhao, Q., Chen, H. and Pan, F. (2022) Surface design with cation and anion dual gradient stabilizes high-voltage $LiCoO_2$. *Advanced Energy Materials*, 12(20), 2200813.

32 Gong, Y., Zhang, J., Jiang, L., Shi, J.-A., Zhang, Q., Yang, Z., Zou, D., Wang, J., Yu, X., Xiao, R., Hu, Y.-S., Gu, L., Li, H. and Chen, L. (2017) In situ atomic-scale observation of electrochemical delithiation induced structure evolution of $LiCoO_2$ cathode in a working all-solid-state battery. *Journal of the American Chemical Society*, 139(12), 4274–4277.

33 Li, J., Lin, C., Weng, M., Qiu, Y., Chen, P., Yang, K., Huang, W., Hong, Y., Li, J., Zhang, M., Dong, C., Zhao, W., Xu, Z., Wang, X., Xu, K., Sun, J. and Pan, F. (2021) Structural origin of the high-voltage instability of lithium cobalt oxide. *Nature Nanotechnology*, 16(5), 599–605.

34 Padhi, A. K., Nanjundaswamy, K. S. and Goodenough, J. B. (1997) Phospho-olivines as positive-electrode materials for rechargeable lithium batteries. *Journal of The Electrochemical Society*, 144(4), 1188.

35 Franger, S., Le Cras, F., Bourbon, C. and Rouault, H. (2002) $LiFePO_4$ synthesis routes for enhanced electrochemical performance. *Electrochemical and Solid-State Letters*, 5(10), A231.

36 Ravet, N., Chouinard, Y., Magnan, J. F., Besner, S., Gauthier, M. and Armand, M. (2001) Electroactivity of natural and synthetic triphylite. *Journal of Power Sources*, 97–98, 503–507.

37 Streltsov, V. A., Belokoneva, E. L., Tsirelson, V. G. and Hansen, N. K. (1993) Multipole analysis of the electron density in triphylite, $LiFePO_4$, using X-ray diffraction data. *Acta Crystallographica Section B: Structural Science, Crystal Engineering and Materials*, 49(2), 147–153.

38 Andersson, A. S., Kalska, B., Häggström, L. and Thomas, J. O. (2000) Lithium extraction/insertion in $LiFePO_4$: an X-ray diffraction and Mössbauer spectroscopy study. *Solid State Ionics*, 130(1), 41–52.

39 Andersson, A. S. and Thomas, J. O. (2001) The source of first-cycle capacity loss in $LiFePO_4$. *Journal of Power Sources*, 97–98, 498–502.

40 Wang, Y., Wang, Y., Hosono, E., Wang, K. and Zhou, H. (2008) The design of a $LiFePO_4$/carbon nanocomposite with a core-shell structure and its synthesis by an in situ polymerization restriction method. *Angewandte Chemie International Edition*, 47(39), 7461–7465.

41 Gibot, P., Casas-Cabanas, M., Laffont, L., Levasseur, S., Carlach, P., Hamelet, S., Tarascon, J.-M. and Masquelier, C. (2008) Room-temperature single-phase Li insertion/extraction in nanoscale Li_xFePO_4. *Nature Materials*, 7(9), 741–747.

42 Chung, S.-Y., Bloking, J. T. and Chiang, Y.-M. (2002) Electronically conductive phospho-olivines as lithium storage electrodes. *Nature Materials*, 1(2), 123–128.

43 Ou, X., Liang, G., Wang, L., Xu, S. and Zhao, X. (2008) Effects of magnesium doping on electronic conductivity and electrochemical properties of $LiFePO_4$ prepared via hydrothermal route. *Journal of Power Sources*, 184(2), 543–547.

44 Wu, J., Rao, Z., Cheng, Z., Yuan, L., Li, Z. and Huang, Y. (2019) Ultrathin, flexible polymer electrolyte for cost-effective fabrication of all-solid-state lithium metal batteries. *Advanced Energy Materials*, 9(46), 1902767.

45 Yang, X., Jiang, M., Gao, X., Bao, D., Sun, Q., Holmes, N., Duan, H., Mukherjee, S., Adair, K., Zhao, C., Liang, J., Li, W., Li, J., Liu, Y., Huang, H., Zhang, L., Lu, S., Lu, Q., Li, R., Singh, C. V. and Sun, X. (2020) Determining the limiting factor of the electrochemical stability window for PEO-based solid polymer electrolytes: main chain or terminal —OH group? *Energy & Environmental Science*, 13(5), 1318–1325.

46 Auvergniot, J., Cassel, A., Ledeuil, J.-B., Viallet, V., Seznec, V. and Dedryvère, R. (2017) Interface stability of argyrodite Li_6PS_5Cl toward $LiCoO_2$, $LiNi_{1/3}Co_{1/3}Mn_{1/3}O_2$, and $LiMn_2O_4$ in bulk all-solid-state batteries. *Chemistry of Materials*, 29(9), 3883–3890.

47 Liang, J., Zhu, Y., Li, X., Luo, J., Deng, S., Zhao, Y., Sun, Y., Wu, D., Hu, Y., Li, W., Sham, T.-K., Li, R., Gu, M. and Sun, X. (2023) A gradient oxy-thiophosphate-coated Ni-rich layered oxide cathode for stable all-solid-state Li-ion batteries. *Nature Communications*, 14(1), 146.

48 Ihrig, M., Finsterbusch, M., Tsai, C.-L., Laptev, A. M., Tu, C.-h., Bram, M., Sohn, Y. J., Ye, R., Sevinc, S., Lin, S.-k., Fattakhova-Rohlfing, D. and Guillon, O. (2021) Low temperature sintering of fully inorganic all-solid-state batteries—Impact of interfaces on full cell performance. *Journal of Power Sources*, 482, 228905.

49 Ni, J.-F., Zhou, H.-H., Chen, J.-T. and Zhang, X.-X. (2007) Molten salt synthesis and electrochemical properties of spherical $LiFePO_4$ particles. *Materials Letters*, 61(4), 1260–1264.

50 Zhang, Z., Chen, S., Yao, X., Cui, P., Duan, J., Luo, W., Huang, Y. and Xu, X. (2020) Enabling high-areal-capacity all-solid-state lithium-metal batteries by tri-layer electrolyte architectures. *Energy Storage Materials*, 24, 714–718.

51 Jung, S.-K., Gwon, H., Lee, S.-S., Kim, H., Lee, J. C., Chung, J. G., Park, S. Y., Aihara, Y. and Im, D. (2019) Understanding the effects of chemical reactions at the cathode-electrolyte interface in sulfide based all-solid-state batteries. *Journal of Materials Chemistry A*, 7(40), 22967–22976.

52 Ryu, H.-H., Park, K.-J., Yoon, C. S. and Sun, Y.-K. (2018) Capacity fading of Ni-rich $Li[Ni_xCo_yMn_{1-x-y}]O_2$ $(0.6 \leq x \leq 0.95)$ cathodes for high-energy-density lithium-ion batteries: bulk or surface degradation? *Chemistry of Materials*, 30(3), 1155–1163.

53 Han, Y., Jung, S. H., Kwak, H., Jun, S., Kwak, H. H., Lee, J. H., Hong, S.-T. and Jung, Y. S. (2021) Single- or poly-crystalline Ni-rich layered cathode, sulfide or halide solid electrolyte: which will be the winners for all-solid-state batteries? *Advanced Energy Materials*, 11(21), 2100126.

54 Ryu, H.-H., Park, K.-J., Yoon, D. R., Aishova, A., Yoon, C. S. and Sun, Y.-K. (2019) $Li[Ni_{0.9}Co_{0.09}W_{0.01}]O_2$: a new type of layered oxide cathode with high cycling stability. *Advanced Energy Materials*, 9(44), 1902698.

55 Li, Y.-C., Xiang, W., Wu, Z.-G., Xu, C.-L., Xu, Y.-D., Xiao, Y., Yang, Z.-G., Wu, C.-J., Lv, G.-P. and Guo, X.-D. (2018) Construction of homogeneously Al^{3+} doped Ni rich Ni-Co-Mn cathode with high stable cycling performance and storage stability via scalable continuous precipitation. *Electrochimica Acta*, 291, 84–94.

56 Chen, Z., Qin, Y., Amine, K. and Sun, Y. K. (2010) Role of surface coating on cathode materials for lithium-ion batteries. *Journal of Materials Chemistry*, 20(36), 7606–7612.

57 Kim, U.-H., Ryu, H.-H., Kim, J.-H., Mücke, R., Kaghazchi, P., Yoon, C. S. and Sun, Y.-K. (2019) Microstructure-controlled Ni-rich cathode material by microscale compositional partition for next-generation electric vehicles. *Advanced Energy Materials*, 9(15), 1803902.

58 Doerrer, C., Capone, I., Narayanan, S., Liu, J., Grovenor, C. R. M., Pasta, M. and Grant, P. S. (2021) High energy density single-crystal NMC/Li_6PS_5Cl cathodes for all-solid-state lithium-metal batteries. *ACS Applied Materials & Interfaces*, 13(31), 37809–37815.

59 Haruyama, J., Sodeyama, K., Han, L., Takada, K. and Tateyama, Y. (2014) Space–charge layer effect at interface between oxide cathode and sulfide electrolyte in all-solid-state lithium-ion battery. *Chemistry of Materials*, 26(14), 4248–4255.

60 Liu, C., Cao, G., Wu, Z., Hu, J., Wang, H. and Shao, G. (2019) Surficial structure retention mechanism for $LiNi_{0.8}Co_{0.15}Al_{0.05}O_2$ in a full gradient cathode. *ACS Applied Materials & Interfaces*, 11(35), 31991–31996.

61 Liu, X., Zheng, B., Zhao, J., Zhao, W., Liang, Z., Su, Y., Xie, C., Zhou, K., Xiang, Y., Zhu, J., Wang, H., Zhong, G., Gong, Z., Huang, J. and Yang, Y. (2021) Electrochemo-mechanical effects on structural integrity of Ni-rich cathodes with different microstructures in all solid-state batteries. *Advanced Energy Materials*, 11(8), 2003583.

62 Fan, Z., Xiang, J., Yu, Q., Wu, X., Li, M., Wang, X., Xia, X. and Tu, J. (2022) High performance single-crystal Ni-rich cathode modification via crystalline LLTO nanocoating for all-solid-state lithium batteries. *ACS Applied Materials & Interfaces*, 14(1), 726–735.

63 Yi, M., Li, J., Wang, M., Fan, X., Hong, B., Zhang, Z., Zhang, Z., Jiang, H., Wang, A. and Lai, Y. (2023) Suppressing structural degradation of single crystal nickel-rich cathodes in PEO-based all-solid-state batteries: mechanistic insight and performance. *Energy Storage Materials*, 54, 579–588.

64 Thackeray, M. M., David, W. I. F., Bruce, P. G. and Goodenough, J. B. (1983) Lithium insertion into manganese spinels. *Materials Research Bulletin*, 18(4), 461–472.

65 Huang, Y., Dong, Y., Li, S., Lee, J., Wang, C., Zhu, Z., Xue, W., Li, Y. and Li, J. (2021) Lithium manganese spinel cathodes for lithium-ion batteries. *Advanced Energy Materials*, 11(2), 2000997.

66 Zhan, C., Wu, T., Lu, J. and Amine, K. (2018) Dissolution, migration, and deposition of transition metal ions in Li-ion batteries exemplified by Mn-based cathodes—a critical review. *Energy & Environmental Science*, 11(2), 243–257.

67 Xu, W., Zheng, Y., Cheng, Y., Qi, R., Peng, H., Lin, H. and Huang, R. (2021) Understanding the effect of Al doping on the electrochemical performance improvement of the $LiMn_2O_4$ cathode material. *ACS Applied Materials & Interfaces*, 13(38), 45446–45454.

68 Zhang, S., Chen, H., Chen, J., Yin, S., Mei, Y., Ni, L., Di, A., Deng, W., Zou, G., Hou, H. and Ji, X. (2022) Mitigating the Jahn-Teller distortion driven by the spin-orbit coupling of lithium manganate cathode. *Journal of Energy Chemistry*, 72, 379–387.

69 Lin, J., Fan, E., Zhang, X., Li, Z., Dai, Y., Chen, R., Wu, F. and Li, L. (2022) Sustainable upcycling of spent lithium-ion batteries cathode materials: stabilization by in situ Li/Mn disorder. *Advanced Energy Materials*, 12(26), 2201174.

70 Lee, S., Yoon, G., Jeong, M., Lee, M.-J., Kang, K. and Cho, J. (2015) Hierarchical surface atomic structure of a manganese-based spinel cathode for lithium-ion batteries. *Angewandte Chemie International Edition*, 54(4), 1153–1158.

71 Xiao, Y., Zhang, X.-D., Zhu, Y.-F., Wang, P.-F., Yin, Y.-X., Yang, X., Shi, J.-L., Liu, J., Li, H., Guo, X.-D., Zhong, B.-H. and Guo, Y.-G. (2019) Suppressing manganese dissolution via exposing stable {111} facets for high-performance lithium-ion oxide cathode. *Advanced Science*, 6(13), 1801908.

72 Bi, Z., Zhao, N., Ma, L., Shi, C., Fu, Z., Xu, F. and Guo, X. (2020) Surface coating of $LiMn_2O_4$ cathodes with garnet electrolytes for improving cycling stability of solid lithium batteries. *Journal of Materials Chemistry A*, 8(8), 4252–4256.

73 Delluva, A. A., Dudoff, J., Teeter, G. and Holewinski, A. (2020) Cathode interface compatibility of amorphous $LiMn_2O_4$ (LMO) and $Li_7La_3Zr_2O_{12}$ (LLZO) characterized with thin-film solid-state electrochemical cells. *ACS Applied Materials & Interfaces*, 12(22), 24992–24999.

74 Weiss, M., Simon, F. J., Busche, M. R., Nakamura, T., Schröder, D., Richter, F. H. and Janek, J. (2020) From liquid- to solid-state batteries: ion transfer kinetics of heteroionic interfaces. *Electrochemical Energy Reviews*, 3(2), 221–238.

75 Zhao, S., Yan, K., Zhang, J., Sun, B. and Wang, G. (2021) Reaction mechanisms of layered lithium-rich cathode materials for high-energy lithium-ion batteries. *Angewandte Chemie International Edition*, 60(5), 2208–2220.

76 Seo, D.-H., Lee, J., Urban, A., Malik, R., Kang, S. and Ceder, G. (2016) The structural and chemical origin of the oxygen redox activity in layered and cation-disordered Li-excess cathode materials. *Nature Chemistry*, 8(7), 692–697.

77 He, W., Guo, W., Wu, H., Lin, L., Liu, Q., Han, X., Xie, Q., Liu, P., Zheng, H., Wang, L., Yu, X. and Peng, D.-L. (2021) Challenges and recent advances in high capacity Li-rich cathode materials for high energy density lithium-ion batteries. *Advanced Materials*, 33(50), 2005937.

78 Lu, Z. and Dahn, J. R. (2002) Understanding the anomalous capacity of $Li/Li[Ni_xLi_{(1/3-2x/3)}Mn_{(2/3-x/3)}]O_2$ cells using in situ X-ray diffraction and electrochemical studies. *Journal of The Electrochemical Society*, 149(7), A815.

79 Zhang, K., Li, B., Zuo, Y., Song, J., Shang, H., Ning, F. and Xia, D. (2019) Voltage decay in layered Li-rich Mn-based cathode materials. *Electrochemical Energy Reviews*, 2(4), 606–623.

80 Gao, D., Zeng, Z., Mi, H., Sun, L., Ren, X., Zhang, P. and Li, Y. (2019) Enhanced structural stability and overall conductivity of Li-rich layered oxide materials achieved by a dual electron/lithium-conducting coating strategy for high-performance lithium-ion batteries. *Journal of Materials Chemistry A*, 7(41), 23964–23972.

81 Sun, G., Yu, F.-D., Zhao, C., Yu, R., Farnum, S., Shao, G., Sun, X. and Wang, Z.-B. (2021) Decoupling the voltage hysteresis of Li-rich cathodes: electrochemical monitoring, modulation anionic redox chemistry and theoretical verifying. *Advanced Functional Materials*, 31(1), 2002643.

82 Rossouw, M. H. and Thackeray, M. M. (1991) Lithium manganese oxides from Li_2MnO_3 for rechargeable lithium battery applications. *Materials Research Bulletin*, 26(6), 463–473.

83 Yamada, A. (1996) Lattice instability in $Li(Li_xMn_{2-x})O_4$. *Journal of Solid State Chemistry*, 122(1), 160–165.

84 Zuo, Y., Shang, H., Hao, J., Song, J., Ning, F., Zhang, K., He, L. and Xia, D. (2023) Regulating the potential of anion redox to reduce the voltage hysteresis of Li-rich cathode materials. *Journal of the American Chemical Society*, 145(9), 5174–5182.

85 Rosenbach, C., Walther, F., Ruhl, J., Hartmann, M., Hendriks, T. A., Ohno, S., Janek, J. and Zeier, W. G. (2023) Visualizing the chemical incompatibility of halide and sulfide-based electrolytes in solid-state batteries. *Advanced Energy Materials*, 13(6), 2203673.

86 Cui, C., Fan, X., Zhou, X., Chen, J., Wang, Q., Ma, L., Yang, C., Hu, E., Yang, X.-Q. and Wang, C. (2020) Structure and interface design enable stable Li-rich cathode. *Journal of the American Chemical Society*, 142(19), 8918–8927.

87 Shi, T., Zhang, Y.-Q., Tu, Q., Wang, Y., Scott, M. C. and Ceder, G. (2020) Characterization of mechanical degradation in an all-solid-state battery cathode. *Journal of Materials Chemistry A*, 8(34), 17399–17404.

88 Sun, J., Cao, X., Yang, W., Yoo, E. and Zhou, H. (2023) Impact of particle size on the kinetics and structure stability of single-crystal Li-rich cathode materials. *Journal of Materials Chemistry A*, 11(26), 13956–13964.

89 Zhao, X., Tian, Y., Lun, Z., Cai, Z., Chen, T., Ouyang, B. and Ceder, G. (2022) Design principles for zero-strain Li-ion cathodes. *Joule*, 6(7), 1654–1671.

90 Yu, R., Wang, C., Duan, H., Jiang, M., Zhang, A., Fraser, A., Zuo, J., Wu, Y., Sun, Y., Zhao, Y., Liang, J., Fu, J., Deng, S., Ren, Z., Li, G., Huang, H., Li, R., Chen, N., Wang, J., Li, X., Singh, C. V. and Sun, X. (2023) Manipulating charge-transfer kinetics of lithium-rich layered oxide cathodes in halide all-solid-state batteries. *Advanced Materials*, 35(5), 2207234.

91 Zhang, J., Chen, Z., Ai, Q., Terlier, T., Hao, F., Liang, Y., Guo, H., Lou, J. and Yao, Y. (2021) Microstructure engineering of solid-state composite cathode via solvent-assisted processing. *Joule*, 5(7), 1845–1859.

92 Little, R. D. and Moeller, K. D. (2018) Introduction: electrochemistry: technology, synthesis, energy, and materials. *Chemical Reviews*, 118(9), 4483–4484.

93 Schmuch, R., Wagner, R., Hörpel, G., Placke, T. and Winter, M. (2018) Performance and cost of materials for lithium-based rechargeable automotive batteries. *Nature Energy*, 3(4), 267–278.

94 Koerver, R., Aygün, I., Leichtweiß, T., Dietrich, C., Zhang, W., Binder, J. O., Hartmann, P., Zeier, W. G. and Janek, J. (2017) Capacity fade in solid-state batteries: interphase formation and chemomechanical processes in nickel-rich layered oxide cathodes and lithium thiophosphate solid electrolytes. *Chemistry of Materials*, 29(13), 5574–5582.

95 Koerver, R., Zhang, W., de Biasi, L., Schweidler, S., Kondrakov, A. O., Kolling, S., Brezesinski, T., Hartmann, P., Zeier, W. G. and Janek, J. (2018) Chemo-mechanical expansion of lithium electrode materials—on the route to mechanically optimized all-solid-state batteries. *Energy & Environmental Science*, 11(8), 2142–2158.

96 Tan, D. H. S., Wu, E. A., Nguyen, H., Chen, Z., Marple, M. A. T., Doux, J.-M., Wang, X., Yang, H., Banerjee, A. and Meng, Y. S. (2019) Elucidating reversible electrochemical redox of Li_6PS_5Cl solid electrolyte. *ACS Energy Letters*, 4(10), 2418–2427.

97 Wang, S., Tang, M., Zhang, Q., Li, B., Ohno, S., Walther, F., Pan, R., Xu, X., Xin, C., Zhang, W., Li, L., Shen, Y., Richter, F. H., Janek, J. and Nan, C. W. (2021) Lithium argyrodite as solid electrolyte and cathode precursor for solid-state batteries with long cycle life. *Advanced Energy Materials*, 11(31), 2101370.

98 Wang, L., Xie, R., Chen, B., Yu, X., Ma, J., Li, C., Hu, Z., Sun, X., Xu, C., Dong, S., Chan, T.-S., Luo, J., Cui, G. and Chen, L. (2020) In-situ visualization of the space-charge-layer effect on interfacial lithium-ion transport in all-solid-state batteries. *Nature Communications*, 11(1), 5889.

99 Ohta, N., Takada, K., Zhang, L., Ma, R., Osada, M. and Sasaki, T. (2006) Enhancement of the high-rate capability of solid-state lithium batteries by nanoscale interfacial modification. *Advanced Materials*, 18(17), 2226–2229.

100 Richards, W. D., Miara, L. J., Wang, Y., Kim, J. C. and Ceder, G. (2015) Interface stability in solid-state batteries. *Chemistry of Materials*, 28(1), 266–273.

101 Sakuda, A., Hayashi, A., Ohtomo, T., Hama, S. and Tatsumisago, M. (2011) All-solid-state lithium secondary batteries using $LiCoO_2$ particles with pulsed laser deposition coatings of $Li_2S–P_2S_5$ solid electrolytes. *Journal of Power Sources*, 196(16), 6735–6741.

102 Banerjee, A., Tang, H., Wang, X., Cheng, J.-H., Nguyen, H., Zhang, M., Tan, D. H. S., Wynn, T. A., Wu, E. A., Doux, J.-M., Wu, T., Ma, L., Sterbinsky, G. E., D'Souza, M. S., Ong, S. P. and Meng, Y. S. (2019) Revealing nanoscale solid-solid interfacial phenomena for long-life and high-energy all-solid-state batteries. *ACS Applied Materials & Interfaces*, 11(46), 43138–43145.

103 Ohta, N., Takada, K., Sakaguchi, I., Zhang, L., Ma, R., Fukuda, K., Osada, M. and Sasaki, T. (2007) $LiNbO_3$-coated $LiCoO_2$ as cathode material for all solid-state lithium secondary batteries. *Electrochemistry Communications*, 9(7), 1486–1490.

104 Wang, C., Liang, J., Hwang, S., Li, X., Zhao, Y., Adair, K., Zhao, C., Li, X., Deng, S., Lin, X., Yang, X., Li, R., Huang, H., Zhang, L., Lu, S., Su, D. and Sun, X. (2020) Unveiling the critical role of interfacial ionic conductivity in all-solid-state lithium batteries. *Nano Energy*, 72, 104686.

105 Deng, S., Li, X., Ren, Z., Li, W., Luo, J., Liang, J., Liang, J., Banis, M. N., Li, M., Zhao, Y., Li, X., Wang, C., Sun, Y., Sun, Q., Li, R., Hu, Y., Huang, H., Zhang, L., Lu, S., Luo, J. and Sun, X. (2020) Dual-functional interfaces for highly stable Ni-rich layered cathodes in sulfide all-solid-state batteries. *Energy Storage Materials*, 27, 117–123.

106 Sakuda, A., Kitaura, H., Hayashi, A., Tadanaga, K. and Tatsumisago, M. (2009) All-solid-state lithium secondary batteries with oxide-coated $LiCoO_2$ electrode and $Li_2S–P_2S_5$ electrolyte. *Journal of Power Sources*, 189(1), 527–530.

107 Ito, S., Fujiki, S., Yamada, T., Aihara, Y., Park, Y., Kim, T. Y., Baek, S.-W., Lee, J.-M., Doo, S. and Machida, N. (2014) A rocking chair type all-solid-state lithium ion battery adopting $Li_2O–ZrO_2$ coated $LiNi_{0.8}Co_{0.15}Al_{0.05}O_2$ and a sulfide based electrolyte. *Journal of Power Sources*, 248, 943–950.

108 Cao, D., Zhang, Y., Nolan, A. M., Sun, X., Liu, C., Sheng, J., Mo, Y., Wang, Y. and Zhu, H. (2020) Stable thiophosphate-based all-solid-state lithium batteries through conformally interfacial nanocoating. *Nano Letters*, 20(3), 1483–1490.

109 Wang, L., Sun, X., Ma, J., Chen, B., Li, C., Li, J., Chang, L., Yu, X., Chan, T. S., Hu, Z., Noked, M. and Cui, G. (2021) Bidirectionally compatible buffering layer enables highly stable and conductive interface for 4.5 V sulfide-based all-solid-state lithium batteries. *Advanced Energy Materials*, 11(32), 2100881.

110 Li, X., Liang, M., Sheng, J., Song, D., Zhang, H., Shi, X. and Zhang, L. (2019) Constructing double buffer layers to boost electrochemical performances of NCA cathode for ASSLB. *Energy Storage Materials*, 18, 100–106.

111 Liang, Y., Liu, H., Wang, G., Wang, C., Li, D., Ni, Y. and Fan, L.-Z. (2022) Heuristic design of cathode hybrid coating for power-limited sulfide-based all-solid-state lithium batteries. *Advanced Energy Materials*, 12(33), 2201555.

112 Zhou, L., Zuo, T.-T., Kwok, C. Y., Kim, S. Y., Assoud, A., Zhang, Q., Janek, J. and Nazar, L. F. (2022) High areal capacity, long cycle life 4 V ceramic all-solid-state Li-ion batteries enabled by chloride solid electrolytes. *Nature Energy*, 7(1), 83–93.

113 Zhang, S., Zhao, F., Wang, S., Liang, J., Wang, J., Wang, C., Zhang, H., Adair, K., Li, W., Li, M., Duan, H., Zhao, Y., Yu, R., Li, R., Huang, H., Zhang, L., Zhao, S., Lu, S., Sham, T. K., Mo, Y. and Sun, X. (2021) Advanced high-voltage all-solid-state Li-ion batteries enabled by a dual-halogen solid electrolyte. *Advanced Energy Materials*, 11(32), 2100836.

114 Manthiram, A. (2020) A reflection on lithium-ion battery cathode chemistry. *Nature Communications*, 11(1), 1550.

115 Kim, S., Thind, A. S., Mohanty, S. K., Hong, J. K., Jung, J. Y., Park, H., Yang, Y., Park, M.-S., Porwal, M. S., Allen, E., Choi, B. K., Jung, H., Sunariwal, N., Kumar, K., Ge, M., Yoo, J., Song, J.-H., Ryu, J. H., Lee, S., Reddy, S. C., Srinivasa, M. K., Pati, A. R., Lee, E., Klie, R. F., Cabana, J. and Yoo, H. D. (2025) High-nickel cathodes with mechanical and interfacial robustness via tailored concentration gradients for stable Li-ion batteries. *ACS Energy Letters*, 10(7), 3600–3609.

4

Preparation and Application of Oxide-based Solid Electrolytes

4.1 Development and Challenges of Oxide-based Solid-state Electrolytes

4.1.1 All-solid-state Lithium Batteries and Solid-state Electrolytes

Lithium-ion batteries (LIBs) have been extensively exploited in recent decades, owing to ever increasing demands in energy storage particularly for powering portable electronic devices and electric vehicles. However, the increase in energy density exacerbates the safety risks associated with flammable organic liquid electrolytes (LEs) in current LIBs, raising the probability of fire incidents. One of the most promising approaches to further enhance the safety and energy density of lithium-based batteries is to replace the conventional "liquid" ion-conducting electrolyte and polymer separator in LIBs with a solid lithium-conducting ceramic electrolyte, thereby forming an all-solid-state lithium battery (ASSLB) configuration (Figure 4.1) [1]. A critical bottleneck impeding the large-scale industrialization and practical application of ASSLBs is the development of solid electrolytes (SEs) that simultaneously achieve high ionic conductivity and low interfacial resistance at electrode–electrolyte interfaces. Replacing organic LEs with solid-state counterparts not only mitigates dendrite growth and side reactions at the metal anode but also addresses leakage, vaporization, and flammability risks, thereby enhancing battery safety by eliminating combustion and explosion hazards inherent to conventional LIBs [2].

As ionic conductors, solid-state electrolytes (SSEs) can be used to construct solid-state batteries. Additionally, all-solid-state lithium metal batteries (ASSLMBs) may serve other purposes. The electronic states, ionic defects, and ionic chemistry in lithium ceramics can be modulated through lithium operations (e.g., intercalation/deintercalation, conversion, and alloying). Such SSEs are applicable in sensors, neuromorphic computing [3], and on-chip energy storage [4], performing diverse functions in chip architectures and electronic devices (Figure 4.2) [5].

Solid-state batteries assembled with SSEs are expected to improve the safety and energy density of LIBs, owing to SSEs' inherent flame retardancy and the

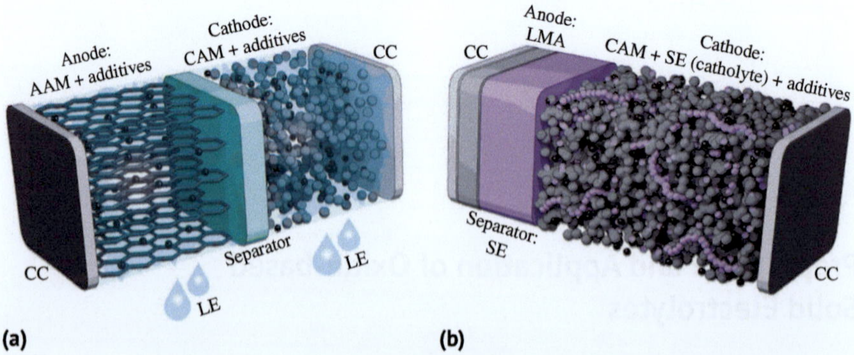

Figure 4.1 Graphical illustrations of (a) a state-of-the-art LIB with LE and (b) an all-solid-state battery with lithium metal anode (CC: current collector; LE: liquid electrolyte; SE: solid electrolyte; AAM/CAM: anode/cathode active material, respectively; and LMA: lithium metal anode). *Source*: [1] / with permission of John Wiley & Sons.

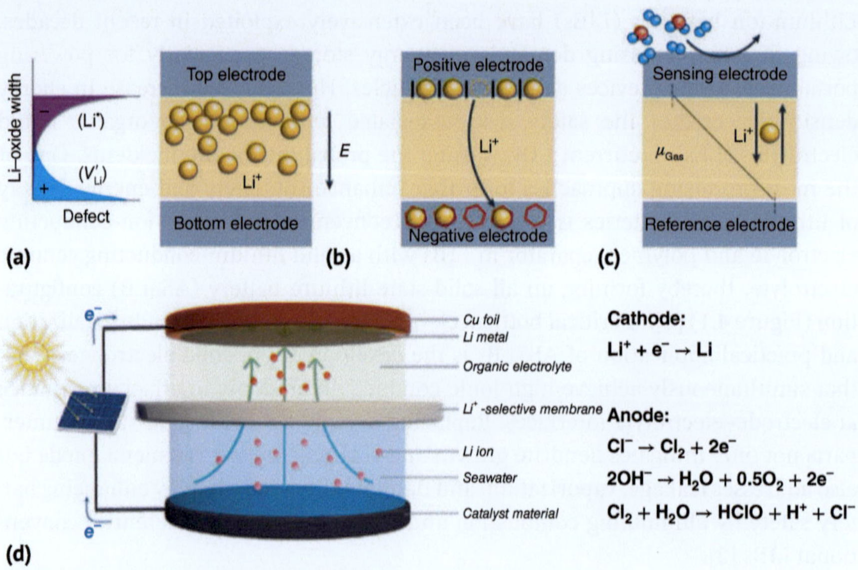

Figure 4.2 Devices using lithium-ion conductors. (a) Neuromorphic-computing memristor. (b) Micro-battery. (c) Electrochemical gas sensor. (d) Schematic diagram of the solar-powered lithium extraction device. *Source*: [5] / John Wiley & Sons / CC BY 4.0.

high capacity of lithium metal anodes. Furthermore, some SSEs exhibit excellent mechanical strength, enabling their use as structural supports to simplify battery design while enhancing safety and durability [6]. Compared to organic LEs, SSEs offer a wider electrochemical window, along with nonflammability, nonvolatility, and high-temperature stability. Certain solid-state lithium-ion conductors also demonstrate higher transference numbers, high elastic modulus, broad electrochemical stability, and leakage-free operation.

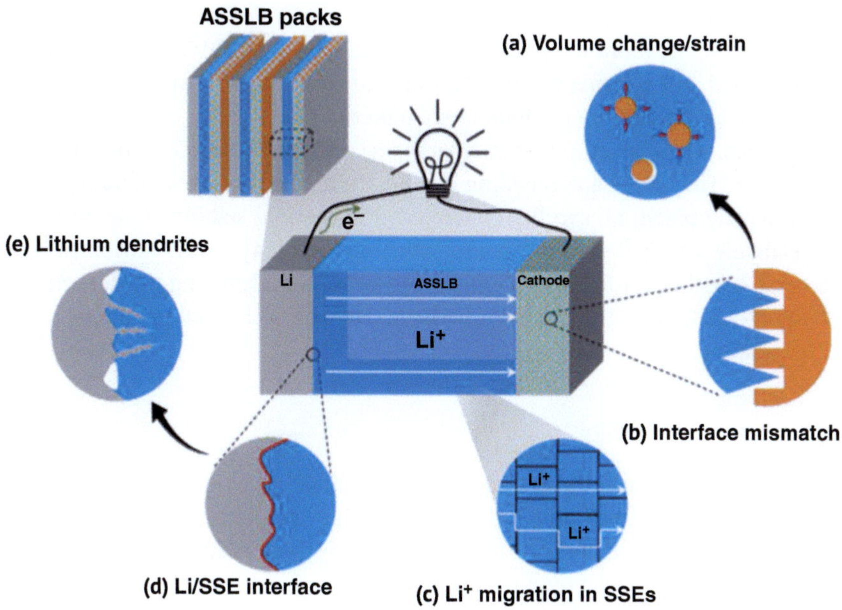

Figure 4.3 Challenges in ASSLMBs: (a, b) interface challenges in cathode/SSE: volume change/strain and interface mismatch; (c) challenges in SSE: Li^+ migration; and (d, e) interface challenges in Li (anode)/SSE: lithium dendrites and Li/SSE interface. *Source*: [8] / with permission of John Wiley & Sons.

Unlike LEs, SSEs are non-fluid and cannot wet electrode materials as effectively. Thus, ASSLMBs face major challenges, including Li^+ migration kinetics within SSEs, volume changes/strains, interfacial mismatches, and Li/SSE interface issues (Figure 4.3) [7, 8].

To address these challenges, ideal SSEs should possess the following characteristics [9]:

1. High ionic conductivity ($\geq 10^{-3}$ S cm^{-1} at room temperature).
2. Low electronic conductivity ($<10^{-10}$ S cm^{-1} to prevent internal short circuits).
3. Minimal interfacial impedance with electrodes (a primary cause of battery failure).
4. Excellent chemical/electrochemical stability (wide operational voltage window) to avoid electrolyte reduction or decomposition.
5. High mechanical strength and compatibility (e.g., matched thermal expansion/compression coefficients) to prevent delamination under thermal/pressure cycling.
6. Ease of processing and cost-effectiveness.

At present, the most extensively studied Li-ion-conducting solid ceramic materials with competitive room-temperature ionic conductivities (10^{-7}–10^{-3} S cm^{-1}) primarily include: polymer electrolytes, oxide-based electrolytes, and sulfide-based electrolytes [10]. Among these, polymer electrolytes exhibit excellent flexibility and can be applied to various battery systems. However, their high flammability significantly limits their practical application [11, 12]. Sulfide-based electrolytes demonstrate higher room-temperature ionic conductivity (> 10^{-3} S cm^{-1}) compared to oxide-based and polymer electrolytes. Yet the polysulfide shuttle effect severely degrades the performance of sulfide-based batteries [13].

In contrast, oxide-based SEs have emerged as a leading research direction in the field of SEs. They offer a compelling combination of advantages, including high energy density potential, excellent chemical and thermal stability, long cycle life, and relatively low material and manufacturing costs [14]. These attributes position oxide-based electrolytes as a critical candidate for advancing the commercial viability of ASSLBs, addressing both performance and scalability requirements for next-generation battery technologies.

4.1.2 Oxide-based SSEs

Compared to polymer electrolytes, oxide- and sulfide-based ceramic SSB electrolytes can achieve ionic conductivities comparable to LEs (approximately 10^{-2}–10^{-3} S cm^{-1}) [15, 16]. In contrast to polymer electrolytes—which suffer from poor oxidative stability and low Li-ion transference numbers—many ceramic SSB electrolytes exhibit a cation transference number approaching unity, thereby eliminating energy losses caused by anion migration [17]. Among the two most promising SSB electrolyte chemistries (oxides and sulfides), only oxides offer a relatively wide electrochemical stability window (ESW), enabling the pairing of high-voltage cathodes (up to 5 V) [18] with lithium metal anodes. This synergy facilitates the development of batteries with higher power and energy densities. Inorganic oxide-based SSEs have been extensively investigated (Figure 4.4), including sodium superionic

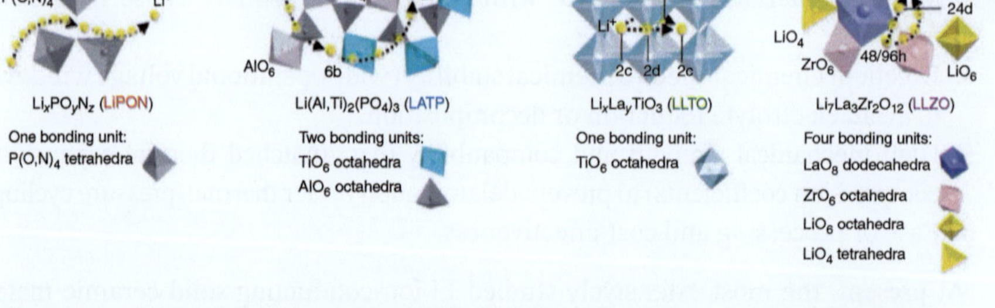

Figure 4.4 Properties of diverse oxide SEs. The figure illustrates the structures, local bonding units, and network facilitating Li-ion migration pathways in oxide-based Li-ion conductors, including a-LiPON, NASICON-type LATP, perovskite-type LLTO, and garnet-type LLZO. Here, Na and Nd denote apical nitrogen and doubly coordinated nitrogen, respectively. *Source*: [22] / with permission of Springer Nature.

conductor (NASICON)-type ($Li_{1+x}Al_xGe_{2-x}(PO_4)_3$ [LAGP]) [19], perovskite-type ($Li_{3x}La_{2/3-x}TiO_3$ [lithium lanthanum titanium oxide, LLTO]) [20], garnet-type ($Li_7La_3Zr_2O_{12}$ [lithium lanthanum zirconium oxide, LLZO]) [21], and oxynitride glasses ($Li_xPO_yN_z$) [22], among others. The LAGP electrolyte is synthesized by substituting Ge^{4+} with Al^{3+} in $LiGe_2(PO_4)_3$ [23, 24]. Compared to pristine $LiGe_2(PO_4)_3$, LAGP exhibits significantly enhanced ionic conductivity, reaching a maximum value of 5×10^{-3} S cm^{-1}. The perovskite-type SE is fabricated through partial replacement of trivalent lanthanum in lanthanum titanate with Li^+ ions [25, 26]. Garnet-type SSEs can be categorized into different systems based on their lithium content, among which LLZO represents the Li_7-rich system and is commonly referred to as lithium-stuffed garnet.

The introduction of SSB oxide electrolytes will enhance stacking density, leading to negligible self-discharge and potentially extending battery lifespan with improved performance exceeding 10^4 cycles. Beyond electrochemical stability during battery operation, SSEs must also be mechanically robust and as thin as possible to maximize the valuable volume for electrodes in SSB design. To compete with the ~20-μm-thick polymer separators in conventional LIBs, ceramic fabrication strategies within this dimensional range are required. However, in the current literature, most reported solid-state lithium electrolyte conductors for SSBs are processed as ceramic bulk materials, with thicknesses ranging from 300 to 1,000 μm (Figure 4.5a). These are fabricated via conventional ceramic processing routes and subsequently densified at relatively high sintering temperatures (around 1,000 °C). Recently, ultrafast (up to 10 seconds) high-temperature (up to 3,000 °C) sintering processes have been strategically proposed and successfully mitigated the common issue of volatile element loss in traditional ceramic sintering [27]. This suggests the potential for high-throughput material screening and synthesis in the future. However, transitioning from bulk-processed SEs—primarily used as model systems—to cost-effective thin-film manufacturing with target thicknesses below 20 μm brings wet-chemical and other methods to the forefront.

To gain deeper insight into the capabilities of current ceramic SSB processing beyond bulk materials, we analyzed the possible thickness ranges of SSB electrolytes based on known lithium oxide solid-state chemistry and ceramic processing routes (Figure 4.5a). Yet, transforming bulk ceramic particles into thin films with thicknesses of several hundred nanometers requires significant effort—on average, an additional decade or more of R&D is needed to understand the chemical properties, deposition processes, and characteristics of thin films when starting from bulk materials (Figure 4.5b, top). Optimizing ceramic processing to control material chemistry and gaining a thorough understanding of the relationship between thin-film structure, phase composition, and lithium-ion transport will directly determine whether these materials can be successfully integrated as "thin-film" electrolytes into next-generation solid-state batteries (Figure 4.5b, bottom).

Consequently, oxide-based electrolytes face three major drawbacks that hinder their current commercialization [22]. First, oxides are highly brittle and exhibit poor mechanical properties (e.g., high Young's modulus, low fracture toughness), which may impede tight interfacial contact and necessitate additional technical solutions

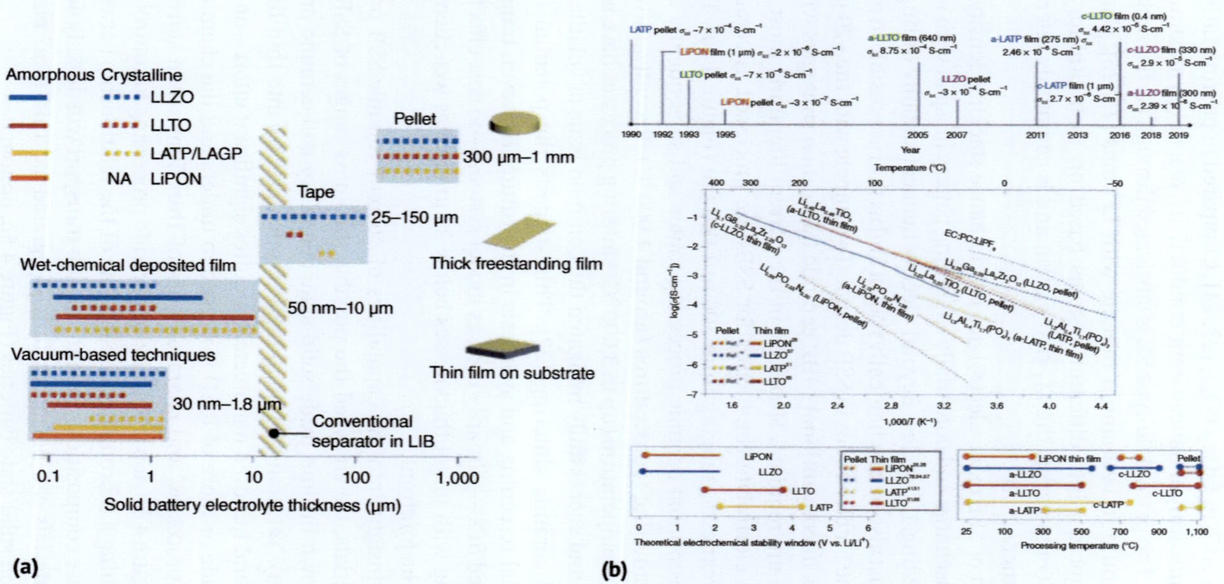

Figure 4.5 Design considerations toward oxide-based SEs. (a) Typical thickness ranges of SSEs reported for pellets, tapes, and wet-chemical and vacuum-based deposited films. (b) Comparative properties of diverse oxide SEs, material processing years and forms, and conductivity comparison. *Source*: [22] / with permission of Springer Nature.

(e.g., buffer layers, electrolyte blending, and additives). Mechanical challenges could become even more pronounced as thickness decreases, warranting further attention to chemoelectromechanical phenomena at interfaces. Second, oxide SEs have limited compatibility with current cathode chemistries, primarily due to the high-temperature processes involved in co-sintering steps between components. Reducing processing temperatures is a prerequisite for ensuring good chemical compatibility. Third, oxides are generally denser than other electrolyte types (sulfides and polymers), which negatively impacts overall gravimetric energy density—thus necessitating the use of lithium metal anodes and high-voltage cathodes.

4.1.2.1 LiPON-type Electrolytes

LiPON (with the general chemical formula $Li_xPO_yN_z$) is one of the earliest developed lithium-based thin-film SEs and has successfully facilitated the commercialization of solid-state micro-batteries [28]. Bulk oxynitride phosphate glasses are typically prepared by melting $LiPO_3$ followed by annealing in a nitrogen atmosphere [29], whereas amorphous LiPON (a-LiPON) thin films with significantly higher ionic conductivity (~10^{-6} S cm^{-1} at room temperature) are primarily deposited via radio-frequency (RF) magnetron sputtering (RF sputtering) [30] using Li_3PO_4 targets under a nitrogen or ammonia atmosphere at room temperature. The typical film thickness is approximately 1 μm, but the inherent limitations of physical deposition necessitate the use of nitrogen-enriched sputtering targets to achieve sufficient nitrogen content [31, 32]. Similarly, pulsed laser deposition (PLD) of LiPON requires nitrogen-rich targets and precise atmosphere control, though the process remains highly challenging due to poor N_2 surface adsorption and difficulties in Li/N transfer [33].

LiPON's moderate ionic conductivity and apparently wide electrochemical window [22] make it an ideal thin-film electrolyte for micro-batteries, enabling short-range Li-ion transport and low internal resistance. Although first-principles calculations suggest a narrow theoretical electrochemical window (0.68–2.63 V vs. Li^+/Li) [34], experimental measurements reveal a stable window of 0–5.5 V, attributed to sluggish decomposition kinetics and the formation of an electronically insulating yet stable interphase at the LiPON/electrode interface. When paired with a lithium metal anode and a high-voltage $LiNi_{0.5}Mn_{1.5}O_4$ cathode, LiPON demonstrates exceptional cycling stability (> 10,000 cycles) with minimal capacity degradation (< 10%) and no lithium dendrite formation. Another key advantage is its low processing temperature range (25–250 °C), which enhances chemical compatibility with other oxides and is advantageous for large-scale manufacturing.

4.1.2.2 NASICON-type Electrolytes

The NASICON-structured ionic conductors, with the general chemical formula $NaM_2(PO_4)_3$ (M = Ge, Ti, and Zr), were first discovered in 1976 [35]. A decade later, the first lithium-ion-conducting NASICON-type electrolyte, $LiZr_2(PO_4)_3$, was reported [36]. Substituting Zr^{4+} with Ti^{4+} and Al^{3+} significantly enhances ionic conductivity—for example, LATP ($Li_{1.3}Al_{0.3}Ti_{1.7}(PO_4)_3$) achieved a room-temperature conductivity of 7×10^{-4} S cm^{-1}. The highest reported bulk conductivity

(1×10^{-3} S cm^{-1}) was achieved for LATP pellets prepared via sol–gel synthesis combined with field-assisted sintering [37].

Despite LATP's high bulk ionic conductivity, structural stability, and oxidation resistance, its integration into batteries is hindered by the reduction of Ti^{4+} when in contact with low-potential anodes (e.g., Li metal or graphite). This issue can be mitigated by introducing an Al_2O_3 protective interlayer [38]. Thin-film LATP has been prepared via sol–gel spin-coating [39], aerosol deposition [40], and RF magnetron sputtering [41]. However, thin films exhibit two to three orders of magnitude lower conductivity than bulk counterparts. Aerosol-deposited films demonstrate room-temperature conductivities of 1.1×10^{-6}–2.7×10^{-6} S cm^{-1}, with the advantage of dense ceramic coating formation without high-temperature sintering [42]. Yet performance is compromised by porosity, amorphous phases, and low crystallinity. In contrast, amorphous LATP films deposited by RF sputtering achieve a higher conductivity of 2.46×10^{-5} S cm^{-1}. A critical limitation is LATP's low reduction potential (2.17 V vs. Li/Li^+), which complicates direct compatibility with lithium metal anodes.

4.1.2.3 Perovskite-type LLTO Electrolytes

Perovskite-type LLTO represents another extensively studied class of SSEs, available in both polycrystalline bulk and thin-film forms. This ABO_3-type perovskite is derived from $LaTiO_3$ through partial substitution of La^{3+} at the A-site with Li^+, where Ti^{4+} coordinates with oxygen to form corner-sharing TiO_6 octahedral frameworks that accommodate mobile Li^+ and La^{3+} ions. Consequently, the ionic conductivity of LLTO exhibits strong dependence on lithium-ion concentration. The Li^+ transport in LLTO occurs via a vacancy-hopping mechanism through quadrilateral bottlenecks formed by four oxygen atoms. The concentration of these bottlenecks critically determines the ionic mobility. In polycrystalline LLTO pellets, the high grain-boundary resistance remains a critical bottleneck, limiting the total ionic conductivity to ~10^{-5} S cm^{-1} [43, 44]. The origin of this high grain-boundary resistance is primarily attributed to: (i) the presence of ion-blocking secondary phases (e.g., Li_2CO_3) at grain boundaries, and (ii) local structural distortions of Ti–O polyhedra, which may lead to La and Li depletion in grain-boundary regions [45, 46]. To address the latter issue, studies have demonstrated that introducing excess Li at grain boundaries can effectively mitigate this problem. For instance, Kwon et al. [45] reported that Li-enriched LLTO ceramic pellets achieved a maximum total ionic conductivity of 4.8×10^{-4} S cm^{-1}.

LLTO thin films have been fabricated via various deposition techniques, including electron-beam evaporation, PLD, RF sputtering, atomic layer deposition (ALD), and sol-gel-based spin-coating. The highest reported conductivity for LLTO films, 8.7×10^{-4} S cm^{-1}, was achieved in grain-boundary-free amorphous LLTO (a-LLTO) films prepared by PLD [47]. Other techniques—such as RF sputtering, sol-gel coating, and PLD—yielded films with total ionic conductivities ranging from 10^{-4} to 10^{-8} S cm^{-1} [47, 48]. Regarding grain-boundary control strategies, epitaxial growth eliminates grain-boundary resistance but requires single-crystal substrates, rendering it impractical for large-scale manufacturing [49]. For sol–gel-derived polycrystalline LLTO films, annealing above 1,150 °C significantly reduces grain-boundary resistance due to decreased grain-boundary density perpendicular to the substrate, achieving a total Li-ion conductivity of 4.42×10^{-5} S cm^{-1} after post-annealing at 1,100 °C. Similar to NASICON-type LATP electrolyte films, the direct application of LLTO thin films with metallic Li anodes may face interfacial instability issues owing to Ti^{4+} reduction at low potentials [50].

4.1.2.4 Garnet-type LLTO Electrolytes

Lithium garnet-type ionic conductors have garnered significant attention over the past decade due to their wide electrochemical window (experimentally ~0–6 V, attributed to kinetically limited surface passivation layers) and superior lithium-ion conductivity compared to other advanced SSEs. These materials adopt the general formula $A_3B_2(XO_4)_3$ (A = La; X = Zr/Nb/Ta; and B = Al/Ga/Ge/Fe), with their lithium-ion conductivity first investigated [51] in 2005. $Li_5La_3X_2O_{12}$ (X = Nb/Ta) exhibited a conductivity of 10^{-8} S cm^{-1}. Subsequent studies identified cubic-phase LLZO with higher room-temperature conductivity (~10^{-3} S cm^{-1}), outperforming its low-temperature sintered tetragonal-phase counterpart [9, 52]. Unlike LLTO [53, 54], LLZO demonstrates low grain-boundary resistance owing to its widened band-gap and distinct space–charge distribution characteristics [55]—critical for high-performance SSB electrolytes.

Stabilization of cubic-phase LLZO is particularly challenging in thin-film forms. Current fabrication methods include:

- Vacuum techniques: RF sputtering [56], PLD [57], and metal–organic chemical vapor deposition (CVD) [58], though these exhibit significant conductivity variations (10^{-8}–10^{-5} S cm^{-1}).
- Chemical solution methods: sol–gel [59] and ALD [60], which face issues like drying cracks and densification, achieving optimal conductivities of only 2.4×10^{-8} S cm^{-1} (one order of magnitude lower than PLD multilayer deposition).

By optimizing stoichiometry via co-deposition and post-lithiation, PLD multilayer deposition combined with annealing can enhance conductivity to 10^{-5}–10^{-4} S cm^{-1}. Solution-based methods allow flexible lithium content adjustment through precursor lithium salts, enabling cost-effective scaling. Similar to LiPON and LLTO, garnets can also form amorphous thin films. A 2018 study revealed that PLD at low temperatures (300 °C) produces amorphous LLZO (a-LLZO), with ion transport properties (e.g., conductivity, activation energy) dependent on processing temperature. As LLZO thin-film stoichiometry control remains nascent, research on amorphous phases is still exploratory, requiring further validation of short-range ordered structures and their conduction mechanisms.

4.1.3 Production Processes of SSEs

4.1.3.1 Fabrication of Thin SSEs

The development of thin SSEs is crucial for achieving high-energy-density/high-power-density output in solid-state lithium batteries (SSLBs). Figure 4.6 illustrates

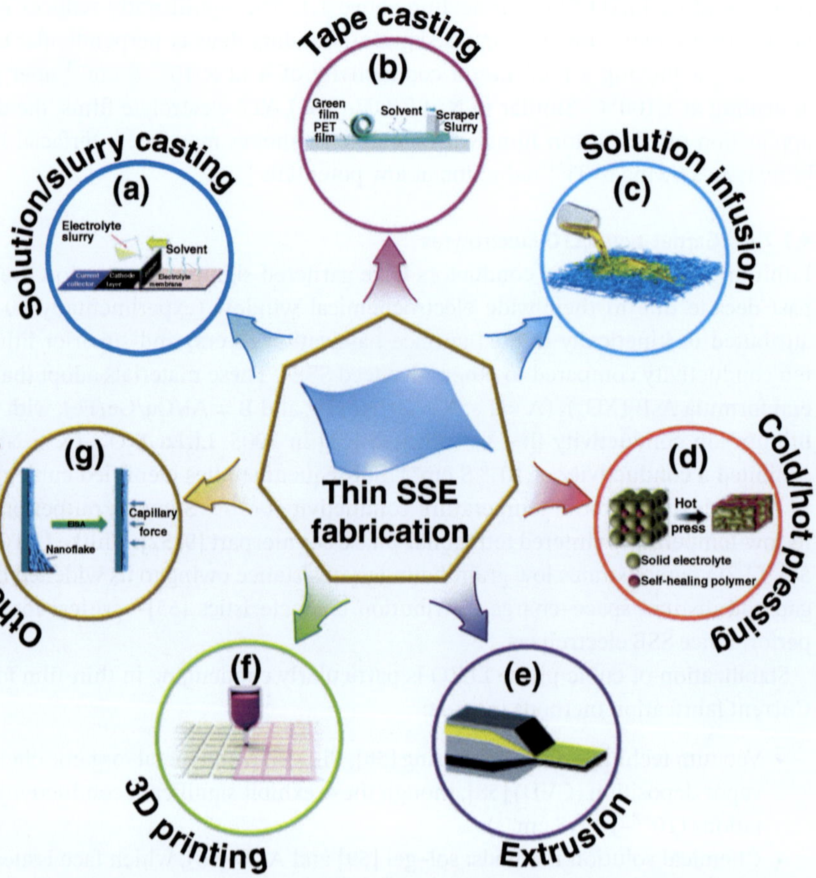

Figure 4.6 Fabrication methods for thin SSEs. (a) Solution/slurry casting method. (b) Tape casting method. (c) Solution infusion method. (d) Cold/hot pressing method. (e) Extrusion method. (f) 3D printing method. (g) Other methods. *Source*: [63] / with permission of Elsevier.

seven common methods: solution/slurry casting, tape casting, solution impregnation, hot/cold pressing, extrusion, 3D printing, and other techniques [61]. Thin SSEs with thicknesses ≤ 100 μm prepared using these methods and their applications are summarized in Table 4.1 [62–121].

4.1.3.1.1 Solution/Slurry Casting

Among the various preparation methods, solution/slurry casting is widely adopted due to its scalability and practicality. Solution casting is suitable for preparing solid polymer electrolytes (SPEs) and polymer-based composite electrolytes [122–124], while slurry casting is used for ceramic electrolytes where SSE particles are bonded by binders. This method typically yields two types of products: (i) freestanding SSEs and (ii) cathode-supported SSEs. The former is achieved by pouring the solution/slurry into an inert

Table 4.1 Fabrication of various SSEs.

Fabrication method	SSE	Thickness (μm)	Application	Source
Slurry casting	Li_6PS_5Cl	30	Ag–C/NMC90.50.5 (Pouch cell+)	[62]
	β-Li_3PS_4	21	Li/NMC622	[63]
	LLZO	5	Li/LCO	[64]
	Li_3PS_4	59	Graphite/NMC111	[65]
	PAN–LAGP–PEGDA	25	Li-NMC811 (Pouch cell+)	[66]
	LLZO	3–5	Li/LFP	[66]
	Li_3PS_4	50	Graphite–NMC111 (Pouch cell+)	[67]
	Li_6PS_5Cl	30	Graphite–NMC622 (Pouch cell+)	[68]
Solution casting	PEO–CDQs	80	Li/LFP	[69]
	PEO–LSPS	70	Li/S	[70]
	PEO–LLTO	15 ± 5	Li/S	[71]
	PEO–$Li_7P_3S_{11}$	50	–	[72]
	PEO	40	Li/LCO (Pouch cell)	[73]
	PEO	9.5	Li/LFP	[74]
	polyether–LiTFSI	29	Li/LFP (Pouch cell+)	[75]
	PEO–MIL-53(Al)	60	Li/S	[76]
	PEO–LLZO	20–75	Li/S	[77]
	MIL-53(Al)	60	Li/LFP	[78]
	PEO–SiO_2	30–150	Li/LFP	[79]
	PAN–$LiClO_4$–BNNF	12	Li/NMC811	[80]
Tape casting	Garnet	15	Li/S (Pouch cell)	[81]
	LAGP	75	–	[82]
	LLTO	25–16	Li/LFP	[83]
	LLZO	14–20	Li/S (Pouch cell+)	[84]
	LLCZNO	35	Li/S	[85]
	LLZO	20	Li/NMC622	[86]
	Garnet	30	–	[87]
	LLZO	< 30	–	[88]
	LLZO	25	–	[89]
	LLZO	20	Li/S	[90]
	LAGP	70	–	[91]

(Continued)

Table 4.1 (Continued)

Fabrication method	SSE	Thickness (μm)	Application	Source
Solution infusion	PEG–Ti–$Li_7P_3S_{11}$	10	Li/LFP	[92]
	PEO–LATP	100	–	[93]
	LLZTO–SN–LiTFSI	100	Li/LFP&Li/NMC532	[94]
	LLZTO–PAN–$LiClO_4$	25	Li/LNMO	[95]
	Polyethylene–PEO	7.5	Li/LFP&Li/S (Pouch cell)	[96]
	PI–PEO	8.6	Li/LFP (Pouch cell)	[97]
	PI–$Li_6PS_5Cl_{0.5}Br_{0.5}$	40–70	Graphite/NMC622	[98]
	PEO–LLZO	40–50	–	[99]
	Celgard 3501–PEGDME	50	Li/LFP&Li/NMC532 (Pouch cell+)	[100]
Cold/hot press	KF–PES–Li_3PS_4	60	Li/NCA	[101]
	KF–Li_3PS_4	100	Li/S	[102]
	[C2mpyr][BF4]–PVDF	30	Li/LFP	[103]
	NW-Li_3PS_4	70	LTO/LCO	[104]
	$77.5Li_2S$–$22.5P_2S_5$	100	Li/FeS_2	[105]
	β-Li_3PS_4	6–35	–	[106]
	PEO–LGPS	100	Li/S	[107]
Extrusion	PEO–LLZO	<100	–	[108]
	PEO–LiBETI	80–90	Li/V_2O_5 (Pouch cell+)	[109]
3D printing	LLZO	5–10	–	[110]
	SBC-LLZO	100	–	[111]
Hydrothermal	LATP	40–90	Graphite/LFP	[112]
	LGTP	~100	–	[113]
Solvent evaporation	β-Li_3PS_4	30	–	[114]
	$75.51Li_2S$–$24.49P_2S_5$	1.5	–	[115]
In situ polymerization	PolyDOL	~25	Li/NMC622&Li/S	[116]
	LLZO–PEGDA	36	Li/LFP	[117]
	PEGDA–PEGDE	90	Li/LFP (Pouch cell)	[118]
Spin-coating	LLTO	19.8	Li/O_2	[119]
	LATP	36	Li/O_2	[120]
	LATP	20	Li/O_2	[121]

mold or blade-coating it onto an inert substrate, followed by solvent evaporation, with thickness controlled by adjusting the solution/slurry volume and concentration. For example, Yang et al. and Wei and coworkers prepared 70 μm and 80 μm polyethylene oxide (PEO)-based composite SPEs via solution casting for Li–S and Li–LFP batteries, respectively [62, 70]. Lee and coworkers fabricated a 40 μm Li_6PS_5Cl SSE using slurry casting [74]. Cathode-supported SSEs can be formed by directly coating the slurry onto the cathode surface. Freestanding SSEs require thicknesses > 30 μm for mechanical integrity, whereas cathode-supported SSEs can be further thinned to 5–10 μm. Wang and coworkers achieved a 9.5 μm ultrathin SSE film on the cathode via repeated solution casting (Figure 4.6a), while another study realized a 5 μm LLZO SSE layer [66], simultaneously reducing cathode/SSE interfacial resistance [125].

Solvent selection is critical: ideal solvents should have low boiling points, chemical inertness, and good solubility for polymers/lithium salts. For instance, acetonitrile is commonly used in PEO systems and is inert to fillers like LLZO/SiO_2 [126], while nonpolar solvents like xylene or hexane are preferred for sulfide SSEs [127]. Binder selection also requires trade-offs; traditional PVDF/CMC are unsuitable due to poor solubility in nonpolar solvents, whereas rubber-based binders like SBR/SR are better for sulfide SSEs [126]. However, nonconductive binders reduce ionic conductivity, necessitating optimization to balance conductivity and adhesion.

4.1.3.1.2 Tape Casting

Tape casting (also called doctor blade-coating) is widely used for large-area thin ceramic layers (Figure 4.6b) [83]. This method disperses ceramic SSE powder and organic binders in a solvent, forming a uniform slurry spread by an adjustable blade. After solvent evaporation, flexible ceramic films of 20–1,000 μm are obtained. Subsequent hot pressing reduces interparticle interfacial resistance, and annealing removes organics to yield dense SSE films [83]. For example, Wang and coworkers reported 25–160-μm-thick LLTO SSEs [85]. This technique can also produce multifunctional multilayer SSEs: Hu and coworkers and Wachsman and coworkers created LLCZNO/LLZO SSEs with "porous–dense–porous" structures by stacking monolayers or introducing PMMA pore-forming agents, where the dense layer was ~ 15 μm and porous layers were 50–70 μm [84, 97].

4.1.3.1.3 Solution Impregnation

Solution impregnation involves immersing a porous substrate into SSE slurry/solution, followed by solvent removal to form thin SSEs [98, 128]. Thickness is determined by the substrate, while ionic conductivity depends on both SSE and substrate (Figure 4.6c). This method has been applied to SPEs and ceramic SSEs: Yu and coworkers designed a 3D nanostructured hydrogel-derived LLTO host impregnated with polymer, achieving a high room-temperature conductivity of 8.8×10^{-5} S cm^{-1} [128]; Hu and coworkers used electrospinning to prepare a 3D LLZO host impregnated with SPE, yielding a 40–50 μm hybrid SSE with 2.5×10^{-4} S cm^{-1} conductivity [128]. For ceramic SSEs, solvent selection must align with substrate thermal stability (e.g., polyimide withstands 400 °C annealing), enabling Jung and coworkers to fabricate 40–70 μm $Li_6PS_5Cl_{0.5}Br_{0.5}$ SSEs with 2.5×10^{-3} S cm^{-1} conductivity [98].

4.1.3.1.4 Cold/Hot Pressing

Cold pressing forms thin layers via mechanical compaction (typically for sulfide SSEs) at room temperature [105], but requires solvent-assisted dispersion (e.g., toluene, heptane) for uniformity. Jung and coworkers cold-pressed Li_3PS_4/LGPS to obtain 70 μm SSEs with 7.3×10^{-4}–4×10^{-3} S cm^{-1} conductivity [105]; Wang and coworkers prepared ~100 μm cathode-supported SSEs [103]. Hot pressing melts polymers/binders without solvents (Figure 4.6d), successfully producing < 100 μm freestanding SPEs and ceramic SSEs [106], though binder content must be optimized to avoid reducing ionic conductivity.

4.1.3.1.5 Extrusion

Extrusion forms films by mixing high-viscosity slurries at high temperatures and pressing them. Passerini and coworkers developed < 100 μm LLZO–$P(EO)_{15}$LiTFSI hybrid SSEs and 90 μm PEO–LiBETI electrolytes (used in 0.6 Ah Li/V_2O_5 pouch cells with only 0.3% per cycle capacity decay over 140 cycles) [110]. Coextrusion synchronously produces cathode-supported SSEs (Figure 4.6e), offering solvent-free, low-porosity advantages over slurry casting but with lower throughput [129].

4.1.3.1.6 3D Printing

3D printing enables precise micron-scale SSE structures (Figure 4.6f). Hu and coworkers printed 50–250 μm flexible electrolyte films with "tile-mortar" patterns, where garnet chips served as ion channels and polymer grids buffered stress [112]; Kim et al. UV-cured 3D-printed 40 μm SPE layers on sulfur cathodes [130].

4.1.3.1.7 Other Methods

Hydrothermal synthesis controls LATP SSE thickness (40–90 μm) [118]; solvent evaporation prepares < 50 μm SSEs; in situ polymerization forms SSEs by injecting pre-polymers into cells and curing them, with thickness determined by electrode gaps, reducing interfacial resistance.

4.1.3.2 Fabrication of Ultrathin SSE Films

Unlike millimeter-scale pellets or solid-state battery electrolyte tapes thicker than 40 μm (which require sintering at approximately two-thirds of the melting temperature under high pressure to achieve densification), lithium ceramic thin-film processing can achieve thicknesses ranging from a few hundred nanometers to 20 μm without traditional sintering. Achieving the ideal thickness for SSEs (i.e., close to or even thinner than the 13 μm thickness of conventional polymer separators) is challenging or even impractical for traditional powder-sintered pellets and poses difficulties for tape processing. In contrast, thin-film processing of lithium oxides enables direct densification from gaseous, liquid, or solid states, even at temperatures as low as one-fifth to one-third of the melting temperature. This approach reduces thickness by up to three orders of magnitude for most solid-state lithium conductors, whether nanocrystalline or amorphous. Additionally, thin-film processing offers greater flexibility in stabilizing attractive new amorphous phases, such as LATP or LLTO, beyond the established LiPON, owing to low-temperature "sinter-free" fabrication [22]. This may have significant implications for lithium dendrite mitigation strategies in grain-boundary-free solid-state battery electrolyte designs. To make informed choices in future solid-state battery designs, we revisit and categorize current ceramic processing options, focusing on opportunities and future prospects for "thin" SSEs. We address critical issues such as lithium loss and phase maintenance and evaluate scalability and mass-manufacturing potential based on chemical synthesis routes.

Vacuum-based thin-film processing techniques, such as physical vapor deposition (PVD) and CVD, enable direct assembly of dense multi-cation thin films for SSEs and electrodes (<~1.5 μm. Among the PVD techniques, PLD and sputtering are commonly used for depositing lithium-based thin films. These methods rely on generating vapor and particle jets from ceramic targets under ultralow pressure, followed by diffusion, nucleation, growth, and condensation on substrates (Figure 4.7a).

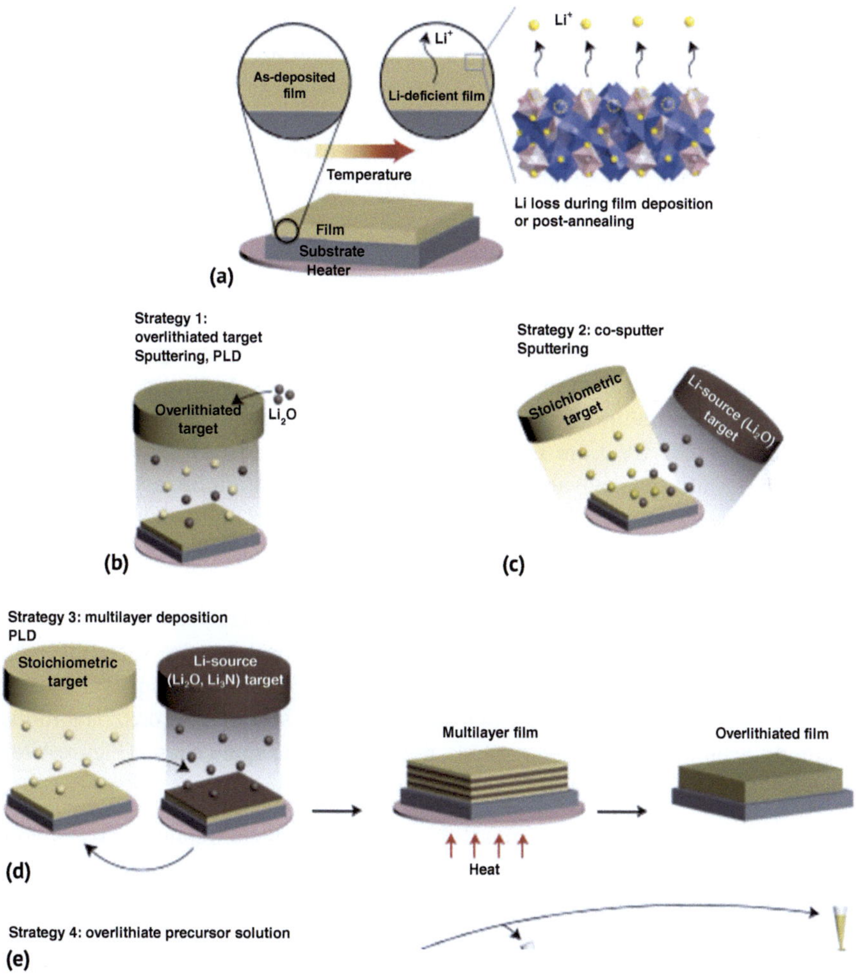

Figure 4.7 Overview of the different lithiation strategies available. (a) Lithium loss mechanism during thin-film annealing at high temperatures. (b) Overlithiation of deposition target for vacuum-based films. (c) Co-deposition with a Li-source target for vacuum-based films. (d) Construction of internal lithiation sources with secondary target for vacuum-based films. (e) Overlithiation of precursor solution for wet-chemical films. *Source*: [22] / with permission of Springer Nature.

This high-energy process typically yields highly oriented microstructures, often columnar grains or amorphous structures. These techniques can stabilize metastable phases, enabling the formation of otherwise thermodynamically unfavorable material phases. Another advantage is precise control over thickness and surface roughness, producing dense films ideal for device integration, where film quality and phase purity outweigh processing costs.

For lithium-based oxides, lithium loss during deposition (Figure 4.7a) often leads to stoichiometric imbalances in films [56]. To compensate, sputtered films typically employ dual-target co-deposition, while PLD films use lithium-overstoichiometric ceramic targets (Figure 4.7b, c). However, even with overlithiated ceramic pellet targets, PLD struggles to stabilize highly lithiated phases of most crystalline SSEs (e.g., LLZO and LLTO) and electrode materials (e.g., spinel and layered compounds). Recently, a novel ceramic processing method was proposed, incorporating an internal lithium source during deposition—using Li_3N and lithium garnet as a second target followed by post-annealing (Figure 4.7d)—successfully mitigating lithium loss. Notably, Li_3N was chosen due to its bandgap matching the PLD laser wavelength, enabling rapid deposition with minimal lithium loss [131].

Similar to PVD, CVD employs gaseous precursors that undergo chemical reactions or decomposition in the vapor phase to deposit thin films. CVD reactors can activate these processes thermally, via plasma, or photochemically. A key advantage of CVD is the lower kinetic energy of deposited particles compared to other vacuum-based techniques (e.g., sputtering or PLD), reducing surface and film degradation during growth and expanding the range of explorable chemistries. ALD, a specialized CVD variant, excels in conformal coatings, even on high-aspect-ratio 3D structures.

While PVD and CVD are more mature, vacuum-based techniques (e.g., PLD, sputtering) continue to advance in conformal film deposition on larger substrates. However, due to the high investment required for scaling PLD, such films remain primarily used as model systems for studying electrode/electrolyte interfacial reactions, not yet commercialized in solid-state batteries.

Wet-chemical synthesis is attractive for reducing manufacturing costs and scaling solid-state battery electrolytes, eliminating the need for expensive vacuum chambers or high energy consumption. Additionally, wet-chemical techniques can easily achieve electrolyte thicknesses from submicrons to several microns. Traditional methods include sol–gel, spray pyrolysis, dip-coating, and spin-coating (Figure 4.7e).

However, wet-chemical deposition faces dual challenges:

- Multi-cation coordination: most lithium oxide conductors comprise two to four cations, requiring precise phase control for fast lithium migration. For example, doped lithium garnets like $Li_{6.25}Al_{0.25}La_3Zr_2O_{12}$ demand careful coordination of metal salt decomposition points and tailored organic chemistry (e.g., adjusting decomposition temperatures and chain lengths) to facilitate chemical reactions, nucleation, and condensation [132].
- Crack prevention: controlling the drying process of lithium ceramic films is critical to avoid cracking during deposition and post-annealing. This may involve single- or multistep temperature/pH-controlled reactions to transfer salts into oxides, or stepwise sub-reactions to manage densification.

Lithium oxide films benefit from a wide range of lithium salts with decomposition temperatures between 100 and 550 °C—broader than those of heavier cations (e.g., La, Ti, and Zr). A key advantage of wet-chemical synthesis is the ease of adjusting lithium content by adding excess lithium salts to precursors, provided they remain soluble and reactive. This enables low-temperature fabrication (< 750 °C) [16] and stabilizes fast-conducting phases during post-annealing. Excess lithium can also act as a grain-boundary modifier, enhancing densification and film density [133].

Developing inorganic synthesis protocols for each material class is complex and time-consuming, but further improvements could make this route cost-effective and scalable for mass production. Moreover, tuning lithium distribution at grain boundaries offers a powerful tool to modify boundary chemistry, including migration numbers, bandgaps, and structural properties, thereby enhancing dendrite resistance and cycle life.

To date, apart from LiPON in solid-state micro-batteries, promising thin-film lithium conductors like LATP, LLTO, and LLZO have not been fully integrated into solid-state batteries. Given cubic LLZO's narrower thermal processing window compared to LLTO and the flexibility of wet-chemical routes for oxide-based LLZO (vs. polyanionic oxides like LiPON and LATP), LLZO is expected to face fewer compatibility issues in device integration. a-LLZO and a-LLTO films are also promising alternatives due to their grain-boundary-free nature and lower processing temperatures. The high ionic conductivity reported for PLD-deposited a-LLTO suggests great potential for SSEs. However, further research is needed to understand the amorphous nature of these materials, optimize their ionic/electronic conductivity, and characterize local bonding units.

4.2 Study of the Grain Boundary and Interface of Oxide-based SEs

ASSLMBs replace LEs with SEs, enabling the use of metallic lithium anodes wetted by SEs [134]. Ideal SEs should possess high ionic conductivity, robust mechanical strength, a wide ESW, and excellent chemical stability [135]. Weppner and coworkers reported that the garnet-type SE LLZO exhibits high room-temperature ionic conductivity (up to 3×10^{-4} S cm^{-1}) and excellent stability against lithium metal [9]. Due to these advantages, LLZO has been widely studied as a promising candidate for solid-state batteries. However, two major challenges hinder its practical application:

1. Poor chemical and electrochemical stability at solid–solid interfaces: the continuous chemical and electrochemical reactions at the cathode–electrolyte interface lead to gradual lithium-ion consumption during cycling, resulting in high interfacial resistance.
2. Inferior mechanical stability: the weak mechanical properties cause lithium dendrite penetration, at the cathode–electrolyte interface, reducing the contact area between active materials, conductive additives, and current collectors. This significantly increases battery impedance, further degrading capacity and cycle life.

The ionic conductivity, mechanical strength, and critical current density (CCD) of garnet electrolytes are closely related to their microstructure. Polycrystalline ceramics typically consist of grains, grain boundaries, and pores. Grain boundaries are inherent structural features of polycrystalline materials and have a profound impact on the electrical behavior of functional materials. In semiconductors, grain boundaries can reduce electron mobility due to the carrier scattering effect, which can be modified through various methods: doping (e.g., Ga^{3+}, Al^{3+} at Li^+ sites; Ta^{5+}, Nb^{5+} at Zr^{4+} sites) adjusts crystal structure and Li^+/vacancy ratios, enhancing bulk conductivity; sintering aids (e.g., Al_2O_3, LiF, and Li_3BO_3) or hot pressing reduces porosity and increases density, improving overall ionic transport.

Despite achieving near-zero porosity via hot-press sintering, LLZO still suffers from Li-dendrite-induced short-circuiting. Studies suggested that CCD is strongly influenced by grain conductivity and grain boundary properties [136]. While widely observed, the exact mechanism of Li dendrite penetration in LLZO remains unclear [137]. Compared to LE batteries, dendrite growth and short-circuiting occur much faster in solid-state systems. Key contributing factors include: poor interfacial contact, leading to uneven Li^+ flux and localized high current density, which promotes Li deposition/dissolution inhomogeneity; high electronic conductivity at grain boundaries, facilitating $Li^+ + e^- \rightarrow Li^0$ reactions and dendrite nucleation [138].

Recent studies have proposed several approaches to suppress dendrite growth (Figure 4.8a): (i) 3D interface engineering: A 3D-LLZTO@ZnO/Li interface [22] was designed to reduce local current density and enhance CCD from 0.1 to 1.4 mA cm^{-2}. (ii) Grain boundary modification: Song and coworkers [138] coated LLZO particles with $LiAlO_2$ forming low-electronic-conductivity grain boundaries, significantly increasing CCD.

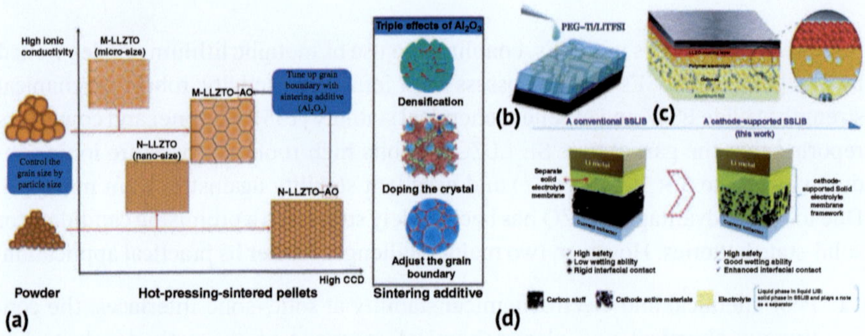

Figure 4.8 Grain boundary and interface engineering. (a) Schematic diagram of the modified grain boundary of LLZTO. *Source*: [61] / with permission of the Royal Society of Chemistry. (b) Schematic illustration of a $Li_7P_3S_{11}$ host for PEG-Ti SPE infusion. *Source*: [140] / with permission of the American Chemical Society. (c) Schematic diagram of the LLZO/LFP interface modified with an in situ polymerized SPE. *Source*: Reprinted with permission from [139]; © 2020 American Chemical Society. (d) Schematic of the construction of a conventional Li-LFP SSLB and a cathode-supported Li-LFP SSLB. *Source*: Reprinted with permission from [74]; © 2019 Royal Society of Chemistry.

Furthermore, a well-designed interface is equally crucial [102]. In situ polymerized polymer precursors or coating the cathode with an SPE solution can form an intimately contacted interface. In situ SPEs serve dual functions: (i) acting as an ion-conductive separator on the electrode surface to transport Li^+ while isolating electrodes; and (ii) constructing Li^+ transport channels within the electrode. This continuous ion-conducting pathway penetrating the electrode significantly reduces interfacial impedance [74, 118]. For instance, Duan et al. [118] polymerized poly (ethylene glycol) methyl ether acrylate in situ at the LLZO/LFP cathode interface, forming a 5.7 μm SPE layer (Figure 4.8b, c), achieving a 94.5% capacity retention after 120 cycles. Wang and coworkers [74] employed a two-step slurry-coating method to prepare a cathode-supported SPE (9.5 μm) on LFP, demonstrating superior rate performance due to reduced interfacial resistance and thin SSE layers (Figure 4.8d). A battery pack composed of three SSLB units exhibited an open-circuit voltage of 9.12 V, confirming its practical potential.

4.3 Application of Thin SSEs

4.3.1 Application of Thin SSEs in LIBs

SSLBs assembled with thin SSEs and lithium metal anodes hold promise for achieving high energy density and enhanced safety. The rational design of SSEs with multi-structures, multifunctionality, and tailored properties to meet the diverse requirements of different SSLB systems is equally critical. Based on battery configurations, they can be classified into coin/model cells and pouch cells.

Li–LFP batteries have emerged as one of the most popular SSLB systems due to their nanosized active materials, which facilitate Li^+/e^- transport, and their relatively low operating voltage range (typically 2.5–4 V), which falls within the ESWs of most SSEs. PEO, the most widely used SPE, exhibits considerable ionic conductivity at operating temperatures above 60 °C, high flexibility, and excellent interfacial wettability in Li–LFP SSLBs. Its stability against lithium metal and oxidation potential of 4 V also make it an ideal choice for Li–LFP SSLBs. When combined with a lithium metal anode and thin SSEs, Li–LFP SSLBs can achieve energy densities exceeding 200 Wh kg^{-1}.

By infusing a PEO and lithium salt solution into a commercial polyethylene separator (labeled PPL), researchers developed an ultrathin PEO SPE with a thickness of 7.5 μm. Benefiting from the shortened Li^+ diffusion distance in the PPL electrolyte, the assembled Li–LFP cell demonstrated superior rate performance: at a high rate of 10C, it delivered a capacity of 50 mAh g^{-1} at 60 °C, twice that of its counterpart with a solution-cast PEO electrolyte (120 μm) at 5C. Additionally, a high-loading Li–LFP SSLB (LFP loading: 7 mg cm^{-2}) exhibited stable cycling over 50 cycles. Notably, the thin PPL had an areal density below 1 mg cm^{-3}, nearly 10 times lower than previously reported SPEs. When paired with a lithium metal anode, the assembled Li–LFP cell achieved gravimetric/volumetric energy densities of 216 Wh kg^{-1} and 317 Wh L^{-1}.

Improving SPE ionic conductivity is a key strategy. Cui and coworkers considered the influence of tortuosity on Li^+ transport and developed a PI membrane with vertically aligned channels for PEO SPE infusion (labeled PI–PEO). Compared to blade-cast pure PEO SPE (130 μm), the PI–PEO SPE (8.6 μm) exhibited a room-temperature ionic conductivity of 2.3×10^{-4} S cm^{-1}, a fourfold improvement [97]. Incorporating inorganic fillers is another effective approach [79, 92, 93]. For instance, Lin and coworkers [79] employed a 3D SiO_2 aerogel scaffold as a PEO SPE host, achieving a composite electrolyte with a room-temperature ionic conductivity of 6×10^{-4} S cm^{-1}. This enabled stable cycling of Li–LFP cells for over 200 cycles at 0.5C (capacity > 100 mAh g^{-1}, LFP loading ≈ 1 mg cm^{-2}). A high-loading cell (13.6 mg cm^{-2}) delivered an areal capacity of 2.1 mAh cm^{-2} (13 cycles, 55 °C).

4.3.2 Application of SSEs in Gas Sensor

Conventional wisdom holds that solid-state ionic conductivity is strongly influenced by the valence state of mobile ions, rendering multivalent cation-conducting solids unsuitable for electrochemical devices, including gas sensors. However, recent discoveries of trivalent cation-conducting solids with 2D or 3D network structures have challenged this paradigm. These materials exhibit exceptionally high ionic conductivity even at moderate temperatures, enabling gas sensors with superior performance for detecting CO_2, NO_x, and NH_3.

SE-type sensors possess the advantages of selective and quantitative detection of target gases, benefiting from the unique properties of SEs—where only a single type of ion migrates within the solid. Additionally, since all sensor components are solid-state materials, compact sensor elements can be easily fabricated, similar to other types of all-solid-state gas sensors mentioned above. Therefore, SE-type sensors are generally considered strong candidates for practical gas sensing devices.

However, reports on multivalent cation-conducting SEs with high ionic conductivity are scarce because cation migration in solid lattices is strongly influenced by interactions (bond strength) with counter-anions (e.g., O^{2-}). The higher the cation valence, the stronger the interaction with anions. Thus, although there are some reports on sensors using divalent Mg^{2+} or Ca^{2+} cation-conducting solids [140], these sensors operate at relatively high temperatures due to the low conductivity of the employed SEs. Owing to this limitation, multivalent cation-conducting SEs cannot serve as key components in sensors. Most cation-conducting SEs applied in sensors are Li^+- and Na^+-ion conductors. While these sensors can reduce electrical noise (critical for achieving high sensitivity and rapid response even at lower temperatures), their gas selectivity and long-term stability remain challenging to achieve.

To develop practically viable SE-type sensors, it is crucial to create novel SEs that combine high ionic conductivity with high chemical stability, which could be achieved using trivalent cation-conducting SEs. The first report on trivalent cation conduction in solids was in 1995, focusing on the layered $Sc_2(WO_4)_3$ solid [141]. The $Sc_2(WO_4)_3$-type structure is shown in Figure 4.9a. Although the trivalent Sc^{3+} conductivity in $Sc_2(WO_4)_3$ remains lower compared to well-known monovalent or divalent cation conductors, its discovery opened a new phase in the field of SEs.

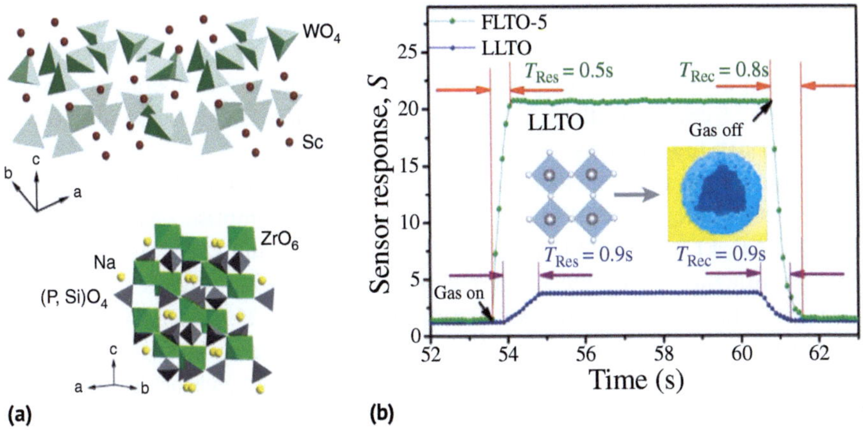

Figure 4.9 SSEs in sensor. (a) $Sc_2(WO_4)_3$-type structure (top) and NASICON ($Na_{1+x}Zr_2P_{3-x}Si_xO_{12}$)-type structure (bottom). *Source*: [142] / with permission of Elsevier. (b) Perovskite LLTO-type structure. *Source*: Reprinted with permission from [144]; © 2020 The Royal Society of Chemistry.

Since 1995, multiple trivalent cations have been reported in layered $Sc_2(WO_4)_3$-type solids [142] and three-dimensional NASICON-type solids [143]. Among the developed trivalent cation-conducting solids, the NASICON-type $(Al_{0.2}Zr_{0.8})_{20/19}Nb(PO_4)_3$ solid exhibits the highest trivalent ion (here, Al^{3+}) conductivity, reaching approximately 5×10^{-4} S cm^{-1} at 600 °C. While the conductivity of $(Al_{0.2}Zr_{0.8})_{20/19}Nb(PO_4)_3$ is lower than that of representative monovalent and divalent cation-conducting SEs used in gas sensors, its high chemical stability suggests potential for achieving sensors with superior selectivity and long-term stability. Expanding the range of candidate materials for sensor SEs may also lead to the application of novel materials in other sensor components.

Furthermore, LLTO based on a perovskite-type hollow porous structure has demonstrated unique advantages in H_2S gas sensing [144]. As illustrated, its distinctive material architecture enables ultrafast response and recovery speeds to H_2S (Figure 4.9b), while optimization of its band structure has improved both response values and operating temperatures [144, 145].

4.4 Study of Oxide-based Composite Electrolytes

In 1998, Scrosati and coworkers [146] incorporated nanosized TiO_2 and Al_2O_3 ceramic powders into a PEO matrix, achieving a one to two orders of magnitude increase in ionic conductivity while enhancing the mechanical properties of the polymer matrix. This novel organic–inorganic composite electrolyte sparked significant research interest. Researchers classified fillers as either "inert fillers" or "active fillers" based on their ability to facilitate rapid Li^+ transport. Early studies primarily focused on inert fillers incapable of Li^+ transport. When added to the polymer matrix, these fillers not only improved mechanical properties but also altered the polymer's crystalline state, thereby enhancing Li^+ transport. Active fillers offer additional advantages: they exhibit intrinsic room-temperature conductivity $\geq 10^{-4}$ S cm^{-1}, directly participate in Li^+ conduction, and possess wide electrochemical stability windows.

Common inert fillers include Al_2O_3 [147], TiO_2 [146], SiO_2 [148, 149], graphene oxide (GO) [150], and MOFs [151], while active fillers comprise LATP [152, 153], LAGP [154], LLTO [155], LLZO [156, 157], LGPS [154], and LPOS [158]. Multiple factors influence composite performance, including ceramic particle properties (size, morphology), filler ratio, and dispersion uniformity. The mechanisms behind conductivity enhancement remain debated. Scrosati and coworkers [146] proposed that surface functional groups on ceramic particles act as Lewis acids, interacting with lithium salt anions to immobilize them and increase free Li^+ concentration. Additionally, ceramic particles may crosslink with PEO chains, reducing crystallinity and promoting Li^+ mobility. Advanced characterization techniques and computational simulations have been employed to elucidate ion transport mechanisms. Finite element simulations [159] revealed accelerated Li^+ migration at polymer–ceramic interfaces. In PEO–LLZO systems, Hu and coworkers [160] used high-resolution solid-state Li NMR and isotopic labeling to demonstrate preferential Li^+ migration through the ceramic phase rather than the polymer matrix or interfaces. The effects of LLZTO particle size and loading content on PEO-based composite electrolytes have been systematically investigated.

Ceramic particle size and filler ratio critically affect performance. The effects of LLZTO particle size and loading on PEO composites have systematically studied [154]. Nanosized LLZTO (vs. micron-scale) significantly improved conductivity (2.1×10^{-4} S cm^{-1} at 30 °C), attributed to percolation effects [161]. The optimal filler ratio decreased with smaller particle size, suggesting lower percolation thresholds for nanoparticles. Enhanced conductivity in nanocomposites likely stems from larger interfacial areas, providing more Li^+ transport pathways. Goodenough and coworkers [161] prepared LLZTO–PEO films across all ratios via hot pressing, identifying the "percolation threshold" for peak conductivity (Figure 4.10). Below this

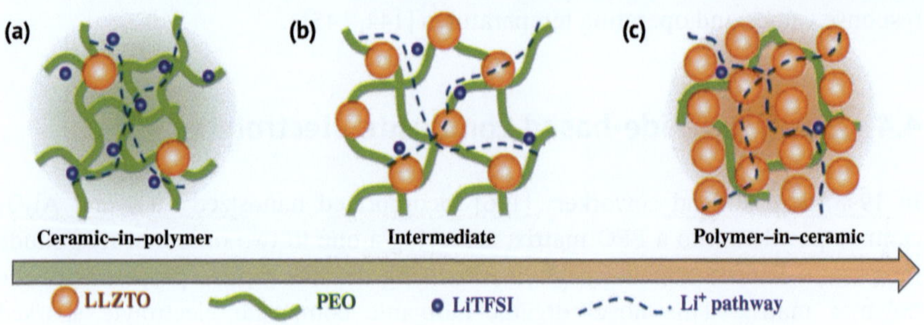

Figure 4.10 Schematic illustration of CPEs with different composite ratios: (a) fewer ceramic particles; (b) permeability threshold; and (c) more ceramic particles. *Source*: [162] / with permission of Elsevier.

threshold, conductivity relies on polymer segmental motion, with fillers reducing crystallinity. At the threshold, dual transport mechanisms (ceramic phase + polymer segments) operate. Exceeding the threshold degrades conductivity due to ceramic agglomeration disrupting Li^+ pathways.

The dispersion degree of ceramic particles is another critical factor influencing the conductivity of composite electrolytes. The primary preparation methods for composite electrolytes include solution casting [157], hot pressing [109, 162], and in situ polymerization [151]. These methods involves dissolving polymers and lithium salts in organic solvents to form a homogeneous solution, which is then cast into a mold. After solvent evaporation, a flexible freestanding polymer membrane is obtained. While solution casting is operationally simple and suitable for large-scale production, it often leads to ceramic particle agglomeration. Hot pressing is applicable to composites with varying ratios but requires specialized equipment and precise temperature/pressure control. In this approach, reactive monomers are introduced into the interlayer spaces of nanomaterials to initiate polymerization. This technique enables uniform dispersion of ceramic particles within the polymer matrix. For instance, Yap et al. [148] reported the in situ hydrolysis synthesis of nanosized SiO_2 spheres in a PEO matrix. Compared to conventional methods, the in situ composite exhibited stronger chemical interactions between SiO_2 and PEO chains, enhancing $LiClO_4$ dissociation and suppressing PEO crystallization. The resulting ionic conductivity reached 4.4×10^{-5} S cm^{-1} at 30 °C.

The morphology of inorganic fillers significantly impacts conductivity. As illustrated in Figure 4.11 [162], beyond ceramic particles, constructing 2D or 3D inorganic architectures (e.g., porous structures [150], nanowires [64, 99], or nanopillars [93]) can

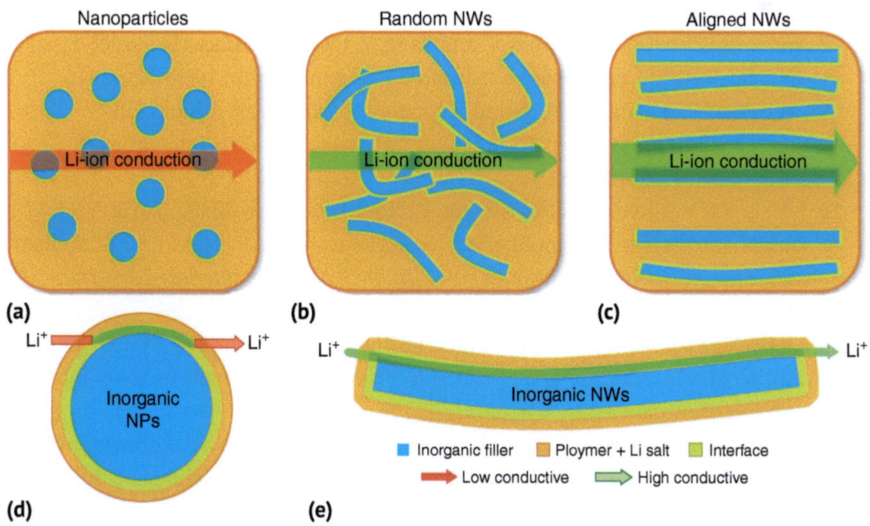

Figure 4.11 Different lithium-ion conduction pathways in composite electrolytes: (a) nanoparticles; (b) random nanowires; (c) aligned nanowires; (d) phase interface of nanoparticles; and (e) phase interface of nanowires. *Source*: [163] / with permission of Springer Nature.

further improve electrochemical performance. Advanced synthesis techniques—such as electrospinning, ice-templating, and hydrogel methods—have enabled diverse inorganic morphologies in composites. For example, Yu and coworkers [163] developed a 3D LLZO scaffold via hydrogel templating, infiltrated with PEO to form a composite electrolyte. This design achieved rapid ion transport, mechanical robustness, and interfacial stability, offering a novel strategy for organic–inorganic hybrid electrolytes.

References

1 Schmaltz, T., Hartmann, F., Wicke, T., Weymann, L., Neef, C. and Janek, J. (2023) A roadmap for solid-state batteries. *Advanced Energy Materials*, 13(43), 2301886.

2 Janek, J. and Zeier, W. G. (2023) Challenges in speeding up solid-state battery development. *Nature Energy*, 8(3), 230–240.

3 Zhu, Y., Gonzalez-Rosillo, J. C., Balaish, M., Hood, Z. D., Kim, K. J. and Rupp, J. L. (2021) Lithium-film ceramics for solid-state lithionic devices. *Nature Reviews Materials*, 6(4), 313–331.

4 Yang, S., Zhang, F., Ding, H., He, P. and Zhou, H. (2018) Lithium metal extraction from seawater. *Joule*, 2(9), 1648–1651.

5 Wei, R., Chen, S., Gao, T. and Liu, W. (2021) Challenges, fabrications and horizons of oxide solid electrolytes for solid-state lithium batteries. *Nano Select*, 2(12), 2256–2274.

6 Nanda, S., Gupta, A. and Manthiram, A. (2021) Anode-free full cells: a pathway to high-energy density lithium-metal batteries. *Advanced Energy Materials*, 11(2), 2000804.

7 Xia, S., Wu, X., Zhang, Z., Cui, Y. and Liu, W. (2019) Practical challenges and future perspectives of all-solid-state lithium-metal batteries. *Chem*, 5(4), 753–785.

8 Wang, D., Zhu, C., Fu, Y., Sun, X. and Yang, Y. (2020) Interfaces in garnet-based all-solid-state lithium batteries. *Advanced Energy Materials*, 10(39), 2001318.

9 Murugan, R., Thangadurai, V. and Weppner, W. (2007) Fast lithium ion conduction in garnet-type $Li_7La_3Zr_2O_{12}$. *Angewandte Chemie International Edition*, 46(41), 7778–7781.

10 Nie, L., Li, Y., Chen, S., Li, K., Huang, Y., Zhu, Y., Sun, Z., Zhang, J., He, Y. and Cui, M. (2019) Biofilm nanofiber-coated separators for dendrite-free lithium metal anode and ultrahigh-rate lithium batteries. *ACS Applied Materials & Interfaces*, 11(35), 32373–32380.

11 Liang, Y., Zhao, C.-Z., Yuan, H., Chen, Y., Zhang, W., Huang, J.-Q., Yu, D., Liu, Y., Titirici, M. M. and Chueh, Y. L. (2019) A review of rechargeable batteries for portable electronic devices. *InfoMat*, 1(1), 6–32.

12 He, Y., Chen, S., Nie, L., Sun, Z., Wu, X. and Liu, W. (2020) Stereolithography three-dimensional printing solid polymer electrolytes for all-solid-state lithium metal batteries. *Nano Letters*, 20(10), 7136–7143.

13 Liu, H., Cheng, X.-B., Huang, J.-Q., Yuan, H., Lu, Y., Yan, C., Zhu, G.-L., Xu, R., Zhao, C.-Z. and Hou, L.-P. (2020) Controlling dendrite growth in solid-state electrolytes. *ACS Energy Letters*, 5(3), 833–843.

14 Banerjee, A., Wang, X., Fang, C., Wu, E. A. and Meng, Y. S. (2020) Interfaces and interphases in all-solid-state batteries with inorganic solid electrolytes. *Chemical Reviews*, 120(14), 6878–6933.

15 Wang, C., Ping, W., Bai, Q., Cui, H., Hensleigh, R., Wang, R., Brozena, A. H., Xu, Z., Dai, J. and Pei, Y. (2020) A general method to synthesize and sinter bulk ceramics in seconds. *Science*, 368(6490), 521–526.

16 Kamaya, N., Homma, K., Yamakawa, Y., Hirayama, M., Kanno, R., Yonemura, M., Kamiyama, T., Kato, Y., Hama, S. and Kawamoto, K. (2011) A lithium superionic conductor. *Nature Materials*, 10(9), 682–686.

17 Zhao, Q., Stalin, S., Zhao, C.-Z. and Archer, L. A. (2020) Designing solid-state electrolytes for safe, energy-dense batteries. *Nature Reviews Materials*, 5(3), 229–252.

18 Li, J., Ma, C., Chi, M., Liang, C. and Dudney, N. J. (2014) Solid electrolyte: the key for high-voltage lithium batteries. *Advanced Energy Materials*, 5(4), 1401408.

19 Xu, T., Zhao, M., Su, Z., Duan, W., Shi, Y., Li, Z., Pol, V. G. and Song, X. (2021) Nanostructured $LiTi_2(PO_4)_3$ anode with superior lithium and sodium storage capability in aqueous electrolytes. *Journal of Power Sources*, 481, 229110.

20 Hayamizu, K., Seki, S. and Haishi, T. (2018) ^{7}Li NMR diffusion studies in micrometre-space for perovskite-type $Li_{0.33}La_{0.55}TiO_3$ (LLTO) influenced by grain boundaries. *Solid State Ionics*, 326, 37–47.

21 Aleksandrov, D., Popovich, A., Qingsheng, W. and Novikov, P. (2020) Air stability of tetragonal solid-state electrolyte $Li_7La_3Zr_2O_{12}$. *Materials Today: Proceedings*, 30, 583–586.

22 Balaish, M., Gonzalez-Rosillo, J. C., Kim, K. J., Zhu, Y., Hood, Z. D. and Rupp, J. L. (2021) Processing thin but robust electrolytes for solid-state batteries. *Nature Energy*, 6(3), 227–239.

23 Arbi, K., Bucheli, W., Jiménez, R. and Sanz, J. (2015) High lithium ion conducting solid electrolytes based on NASICON $Li_{1+x}Al_xM_{2-x}(PO_4)_3$ materials (M = Ti, Ge and $0 \leq x \leq 0.5$). *Journal of the European Ceramic Society*, 35(5), 1477–1484.

24 Anantharamulu, N., Koteswara Rao, K., Rambabu, G., Vijaya Kumar, B., Radha, V. and Vithal, M. (2011) A wide-ranging review on NASICON-type materials. *Journal of Materials Science*, 46(9), 2821–2837.

25 Abhilash, K., Sivaraj, P., Selvin, P. C., Nalini, B. and Somasundaram, K. (2015) Investigation on spin-coated LLTO thin-film nano-electrolytes for rechargeable lithium-ion batteries. *Ceramics International*, 41(10), 13823–13829.

26 Abhilash, K., Selvin, P. C., Nalini, B., Nithyadharseni, P. and Pillai, B. (2013) Investigations on pure and Ag-doped lithium lanthanum titanate (LLTO) nanocrystalline ceramic electrolytes for rechargeable lithium-ion batteries. *Ceramics International*, 39(2), 947–952.

27 Ping, W., Wang, C., Wang, R., Dong, Q., Lin, Z., Brozena, A. H., Dai, J., Luo, J. and Hu, L. (2020) Printable, high-performance solid-state electrolyte films. *Science Advances*, 6(47), eabc8641.

28 Bates, J., Dudney, N., Gruzalski, G., Zuhr, R., Choudhury, A., Luck, C. and Robertson, J. (1993) Fabrication and characterization of amorphous lithium electrolyte thin films and rechargeable thin-film batteries. *Journal of Power Sources*, 43 (1–3), 103–110.

29 Wang, C., Fu, K., Kammampata, S. P., McOwen, D. W., Samson, A. J., Zhang, L., Hitz, G. T., Nolan, A. M., Wachsman, E. D. and Mo, Y. (2020) Garnet-type solid-state electrolytes: materials, interfaces, and batteries. *Chemical Reviews*, 120(10), 4257–4300.

30 Su, Y., Falgenhauer, J., Polity, A., Leichtweiß, T., Kronenberger, A., Obel, J., Zhou, S., Schlettwein, D., Janek, J. and Meyer, B. K. (2015) LiPON thin films with high nitrogen content for application in lithium batteries and electrochromic devices prepared by RF magnetron sputtering. *Solid State Ionics*, 282, 63–69.

31 Ruzmetov, D., Oleshko, V. P., Haney, P. M., Lezec, H. J., Karki, K., Baloch, K. H., Agrawal, A. K., Davydov, A. V., Krylyuk, S. and Liu, Y. (2012) Electrolyte stability determines scaling limits for solid-state 3D Li-ion batteries. *Nano Letters*, 12(1), 505–511.

32 Kozen, A. C., Pearse, A. J., Lin, C.-F., Noked, M. and Rubloff, G. W. (2015) Atomic layer deposition of the solid electrolyte LiPON. *Chemistry of Materials*, 27(15), 5324–5331.

33 West, W. C., Hood, Z. D., Adhikari, S. P., Liang, C., Lachgar, A., Motoyama, M. and Iriyama, Y. (2016) Reduction of charge-transfer resistance at the solid electrolyte–electrode interface by pulsed laser deposition of films from a crystalline Li_2PO_2N source. *Journal of Power Sources*, 312, 116–122.

34 Zhu, Y., He, X. and Mo, Y. (2016) First-principles study on electrochemical and chemical stability of solid electrolyte–electrode interfaces in all-solid-state Li-ion batteries. *Journal of Materials Chemistry A*, 4(9), 3253–3266.

35 Goodenough, J. B., Hong, H.-P. and Kafalas, J. (1976) Fast Na^+-ion transport in skeleton structures. *Materials Research Bulletin*, 11(2), 203–220.

36 Petit, D., Colomban, P., Collin, G. and Boilot, J. (1986) Fast ion transport in $LiZr_2(PO_4)_3$: structure and conductivity. *Materials Research Bulletin*, 21(3), 365–371.

37 Bucharsky, E., Schell, K., Hintennach, A. and Hoffmann, M. (2015) Preparation and characterization of sol–gel derived high lithium-ion conductive NZP-type ceramics $Li_{1+x}Al_xTi_{2-x}(PO_4)_3$. *Solid State Ionics*, 274, 77–82.

38 Xiao, W., Wang, J., Fan, L., Zhang, J. and Li, X. (2019) Recent advances in $Li_{1+x}Al_xTi_{2-x}(PO_4)_3$ solid-state electrolyte for safe lithium batteries. *Energy Storage Materials*, 19, 379–400.

39 Wu, X. M., Chen, S., Mai, F. R., Zhao, J. H. and He, Z. Q. (2013) Influence of the annealing technique on the properties of Li-ion-conductive $Li_{1.3}Al_{0.3}Ti_{1.7}(PO_4)_3$ films. *Ionics*, 19(4), 589–593.

40 Popovici, D., Nagai, H., Fujishima, S. and Akedo, J. (2011) Preparation of lithium aluminum titanium phosphate electrolyte thick films by aerosol deposition method. *Journal of the American Ceramic Society*, 94(11), 3847–3850.

41 Ling, Q., Yu, Z., Xu, H., Zhu, G., Zhang, X., Zhao, Y. and Yu, A. (2016) Preparation and electrical properties of amorphous Li–Al–Ti–PO thin film electrolyte. *Materials Letters*, 169, 42–45.

42 Akedo, J. (2008) Room temperature impact consolidation (RTIC) of fine ceramic powder by aerosol deposition method and applications to microdevices. *Journal of Thermal Spray Technology*, 17(2), 181–198.

43 Takada, K. (2013) Progress and prospective of solid-state lithium batteries. *Acta Materialia*, 61(3), 759–770.

44 García-Martín, S., Amador, U., Morata-Orrantia, A., Rodríguez-Carvajal, J. and Alario-Franco, M. Á. (2009) Structure, microstructure, composition and properties of lanthanum lithium titanates and some substituted analogues. *Research Report*, 635(15), 2363–2373.

45 Kwon, W. J., Kim, H., Jung, K.-N., Cho, W., Kim, S. H., Lee, J.-W. and Park, M.-S. (2017) Enhanced Li^+ conduction in perovskite $Li_{3x}La_{2/3-x}\square_{1/3-2x}TiO_3$ solid electrolytes via microstructural engineering. *Journal of Materials Chemistry A*, 5(13), 6257–6262.

46 Ma, C., Chen, K., Liang, C., Nan, C.-W., Ishikawa, R., More, K. and Chi, M. (2014) Atomic-scale origin of the large grain-boundary resistance in perovskite Li-ion-conducting solid electrolytes. *Energy & Environmental Science*, 7(5), 1638–1642.

47 Furusawa, S.-I., Tabuchi, H., Sugiyama, T., Tao, S. and Irvine, J. T. (2005) Ionic conductivity of amorphous lithium lanthanum titanate thin film. *Solid State Ionics*, 176 (5–6), 553–558.

48 Ulusoy, S., Gulen, S., Aygun, G., Ozyuzer, L. and Ozdemir, M. (2018) Characterization of thin film $Li_{0.5}La_{0.5}Ti_{1-x}Al_xO_3$ electrolyte for all-solid-state Li-ion batteries. *Solid State Ionics*, 324, 226–232.

49 Kim, S., Hirayama, M., Cho, W., Kim, K., Kobayashi, T., Kaneko, R., Suzuki, K. and Kanno, R. (2014) Low-temperature synthesis and ionic conductivity of the epitaxial $Li_{0.17}La_{0.61}TiO_3$ film electrolyte. *CrystEngComm*, 16(6), 1044–1049.

50 Le, H. T., Ngo, D. T., Kim, Y.-J., Park, C.-N. and Park, C.-J. (2017) A perovskite-structured aluminium-substituted lithium lanthanum titanate as a potential artificial solid-electrolyte interface for aqueous rechargeable lithium-metal-based batteries. *Electrochimica Acta*, 248, 232–242.

51 Thangadurai, V. and Weppner, W. (2005) $Li_6ALa_2Ta_2O_{12}$ (A = Sr, Ba): novel garnet-like oxides for fast lithium-ion conduction. *Advanced Functional Materials*, 15(1), 107–112.

52 Bernuy-Lopez, C., Manalastas, Jr., W., del Amo, J. M. L., Aguadero, A., Aguesse, F. and Kilner, J. A. (2014) Atmosphere-controlled processing of Ga-substituted garnets for high Li-ion conductivity ceramics. *Chemistry of Materials*, 26(12), 3610–3617.

53 Tenhaeff, W. E., Wang, Y., Sokolov, A. P., Wolfenstine, J., Sakamoto, J., Dudney, N. J. and Rangasamy, E. (2013) Resolving the grain boundary and lattice impedance of hot-pressed $Li_7La_3Zr_2O_{12}$ garnet electrolytes. *ChemElectroChem*, 1(2), 375–378.

54 Knauth, P. and Tuller, H. L. (2002) Solid-state ionics: roots, status, and future prospects. *Journal of the American Ceramic Society*, 85(7), 1654–1680.

55 Rawlence, M., Filippin, A. N., Wackerlin, A., Lin, T.-Y., Cuervo-Reyes, E., Remhof, A., Battaglia, C., Rupp, J. L. and Buecheler, S. (2018) Effect of gallium substitution on lithium-ion conductivity and phase evolution in sputtered $Li_{7-3x}Ga_xLa_3Zr_2O_{12}$ thin films. *ACS Applied Materials & Interfaces*, 10(16), 13720–13728.

56 Garbayo, I., Struzik, M., Bowman, W. J., Pfenninger, R., Stilp, E. and Rupp, J. L. (2018) Glass-type polyamorphism in Li-garnet thin film solid-state battery conductors. *Advanced Energy Materials*, 8(12), 1702265.

57 Katsui, H. and Goto, T. (2015) Preparation of cubic and tetragonal $Li_7La_3Zr_2O_{12}$ film by metal organic chemical vapor deposition. *Thin Solid Films*, 584, 130–134.

58 Bitzer, M., Van Gestel, T. and Uhlenbruck, S. (2016) Sol–gel synthesis of thin solid $Li_7La_3Zr_2O_{12}$ electrolyte films for Li-ion batteries. *Thin Solid Films*, 615, 128–134.

59 Kazyak, E., Chen, K.-H., Wood, K. N., Davis, A. L., Thompson, T., Bielinski, A. R., Sanchez, A. J., Wang, X., Wang, C. and Sakamoto, J. (2017) Atomic layer deposition of the solid electrolyte garnet $Li_7La_3Zr_2O_{12}$. *Chemistry of Materials*, 29(8), 3785–3792.

60 Wachtel, E. and Lubomirsky, I. (2010) Quasi-amorphous inorganic thin films: non-crystalline polar phases. *Advanced Materials*, 22(23), 2485–2493.

61 Yang, X., Adair, K. R., Gao, X. and Sun, X. (2021) Recent advances and perspectives on thin electrolytes for high-energy-density solid-state lithium batteries. *Energy & Environmental Science*, 14(2), 643–671.

62 Lee, Y.-G., Fujiki, S., Jung, C., Suzuki, N., Yashiro, N., Omoda, R., Ko, D.-S., Shiratsuchi, T., Sugimoto, T. and Ryu, S. (2020) High-energy long-cycling all-solid-state lithium metal batteries enabled by silver–carbon composite anodes. *Nature Energy*, 5(4), 299–308.

63 Ates, T., Keller, M., Kulisch, J., Adermann, T. and Passerini, S. (2019) Development of an all-solid-state lithium battery by slurry-coating procedures using a sulfidic electrolyte. *Energy Storage Materials*, 17, 204–210.

64 Yan, X., Li, Z., Ying, H., Nie, F., Xue, L., Wen, Z. and Han, W.-Q. (2018) A novel thin solid electrolyte film and its application in all-solid-state battery at room temperature. *Ionics*, 24(5), 1545–1551.

65 Yamamoto, M., Terauchi, Y., Sakuda, A. and Takahashi, M. (2018) Binder-free sheet-type all-solid-state batteries with enhanced rate capabilities and high energy densities. *Scientific Reports*, 8(1), 1212.

66 Duan, H., Fan, M., Chen, W. P., Li, J. Y., Wang, P. F., Wang, W. P., Shi, J. L., Yin, Y. X., Wan, L. J. and Guo, Y. G. (2019) Extended electrochemical window of solid electrolytes via heterogeneous multilayered structure for high-voltage lithium metal batteries. *Advanced Materials*, 31(12), e1807789.

67 Yan, X., Li, Z., Wen, Z. and Han, W. (2017) $Li/Li_7La_3Zr_2O_{12}/LiFePO_4$ all-solid-state battery with ultrathin nanoscale solid electrolyte. *The Journal of Physical Chemistry C*, 121(3), 1431–1435.

68 Sakuda, A., Kuratani, K., Yamamoto, M., Takahashi, M., Takeuchi, T. and Kobayashi, H. (2017) All-solid-state battery electrode sheets prepared by a slurry coating process. *Journal of The Electrochemical Society*, 164(12), A2474.

69 Nam, Y. J., Oh, D. Y., Jung, S. H. and Jung, Y. S. (2018) Toward practical all-solid-state lithium-ion batteries with high energy density and safety: comparative study for electrodes fabricated by dry- and slurry-mixing processes. *Journal of Power Sources*, 375, 93–101.

70 Ma, C., Dai, K., Hou, H., Ji, X., Chen, L., Ivey, D. G. and Wei, W. (2018) High ion-conducting solid-state composite electrolytes with carbon quantum dot nanofillers. *Advanced Science*, 5(5), 1700996.

71 Li, X., Wang, D., Wang, H., Yan, H., Gong, Z. and Yang, Y. (2019) Poly(ethylene oxide)–$Li_{10}SnP_2S_{12}$ composite polymer electrolyte enables high-performance all-solid-state lithium sulfur battery. *ACS Applied Materials & Interfaces*, 11(25), 22745–22753.

72 Zhu, P., Yan, C., Zhu, J., Zang, J., Li, Y., Jia, H., Dong, X., Du, Z., Zhang, C., Wu, N., Dirican, M. and Zhang, X. (2019) Flexible electrolyte-cathode bilayer framework with stabilized interface for room-temperature all-solid-state lithium-sulfur batteries. *Energy Storage Materials*, 17, 220–225.

73 Tan, D. H. S., Banerjee, A., Deng, Z., Wu, E. A., Nguyen, H., Doux, J.-M., Wang, X., Cheng, J.-H., Ong, S. P., Meng, Y. S. and Chen, Z. (2019) Enabling thin and flexible solid-state composite electrolytes by the scalable solution process. *ACS Applied Energy Materials*, 2(9), 6542–6550.

74 Chen, X., He, W., Ding, L.-X., Wang, S. and Wang, H. (2019) Enhancing interfacial contact in all solid-state batteries with a cathode-supported solid electrolyte membrane framework. *Energy & Environmental Science*, 12(3), 938–944.

75 Hovington, P., Lagacé, M., Guerfi, A., Bouchard, P., Mauger, A., Julien, C. M., Armand, M. and Zaghib, K. (2015) New lithium metal polymer solid-state battery for an ultrahigh energy: nano C–$LiFePO_4$ versus nano $Li_{1.2}V_3O_8$. *Nano Letters*, 15(4), 2671–2678.

76 Zhang, C., Lin, Y., Zhu, Y., Zhang, Z. and Liu, J. (2017) Improved lithium-ion and electrically conductive sulfur cathode for all-solid-state lithium–sulfur batteries. *RSC Advances*, 7(31), 19231–19236.

77 Tao, X., Liu, Y., Liu, W., Zhou, G., Zhao, J., Lin, D., Zu, C., Sheng, O., Zhang, W., Lee, H.-W. and Cui, Y. (2017) Solid-state lithium–sulfur batteries operated at 37 °C with composites of nanostructured $Li_7La_3Zr_2O_{12}$/carbon foam and polymer. *Nano Letters*, 17(5), 2967–2972.

78 Zhu, K., Liu, Y. and Liu, J. (2014) A fast charging/discharging all-solid-state lithium ion battery based on PEO–MIL-53(Al)–LiTFSI thin film electrolyte. *RSC Advances*, 4(80), 42278–42284.

79 Lin, D., Yuen, P. Y., Liu, Y., Liu, W., Liu, N., Dauskardt, R. H. and Cui, Y. (2018) A silica-aerogel-reinforced composite polymer electrolyte with high ionic conductivity and high modulus. *Advanced Materials*, 30(32), 1802661.

80 Liu, K., Wu, M., Jiang, H., Lin, Y. and Zhao, T. (2020) An ultrathin, strong, flexible composite solid electrolyte for high-voltage lithium metal batteries. *Journal of Materials Chemistry A*, 8(36), 18802–18809.

81 Xu, S., McOwen, D. W., Zhang, L., Hitz, G. T., Wang, C., Ma, Z., Chen, C., Luo, W., Dai, J., Kuang, Y., Hitz, E. M., Fu, K., Gong, Y., Wachsman, E. D. and Hu, L. (2018) All-in-one lithium-sulfur battery enabled by a porous-dense-porous garnet architecture. *Energy Storage Materials*, 15, 458–464.

82 Zhang, M., Huang, Z., Cheng, J., Yamamoto, O., Imanishi, N., Chi, B., Pu, J. and Li, J. (2014) Solid-state lithium ionic conducting thin film $Li_{1.4}Al_{0.4}Ge_{1.6}(PO_4)_3$ prepared by tape casting. *Journal of Alloys and Compounds*, 590, 147–152.

83 Jiang, Z., Wang, S., Chen, X., Yang, W., Yao, X., Hu, X., Han, Q. and Wang, H. (2020) Tape-casting $Li_{0.34}La_{0.56}TiO_3$ ceramic electrolyte films permit high energy density of lithium-metal batteries. *Advanced Materials*, 32(6), 1906221.

84 Hitz, G. T., McOwen, D. W., Zhang, L., Ma, Z., Fu, Z., Wen, Y., Gong, Y., Dai, J., Hamann, T. R., Hu, L. and Wachsman, E. D. (2019) High-rate lithium cycling in a scalable trilayer Li–garnet–electrolyte architecture. *Materials Today*, 22, 50–57.

85 Fu, K., Gong, Y., Hitz, G. T., McOwen, D. W., Li, Y., Xu, S., Wen, Y., Zhang, L., Wang, C., Pastel, G., Dai, J., Liu, B., Xie, H., Yao, Y., Wachsman, E. D. and Hu, L. (2017) Three-dimensional bilayer garnet solid electrolyte-based high energy density lithium metal–sulfur batteries. *Energy & Environmental Science*, 10(7), 1568–1575.

86 Shen, H., Yi, E., Amores, M., Cheng, L., Tamura, N., Parkinson, D. Y., Chen, G., Chen, K. and Doeff, M. (2019) Oriented porous LLZO 3D structures obtained by freeze casting for battery applications. *Journal of Materials Chemistry A*, 7(36), 20861–20870.

87 Xu, S., McOwen, D. W., Wang, C., Zhang, L., Luo, W., Chen, C., Li, Y., Gong, Y., Dai, J., Kuang, Y., Yang, C., Hamann, T. R., Wachsman, E. D. and Hu, L. (2018) Three-dimensional, solid-state mixed electron–ion conductive framework for lithium metal anode. *Nano Letters*, 18(6), 3926–3933.

88 Yi, E., Wang, W., Kieffer, J. and Laine, R. M. (2016) Flame-made nanoparticles permit processing of dense, flexible, Li^+-conducting ceramic electrolyte thin films of cubic $Li_7La_3Zr_2O_{12}$ (c-LLZO). *Journal of Materials Chemistry A*, 4(33), 12947–12954.

89 Yi, E., Wang, W., Kieffer, J. and Laine, R. M. (2017) Key parameters governing the densification of cubic $Li_7La_3Zr_2O_{12}$ Li^+ conductors. *Journal of Power Sources*, 352, 156–164.

90 Gong, Y., Fu, K., Xu, S., Dai, J., Hamann, T. R., Zhang, L., Hitz, G. T., Fu, Z., Ma, Z., McOwen, D. W., Han, X., Hu, L. and Wachsman, E. D. (2018) Lithium-ion conductive ceramic textile: a new architecture for flexible solid-state lithium metal batteries. *Materials Today*, 21(6), 594–601.

91 Paolella, A., Zhu, W., Xu, G.-L., La Monaca, A., Savoie, S., Girard, G., Vijh, A., Demers, H., Perea, A., Delaporte, N., Guerfi, A., Liu, X., Ren, Y., Sun, C.-J., Lu, J., Amine, K. and Zaghib, K. (2020) Understanding the reactivity of a thin $Li_{1.5}Al_{0.5}Ge_{1.5}(PO_4)_3$ solid-state electrolyte toward metallic lithium anode. *Advanced Energy Materials*, 10(32), 2001497.

92 Cai, T.-J., Lo, Y.-H. and Wu, J.-J. (2019) Porous sulfide scaffolded solvent-free PEG–Ti hybrid polymer: all-solution-processed thin film composite polymer electrolytes directly on electrodes for lithium-ion batteries. *Materials Today Energy*, 13, 119–124.

93 Zhai, H., Xu, P., Ning, M., Cheng, Q., Mandal, J. and Yang, Y. (2017) A flexible solid composite electrolyte with vertically aligned and connected ion-conducting nanoparticles for lithium batteries. *Nano Letters*, 17(5), 3182–3187.

94 Jiang, T., He, P., Wang, G., Shen, Y., Nan, C.-W. and Fan, L.-Z. (2020) Solvent-free synthesis of thin, flexible, nonflammable garnet-based composite solid electrolyte for all-solid-state lithium batteries. *Advanced Energy Materials*, 10(12), 1903376.

95 Hu, C., Shen, Y., Shen, M., Liu, X., Chen, H., Liu, C., Kang, T., Jin, F., Li, L., Li, J., Li, Y., Zhao, N., Guo, X., Lu, W., Hu, B. and Chen, L. (2020) Superionic conductors via bulk interfacial conduction. *Journal of the American Chemical Society*, 142(42), 18035–18041.

96 Wu, J., Rao, Z., Cheng, Z., Yuan, L., Li, Z. and Huang, Y. (2019) Ultrathin, flexible polymer electrolyte for cost-effective fabrication of all-solid-state lithium metal batteries. *Advanced Energy Materials*, 9(46), 1902767.

97 Wan, J., Xie, J., Kong, X., Liu, Z., Liu, K., Shi, F., Pei, A., Chen, H., Chen, W., Chen, J., Zhang, X., Zong, L., Wang, J., Chen, L.-Q., Qin, J. and Cui, Y. (2019) Ultrathin, flexible, solid polymer composite electrolyte enabled with aligned nanoporous host for lithium batteries. *Nature Nanotechnology*, 14(7), 705–711.

98 Kim, D. H., Lee, Y.-H., Song, Y. B., Kwak, H., Lee, S.-Y. and Jung, Y. S. (2020) Thin and flexible solid electrolyte membranes with ultrahigh thermal stability derived from solution-processable Li argyrodites for all-solid-state Li-ion batteries. *ACS Energy Letters*, 5(3), 718–727.

99 Fu, K., Gong, Y., Dai, J., Gong, A., Han, X., Yao, Y., Wang, C., Wang, Y., Chen, Y., Yan, C., Li, Y., Wachsman, E. D. and Hu, L. (2016) Flexible, solid-state, ion-conducting membrane with 3D garnet nanofiber networks for lithium batteries. *Proceedings of the National Academy of Sciences of the United States of America*, 113(26), 7094–7099.

100 Yang, X., Jiang, M., Gao, X., Bao, D., Sun, Q., Holmes, N., Duan, H., Mukherjee, S., Adair, K., Zhao, C., Liang, J., Li, W., Li, J., Liu, Y., Huang, H., Zhang, L., Lu, S., Lu, Q., Li, R., Singh, C. V. and Sun, X. (2020) Determining the limiting factor of the electrochemical stability window for PEO-based solid polymer electrolytes: main chain or terminal—OH group? *Energy & Environmental Science*, 13(5), 1318–1325.

101 Li, Y., Wang, X., Zhou, H., Xing, X., Banerjee, A., Holoubek, J., Liu, H., Meng, Y. S. and Liu, P. (2020) Thin solid electrolyte layers enabled by nanoscopic polymer binding. *ACS Energy Letters*, 5(3), 955–961.

102 Xu, H., Xuan, M., Xiao, W., Shen, Y., Li, Z., Wang, Z., Hu, J. and Shao, G. (2019) Lithium ion conductivity in double antiperovskite $Li_{6.5}OS_{1.5}I_{1.5}$: alloying and boundary effects. *ACS Applied Energy Materials*, 2(9), 6288–6294.

103 Xu, R., Yue, J., Liu, S., Tu, J., Han, F., Liu, P. and Wang, C. (2019) Cathode-supported all-solid-state lithium–sulfur batteries with high cell-level energy density. *ACS Energy Letters*, 4(5), 1073–1079.

104 Howlett, P. C., Ponzio, F., Fang, J., Lin, T., Jin, L., Iranipour, N. and Efthimiadis, J. (2013) Thin and flexible solid-state organic ionic plastic crystal–polymer nanofibre composite electrolytes for device applications. *Physical Chemistry Chemical Physics*, 15(33), 13784–13789.

105 Nam, Y. J., Cho, S.-J., Oh, D. Y., Lim, J.-M., Kim, S. Y., Song, J. H., Lee, Y.-G., Lee, S.-Y. and Jung, Y. S. (2015) Bendable and thin sulfide solid electrolyte film: a new electrolyte opportunity for free-standing and stackable high-energy all-solid-state lithium-ion batteries. *Nano Letters*, 15(5), 3317–3323.

106 Whiteley, J. M., Taynton, P., Zhang, W. and Lee, S.-H. (2015) Ultra-thin solid-state Li-ion electrolyte membrane facilitated by a self-healing polymer matrix. *Advanced Materials*, 27(43), 6922–6927.

107 Hood, Z. D., Wang, H., Pandian, A. S., Peng, R., Gilroy, K. D., Chi, M., Liang, C. and Xia, Y. (2018) Fabrication of sub-micrometer-thick solid electrolyte membranes of β-Li_3PS_4 via tiled assembly of nanoscale, plate-like building blocks. *Advanced Energy Materials*, 8(21), 1800014.

108 Li, M., Frerichs, J. E., Kolek, M., Sun, W., Zhou, D., Huang, C. J., Hwang, B. J., Hansen, M. R., Winter, M. and Bieker, P. (2020) Solid-state lithium–sulfur battery enabled by thio-LiSICON/polymer composite electrolyte and sulfurized polyacrylonitrile cathode. *Advanced Functional Materials*, 30(14), 1910123.

109 Keller, M., Appetecchi, G. B., Kim, G.-T., Sharova, V., Schneider, M., Schuhmacher, J., Roters, A. and Passerini, S. (2017) Electrochemical performance of a solvent-free hybrid ceramic–polymer electrolyte based on $Li_7La_3Zr_2O_{12}$ in $P(EO)_{15}LiTFSI$. *Journal of Power Sources*, 353, 287–297.

110 Appetecchi, G. B., Shin, J. H., Alessandrini, F. and Passerini, S. (2005) 0.6 Ah Li/V_2O_5 battery prototypes based on solvent-free PEO–$LiN(SO_2CF_2CF_3)_2$ polymer electrolytes. *Journal of Power Sources*, 143(1), 236–242.

111 McOwen, D. W., Xu, S., Gong, Y., Wen, Y., Godbey, G. L., Gritton, J. E., Hamann, T. R., Dai, J., Hitz, G. T., Hu, L. and Wachsman, E. D. (2018) 3D-printing electrolytes for solid-state batteries. *Advanced Materials*, 30(18), 1707132.

112 Xie, H., Bao, Y., Cheng, J., Wang, C., Hitz, E. M., Yang, C., Liang, Z., Zhou, Y., He, S., Li, T. and Hu, L. (2019) Flexible garnet solid-state electrolyte membranes enabled by tile-and-grout design. *ACS Energy Letters*, 4(11), 2668–2674.

113 Peng, C., Kamiike, Y., Liang, Y., Kuroda, K. and Okido, M. (2019) Thin-film NASICON-type $Li_{1+x}Al_xTi_{2-x}(PO_4)_3$ solid electrolyte directly fabricated on a graphite substrate with a hydrothermal method based on different Al sources. *ACS Sustainable Chemistry & Engineering*, 7(12), 10751–10762.

114 Liang, Y., Peng, C., Kamiike, Y., Kuroda, K. and Okido, M. (2019) Gallium-doped NASICON-type $LiTi_2(PO_4)_3$ thin film grown on graphite anode as solid electrolyte for all-solid-state lithium batteries. *Journal of Alloys and Compounds*, 775, 1147–1155.

115 Wang, H., Hood, Z. D., Xia, Y. and Liang, C. (2016) Fabrication of ultrathin solid electrolyte membranes of β-Li_3PS_4 nanoflakes by evaporation-induced self-assembly for all-solid-state batteries. *Journal of Materials Chemistry A*, 4(21), 8091–8096.

116 Woo, S. P., Kakati, N., Kim, I. Y., Lee, S. H. and Yoon, Y. S. (2016) A new approach: Li_2S–P_2S_5 thin films prepared by thermal evaporation as solid electrolytes. *Journal of the Korean Physical Society*, 69(4), 617–622.

117 Zhao, Q., Liu, X., Stalin, S., Khan, K. and Archer, L. A. (2019) Solid-state polymer electrolytes with in-built fast interfacial transport for secondary lithium batteries. *Nature Energy*, 4(5), 365–373.

118 Duan, H., Yin, Y.-X., Shi, Y., Wang, P.-F., Zhang, X.-D., Yang, C.-P., Shi, J.-L., Wen, R., Guo, Y.-G. and Wan, L.-J. (2018) Dendrite-free Li-metal battery enabled by a thin asymmetric solid electrolyte with engineered layers. *Journal of the American Chemical Society*, 140(1), 82–85.

119 Duan, H., Yin, Y.-X., Zeng, X.-X., Li, J.-Y., Shi, J.-L., Shi, Y., Wen, R., Guo, Y.-G. and Wan, L.-J. (2018) In situ plasticized polymer electrolyte with double network for flexible solid-state lithium-metal batteries. *Energy Storage Materials*, 10, 85–91.

120 Le, H. T. T., Ngo, D. T., Didwal, P. N., Fisher, J. G., Park, C.-N., Kim, I.-D. and Park, C.-J. (2019) Highly efficient and stable solid-state Li–O_2 batteries using a perovskite solid electrolyte. *Journal of Materials Chemistry A*, 7(7), 3150–3160.

121 Zhu, X. B., Zhao, T. S., Wei, Z. H., Tan, P. and An, L. (2015) A high-rate and long cycle life solid-state lithium–air battery. *Energy & Environmental Science*, 8(12), 3745–3754.

122 Zhu, X., Zhao, T., Tan, P., Wei, Z. and Wu, M. (2016) A high-performance solid-state lithium–oxygen battery with a ceramic–carbon nanostructured electrode. *Nano Energy*, 26, 565–576.

123 Yang, X., Sun, Q., Zhao, C., Gao, X., Adair, K., Zhao, Y., Luo, J., Lin, X., Liang, J., Huang, H., Zhang, L., Lu, S., Li, R. and Sun, X. (2019) Self-healing electrostatic shield enabling uniform lithium deposition in all-solid-state lithium batteries. *Energy Storage Materials*, 22, 194–199.

124 Yang, X., Sun, Q., Zhao, C., Gao, X., Adair, K. R., Liu, Y., Luo, J., Lin, X., Liang, J., Huang, H., Zhang, L., Yang, R., Lu, S., Li, R. and Sun, X. (2019) High-areal-capacity all-solid-state lithium batteries enabled by rational design of fast ion transport channels in vertically aligned composite polymer electrodes. *Nano Energy*, 61, 567–575.

125 Yuan, H., Nan, H.-X., Zhao, C.-Z., Zhu, G.-L., Lu, Y., Cheng, X.-B., Liu, Q.-B., He, C.-X., Huang, J.-Q. and Zhang, Q. (2020) Slurry-coated sulfur/sulfide cathode with Li metal anode for all-solid-state lithium–sulfur pouch cells. *Batteries & Supercaps*, 3(7), 596–603.

126 Liang, J., Luo, J., Sun, Q., Yang, X., Li, R. and Sun, X. (2019) Recent progress on solid-state hybrid electrolytes for solid-state lithium batteries. *Energy Storage Materials*, 21, 308–334.

127 Jabbari, M., Bulatova, R., Tok, A. I. Y., Bahl, C. R. H., Mitsoulis, E. and Hattel, J. H. (2016) Ceramic tape casting: a review of current methods and trends with emphasis on rheological behaviour and flow analysis. *Materials Science and Engineering: B*, 212, 39–61.

128 Bae, J., Li, Y., Zhang, J., Zhou, X., Zhao, F., Shi, Y., Goodenough, J. B. and Yu, G. (2018) Cover picture: a 3D nanostructured hydrogel-framework-derived high-performance composite polymer lithium-ion electrolyte (Angew. Chem. Int. Ed. 8/2018). *Angewandte Chemie International Edition*, 57(8), 2007.

129 Schnell, J., Günther, T., Knoche, T., Vieider, C., Köhler, L., Just, A., Keller, M., Passerini, S. and Reinhart, G. (2018) All-solid-state lithium-ion and lithium metal batteries—paving the way to large-scale production. *Journal of Power Sources*, 382, 160–175.

130 Kim, S. H., Kim, J. H., Cho, S. J. and Lee, S. Y. (2019) All-solid-state printed bipolar Li–S batteries. *Advanced Energy Materials*, 9(40), 1901841.

131 Afyon, S., Krumeich, F. and Rupp, J. L. M. (2015) A shortcut to garnet-type fast Li-ion conductors for all-solid-state batteries. *Journal of Materials Chemistry A*, 3(36), 18636–18648.

132 Pfenninger, R., Struzik, M., Garbayo, I., Stilp, E. and Rupp, J. L. M. (2019) A low firing temperature for fast lithium conduction in garnet solid-state battery films. *Nature Energy*, 4(6), 475–483.

133 Li, Y., Cao, Y. and Guo, X. (2013) Influence of lithium oxide additives on densification and ionic conductivity of garnet-type $Li_{6.75}La_3Zr_{1.75}Ta_{0.25}O_{12}$ solid electrolytes. *Solid State Ionics*, 253, 76–80.

134 Wan, J., Xie, J., Mackanic, D. G., Burke, W., Bao, Z. and Cui, Y. (2018) Status, promises, and challenges of nanocomposite solid-state electrolytes for safe and high-performance lithium batteries. *Materials Today Nano*, 4, 1–16.

135 Zhao, W., Yi, J., He, P. and Zhou, H. (2019) Solid-state electrolytes for lithium-ion batteries: fundamentals, challenges and perspectives. *Electrochemical Energy Reviews*, 2(4), 574–605.

136 Ping, W., Wang, C., Lin, Z., Hitz, E., Yang, C., Wang, H. and Hu, L. (2020) Reversible short-circuit behaviors in garnet-based solid-state batteries. *Advanced Energy Materials*, 10(25), 2000702.

137 Yu, J., Wang, Y., Shen, L., Liu, J., Wang, Z., Xu, S., Law, H. M. and Ciucci, F. (2025) Fast-charging solid-state Li batteries: materials, strategies, and prospects. *Advanced Materials*, 37(23), 2417796.

138 Xu, M., Wang, X., Jin, Z., Jiao, X., Gao, Y., Song, Z., Xu, X., Xi, K. and Liu, Y. (2025) Decoupling the roles of grain boundary strength and grain size hidden in grain-level electro-chemo-mechanical failure of solid-state electrolyte. *Journal of Energy Chemistry*, 101, 685–691.

139 Huang, Z., Chen, L., Huang, B., Xu, B., Shao, G., Wang, H., Li, Y. and Wang, C.-A. (2020) Enhanced performance of $Li_{6.4}La_3Zr_{1.4}Ta_{0.6}O_{12}$ solid electrolyte by regulation of grain and grain boundary phases. *ACS Applied Materials & Interfaces*, 12(50), 56118–56125.

140 Ikeda, S., Kato, S., Nomura, K., Ito, K., Einaga, H., Saito, S. and Fujita, Y. (1994) Carbon dioxide sensor using solid electrolytes with zirconium phosphate framework. *Solid State Ionics*, 70–71, 569–571.

141 Tamura, S. and Imanaka, N. (2022) Solid electrolyte-type gas sensors using trivalent cation conducting solid electrolytes. *Sensors and Actuators B: Chemical*, 368, 132252.

142 Imanaka, N., Kobayashi, Y., Fujiwara, K., Asano, T., Okazaki, Y. and Adachi, G.-Y. (1998) Trivalent rare earth ion conduction in the rare earth tungstates with the $Sc_2(WO_4)_3$-type structure. *Chemistry of Materials*, 10(7), 2006–2012.

143 Tamura, S., Araki, K.-i. and Imanaka, N. (2016) Trivalent gallium ion conduction in NASICON-type solid electrolytes. *Journal of Asian Ceramic Societies*, 4(4), 390–393.

144 Zheng, N., Li, X., Yan, S., Wang, Q., Qiao, R., Hu, J., Fan, J., Cao, G. and Shao, G. (2020) Nano-porous hollow $Li_{0.5}La_{0.5}TiO_3$ spheres and electronic structure modulation for ultra-fast H_2S detection. *Journal of Materials Chemistry A*, 8(5), 2376–2386.

145 Li, X., Hu, J., Ban, J., He, S., Zheng, N., Shao, G. and Cao, G. (2022) Mechanism of enhanced H_2S sensing ability based on emerging $Li_{0.5}La_{0.5}TiO_3$–SnO_2 core–shell structure. *Sensors and Actuators B: Chemical*, 352, 131054.

146 Croce, F., Appetecchi, G. B., Persi, L. and Scrosati, B. (1998) Nanocomposite polymer electrolytes for lithium batteries. *Nature*, 394(6692), 456–458.

147 Yang, L., Wang, Z., Feng, Y., Tan, R., Zuo, Y., Gao, R., Zhao, Y., Han, L., Wang, Z. and Pan, F. (2017) Flexible composite solid electrolyte facilitating highly stable "soft contacting" Li–electrolyte interface for solid-state lithium-ion batteries. *Advanced Energy Materials*, 7(22), 1701437.

148 Yap, Y. L., You, A. H., Teo, L. L. and Hanapei, H. (2013) Inorganic filler size effect on ionic conductivity in polyethylene oxide (PEO) composite polymer electrolyte. *International Journal of Electrochemical Science*, 8(2), 2154–2163.

149 Fan, L., Nan, C.-W. and Zhao, S. (2003) Effect of modified SiO_2 on the properties of PEO-based polymer electrolytes. *Solid State Ionics*, 164(1), 81–86.

150 Lin, D., Liu, W., Liu, Y., Lee, H. R., Hsu, P.-C., Liu, K. and Cui, Y. (2016) High ionic conductivity of composite solid polymer electrolyte via in situ synthesis of monodispersed SiO_2 nanospheres in poly(ethylene oxide). *Nano Letters*, 16(1), 459–465.

151 Yuan, M., Erdman, J., Tang, C. and Ardebili, H. (2014) High-performance solid polymer electrolyte with graphene oxide nanosheets. *RSC Advances*, 4(103), 59637–59642.

152 Yuan, C., Li, J., Han, P., Lai, Y., Zhang, Z. and Liu, J. (2013) Enhanced electrochemical performance of poly(ethylene oxide)-based composite polymer electrolyte by incorporation of nano-sized metal–organic framework. *Journal of Power Sources*, 240, 653–658.

153 Wang, Y.-J., Pan, Y. and Kim, D. (2006) Conductivity studies on ceramic $Li_{1.3}Al_{0.3}Ti_{1.7}(PO_4)_3$-filled PEO-based solid composite polymer electrolytes. *Journal of Power Sources*, 159(1), 690–701.

154 Zhao, Y., Wu, C., Peng, G., Chen, X., Yao, X., Bai, Y., Wu, F., Chen, S. and Xu, X. (2016) A new solid polymer electrolyte incorporating $Li_{10}GeP_2S_{12}$ into a polyethylene oxide matrix for all-solid-state lithium batteries. *Journal of Power Sources*, 301, 47–53.

155 Le, H. T. T., Ngo, D. T., Kalubarme, R. S., Cao, G., Park, C.-N. and Park, C.-J. (2016) Composite gel polymer electrolyte based on poly(vinylidene fluoride-hexafluoropropylene) (PVDF-HFP) with modified aluminum-doped lithium lanthanum titanate for high-performance lithium rechargeable batteries. *ACS Applied Materials & Interfaces*, 8(32), 20710–20719.

156 Zhang, X., Liu, T., Zhang, S., Huang, X., Xu, B., Lin, Y., Xu, B., Li, L., Nan, C.-W. and Shen, Y. (2017) Synergistic coupling between $Li_{6.75}La_3Zr_{1.75}Ta_{0.25}O_{12}$ and poly(vinylidene fluoride) induces high ionic conductivity, mechanical strength, and thermal stability of solid composite electrolytes. *Journal of the American Chemical Society*, 139(39), 13779–13785.

157 Zhang, J., Zhao, N., Zhang, M., Li, Y., Chu, P. K., Guo, X., Di, Z., Wang, X. and Li, H. (2016) Flexible and ion-conducting membrane electrolytes for solid-state lithium batteries: dispersion of garnet nanoparticles in insulating polyethylene oxide. *Nano Energy*, 28, 447–454.

158 Zhang, Z., Zhao, Y., Chen, S., Xie, D., Yao, X., Cui, P. and Xu, X. (2017) An advanced construction strategy of all-solid-state lithium batteries with excellent interfacial compatibility and ultralong cycle life. *Journal of Materials Chemistry A*, 5(32), 16984–16993.

159 Jiang, S. and Wagner, J. B. (1995) A theoretical model for composite electrolytes—II. Percolation model for ionic conductivity enhancement. *Journal of Physics and Chemistry of Solids*, 56(8), 1113–1124.

160 Zheng, J., Tang, M. and Hu, Y.-Y. (2016) Lithium ion pathway within $Li_7La_3Zr_2O_{12}$–polyethylene oxide composite electrolytes. *Angewandte Chemie International Edition*, 55(40), 12538–12542.

161 Chen, L., Li, Y., Li, S.-P., Fan, L.-Z., Nan, C.-W. and Goodenough, J. B. (2018) PEO/garnet composite electrolytes for solid-state lithium batteries: from “ceramic-in-polymer” to “polymer-in-ceramic”. *Nano Energy*, 46, 176–184.

162 Liu, W., Lee, S. W., Lin, D., Shi, F., Wang, S., Sendek, A. D. and Cui, Y. (2017) Enhancing ionic conductivity in composite polymer electrolytes with well-aligned ceramic nanowires. *Nature Energy*, 2(5), 17035.

163 Bae, J., Li, Y., Zhao, F., Zhou, X., Ding, Y. and Yu, G. (2018) Designing 3D nanostructured garnet frameworks for enhancing ionic conductivity and flexibility in composite polymer electrolytes for lithium batteries. *Energy Storage Materials*, 15, 46–52.

5

Research on the Development of Sulfide Solid Electrolyte

5.1 Categories of Advanced Sulfide Solid Electrolytes

This section reviews the structures and ionic conduction mechanisms of several sulfide electrolytes. Based on crystal structure, sulfide electrolytes can be classified into Li_2S–M_xS_y (M = P, Si, and Ge) glassy phases, $Li_7P_3S_{11}$ glass-ceramics, thio-lithium superionic conductors (thio-LISICONs), argyrodite-type Li_6PS_5X (X = Cl, Br, or I), and $Li_{11-x}M_{2-x}P_{1+x}S_{12}$ (M = Ge, Sn, or Si).

Li_2S–M_xS_y (M = P, Si, Ge, etc.) glassy phases were among the earliest materials used as lithium-ion conductors. Because the glassy matrix lacks grain boundary impedance, ion transport is isotropic [1]. By adjusting the ratio of Li_2S to glass formers (e.g., GeS_2, SiS_2, and P_2S_5), their room-temperature ionic conductivity falls within the range of 0.1–1 mS cm−1, which is approximately one to two orders of magnitude higher than that of their crystalline counterparts with the same composition [2]. The incorporation of lithium halides (LiI or LiCl) or ortho-oxosalts (Li_4SiO_4 or Li_3PO_4) into the glassy matrix can further increase the lithium-ion concentration, thereby enhancing the ionic conductivity of these materials [3, 4].

Sulfide electrolytes with a glass-ceramic structure share the same precursor composition as the glassy phase but undergo heat treatment that causes the precipitation of high-ionic-conductivity crystalline phases from the amorphous glass structure. Tatsumisago et al. reported the $Li_7P_3S_{11}$ glass-ceramic precipitated from the $70Li_2S$–$30P_2S_5$ matrix, exhibiting a high room-temperature ionic conductivity of 3.2 mS cm^{-1} [5]. $Li_7P_3S_{11}$ crystallizes in a triclinic unit cell belonging to the *P*-1 space group. Its crystal structure contains numerous interstitial sites between $P_2S_7^{4-}$ ditetrahedra and PS_4^{3-} tetrahedra, which provide fast pathways for Li^+ diffusion [6]. However, these superionic $Li_7P_3S_{11}$ phases precipitated from the glass are typically metastable and cannot be directly synthesized via solid-state methods. Seino et al. optimized the hot-pressing process, reducing porosity and grain boundary impedance in the $Li_7P_3S_{11}$ glass-ceramic, achieving an ionic conductivity of 17 mS cm^{-1} [7].

The crystal structure of $Li_{14}Zn(GeO_4)_4$ (LISICON) consists of LiO_4, ZnO_4, GeO_4, and LiO_6 tetrahedra arranged based on the γ-Li_3PO_4 structure, with lithium-ion transport occurring along the *c*-axis. However, the room-temperature ionic conductivity of LISICON is only ~10^{-4} mS cm^{-1}, and high ionic conductivity (≈100 mS cm^{-1}) is only observed at elevated temperatures (~300 °C), limiting its practical application [8]. Substituting O^{2-} ions with larger, more polarizable S^{2-} ions provides wider channels for Li^+ transport within the lattice, significantly enhancing the ionic conductivity of the LISICON structure. Kanno et al. synthesized $Li_{3.25}Ge_{0.25}P_{0.75}S_4$, whose crystal structure is analogous to LISICON (γ-Li_3PO_4 structure), thus naming it thio-LISICON. This electrolyte exhibited a high room-temperature ionic conductivity of 2.2 mS cm^{-1}, attracting widespread attention from researchers [9].

Due to its highly disordered cation arrangement, cubic argyrodite Ag_8GeS_6 (space group *F*-43*m*) exhibits high Ag^+ conductivity. The argyrodite-type structure is retained when Cu^+ substitutes for Ag^+ [10]. Given the similar ionic radii of Li^+ and Cu^+, Deiseroth et al. proposed substituting Li^+ for Ag^+ and halogen atoms for S, synthesizing a series of cubic argyrodite-type electrolyte materials Li_6PS_5X (X = Cl, Br, and I), belonging to the *F*-43*m* space group [11]. X^- anions occupy the face-centered positions (4a and 4c sites), while PS_4^{3-} units reside in the octahedral voids. Li^+ ions potentially occupy the 48h and 24g sites, forming cage-like local polyhedra. Li^+ diffusion involves three distinct hopping modes: adjacent hopping Li_1 (48h)–Li_2 (24g)–Li_1 (48h), intracage hopping, and intercage hopping. Intercage hopping is dominant for long-range Li^+ transport, enabling a three-dimensional long-range ion conduction network within the cubic lattice. The room-temperature ionic conductivities of Li_6PS_5Cl (LPSCl), Li_6PS_5Br, and Li_6PS_5I are 1.9, 6.8, and 4.6 × 10^{-4} mS cm^{-1}, respectively [12]. Studies indicated that the degree of X^-/S^{2-} site disorder significantly influences the ionic transport rate in lithium argyrodite solid electrolytes (SEs) [13]. In Li_6PS_5Cl and Li_6PS_5Br structures, X^-/S^{2-} site mixing increases disorder, enhancing lithium vacancy concentration and thereby increasing Li^+ diffusion rates. In contrast, in Li_6PS_5I, the large ionic radius of I^- (2.20 Å) hinders exchange with the smaller S^{2-} (1.84 Å), resulting in a more ordered anion framework and reduced ion migration rate [14]. Nazar et al. reported a new chlorine-rich argyrodite electrolyte, $Li_{5.5}PS_{4.5}Cl_{1.5}$. The introduction of Cl^- increased Cl^-/S^{2-} site disorder and modified the lattice softness, leading to a high room-temperature ionic conductivity of 12 mS cm^{-1} [13]. Kraft et al. investigated the effect of covalent Ge^{4+} doping in Li_6PS_5I, as the large ionic radius of Ge^{4+} makes incorporation into Li_6PS_5Br or Li_6PS_5Cl difficult [15]. However, in the Li_6PS_5I structure, X^-/S^{2-} site mixing was observed with 20% Ge^{4+} substitution, and $Li_{6.6}Ge_{0.6}P_{0.4}S_5I$ achieved an ionic conductivity of 18.4 mS cm^{-1}.

In 2011, Kamaya et al. reported a tetragonal sulfide superionic conductor, $Li_{10}GeP_2S_{12}$ (LGPS, space group *P*4_2/*nmc*), with a remarkably high room-temperature ionic conductivity of 12 mS cm^{-1}, comparable to liquid electrolytes [16]. The LGPS structure comprises one-dimensional chains of $(Ge_{0.5}P_{0.5})S_4$ tetrahedra sharing edges with LiS_6 octahedra, running parallel to the *c*-axis, interconnected by PS_4^{3-} tetrahedra. This arrangement creates a zigzag conduction pathway along the *c*-axis, facilitating rapid Li^+ hopping. Due to the high cost of Ge in LGPS, Ong theoretically designed substitutions using safer and more abundant metal elements (e.g., Si, Sn, and Al) for Ge^{4+}, also yielding SEs with high ionic conductivity [17]. The grain ionic conductivity of $Li_{10}SnP_2S_{12}$ was 7 mS cm^{-1}, in good agreement with the 6 mS cm^{-1}

predicted by molecular dynamics (MD) simulations. Subsequent studies reported room-temperature total ionic conductivities of 4 mS cm^{-1} for $Li_{10}SnP_2S_{12}$ [18] and 2.3 mS cm^{-1} for $Li_{10}Si_2PS_{12}$ [19]. The Kanno group synthesized an LGPS-type electrolyte, $Li_{9.54}Si_{1.74}P_{1.44}S_{11.7}Cl_{0.3}$, where Li^+ ions occupy interstitial 16h, 8f, and 4c sites, forming a three-dimensional ion transport network, achieving a further enhanced room-temperature ionic conductivity of 25 mS cm^{-1} [20]. Recently, the Kanno group reported a new LGPS-structured sulfide electrolyte, $Li_{9.54}(Si_{0.6}Ge_{0.4})_{1.74}P_{1.44}S_{11.1}Br_{0.3}O_{0.6}$, exhibiting an exceptionally high room-temperature ionic conductivity of 32 mS cm^{-1}, approximately three times that of the original LGPS, representing the highest reported value for sulfide electrolytes to date [21]. The ionic conductivity of LGPS is highly dependent on the anions in its structure; anions with larger radii and higher polarizability can widen the Li^+ diffusion channels and facilitate Li^+ dissociation from the anion sublattice.

5.2 Theoretical Design of Novel Sulfide SEs

This book discusses the interaction between theory and experiment in the design of new materials for batteries, accelerating the prediction and development of advanced materials. Theoretical simulations in the field of energy storage materials and the calculation methods are mostly microscopic simulations at the atomic or molecular level. In order to accurately evaluate the overall performance of the battery, macro-scale simulation is essential to describe all components of the battery as a whole, which is conducive to promoting a new field of high-throughput battery system design. The properties of battery materials that can be calculated at present and the corresponding calculation methods for simulation are shown in Figure 5.1.

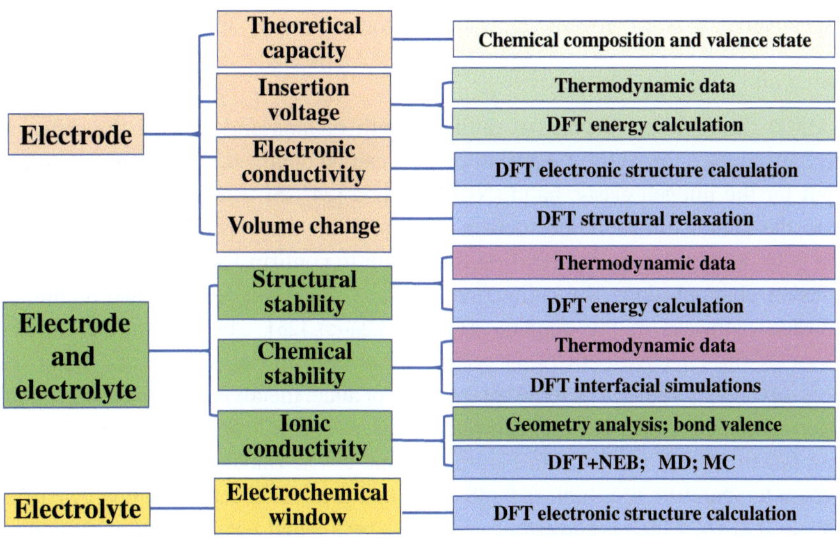

Figure 5.1 Calculation method of key materials in solid-state batteries. *Source*: Reprinted with permission [22].

5.2.1 Ion Diffusion Mechanism

The ionic conductivity is dictated by long-range diffusion, which usually follows an Arrhenius form of relationship. First-principles ab initio molecular dynamics (AIMD) [22, 23] is then carried out to study the ionic transport behavior of Li ions at elevated temperatures, thus providing further insights through statistical processes. In recent works, AIMD explains the transport mechanism of lithium ions by counting the diffusion distance of lithium ions in the electrolyte system at different times. In the process of data statistics, the influence of errors caused by periodic boundary conditions on the actual mean square displacement (MSD) is considered and corrected. Thus, the assessment of diffusion coefficient, D, for Li^+ ions by AIMD simulations at elevated temperatures can be extracted via Equation (5.1),

$$D = \frac{1}{2dt} \langle \Delta r(t) \rangle^2 \tag{5.1}$$

which is obtained from the average MSD $\left\langle [\Delta r(t)]^2 \right\rangle$, calculated from MD runs at each temperature T over a period of time t. Here, d is the dimensionality factor. Using the Arrhenius relation,

$$D = D_0 \exp\left(-\frac{E_a}{k_B T}\right) \tag{5.2}$$

The ionic conductivity σ can be derived using the Nernst–Einstein equation:

$$\sigma = \frac{\rho z^2 F^2}{RT} D = \frac{\rho z^2 F^2}{RT} D_0 \exp\left(-\frac{E_a}{k_B T}\right) = \frac{A_0}{T} \exp\left(-\frac{E_a}{k_B T}\right) \tag{5.3}$$

where ρ is the molar density of diffusing alkali ions in the unit cell, z is the charge of alkali ions (+1 for Li^+), F and R are the Faraday's constant and the gas constant, respectively, D_0 is a constant, E_a is the activation energy for diffusion, k_B is the Boltzmann constant, and $A_0 = \dfrac{\rho z^2 F^2 D_0}{R}$.

AIMD can simulate the actual diffusion path of ions and analyze the diffusion mechanism, so a visual ion migration process can be obtained, which can be used as input for the nudged elastic band (NEB) method [24, 25]. AIMD can not only simulate the actual diffusion mechanism of atoms but also visualize the ion migration path. Mo et al. employed AIMD simulation to confirm fast Li diffusion in a 1D diffusion channel along the *c*-direction of LGPS but also predicted two additional diffusion pathways in the *ab*-plane (Figure 5.2a–c) [26]. The diffusion coefficient D reveals the effective path of Li^+ diffusion, so the diffusion efficiency along the *c*-axis direction is higher. The ionic conductivity of alkali metals is obtained from the diffusion coefficient by the Nernst–Einstein equation. The diffusion coefficients of LGPS from 600 to 1,500 K are given in Figure 5.2d and e. The calculated ionic conductivity at room temperature is as high as 9 mS cm^{-1}, and the overall activation energy barrier is about 0.21 eV, which is consistent with the experimental results (Figure 5.2f and g). From the perspective of materials design, AIMD is an effective means to predict the

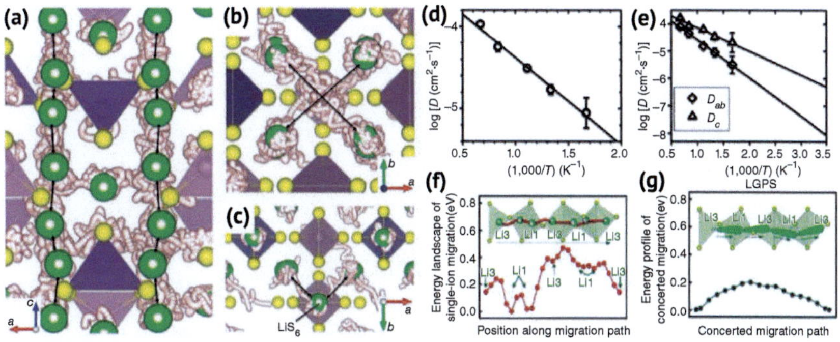

Figure 5.2 (a–e) The Arrhenius plots of the overall diffusion coefficient, D, the diffusion coefficient in the c-direction, D_c, and that in the ab-plane, D_{ab}. *Source:* [26] / with permission of the American Chemical Society. (f) The energy change of a single Li^+ diffusion along the c-axis direction. (g) The migration energy of Li^+ in LGPS diffuses to the next position along the c-axis direction. *Source:* [24] / Springer Nature / CC BY 4.0.

transport properties of lithium ions below room temperature by high-temperature diffusion data. The emergence of LGPS has aroused great interest among energy storage researchers, so that more and more efforts have been made to study the replacement of expensive Ge elements so as to obtain better ionic conductivity [18, 27, 28].

5.2.2 Electrochemical Window

For an inorganic electrolyte, it is very essential to have a wide electrochemical window. It can be obtained not only by testing but also by density functional theory (DFT) analyzing the thermodynamic properties and then calculating the electrochemical window value [29]. An overall assessment of the change in configurational energy for a SE with respect to the content of lithium ion, x, can be established through energy minimization, for example, using the Alloy Theoretic Automated Toolkit (ATAT) [30, 31]. Ceder et al. calculated the electrochemical windows of different components of SE (Figure 5.3a) [32]. The results show that the stability of different types of electrolytes is determined by their binary phase. Therefore, $Li_{1.3}Al_{0.3}Ti_{1.7}(PO_4)_3$ exhibits high oxidation performance due to the formation of strong covalent bonds between P and O [33]. The oxidation potential of the sulfide electrolyte is poor, which is mainly related to the S^{2-} in the binary phase Li_2S. The electrochemical window of LPSCl is shown in Figure 5.3b [33]. It can be seen that LPSCl is about at 1.7 V, while LPSCl is oxidized above 2.3 V, and its stable electrochemical window is only 1.7–2.3 V.

The voltage window calculated by the theory is not completely consistent with the experimental results (Figure 5.3c) [24]. In fact, the difference between the initial potential and the electrochemical window observed in the experiment is mainly due to the difference in the chemical potential of Li between the intermediate phase as

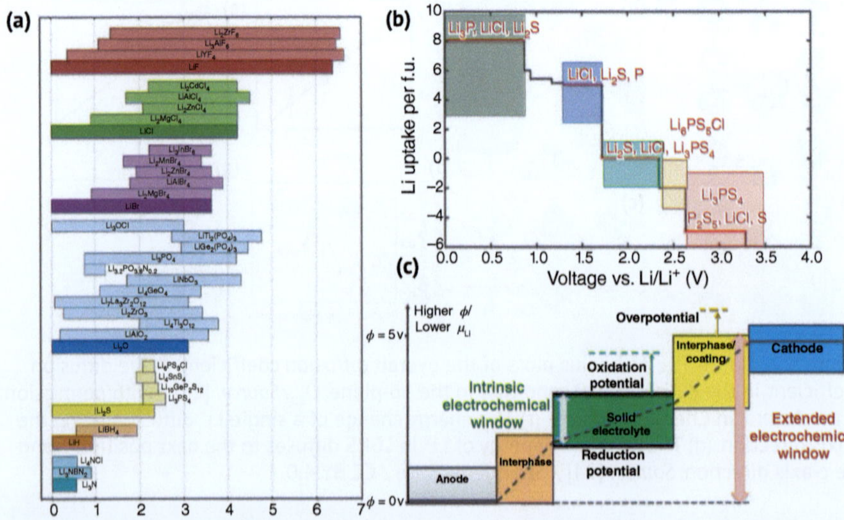

Figure 5.3 (a) The electrochemical window of SE with different components. *Source:* [32] / with permission of the American Chemical Society. (b) The phase diagram of LPSCl at different voltages shows a stable voltage range of 1.7–2.3 V. *Source:* [33] / with permission of the American Chemical Society. (c) Diagram of the nature of the electrochemical window and its practical testing in SE. *Source:* [24] / Springer Nature / CC BY 4.0.

well as the electrolyte and the electrode. The valid electrochemical window of SE not only includes the intrinsic electrochemical window of SE but also the electrochemical window of the interphases. Therefore, in the design process of new electrolytes, we should pay attention to the delithiation redox potential of SSEs.

5.2.3 Electrochemical Stability

The electrochemical stability of the SE is directly related to the integrative performance of the battery. It was verified with GITT analysis, electrochemical charge/discharge profiles, cyclic voltammetry (CV) and other electrochemical techniques in the experiment. DFT can explain the thermodynamic stability of the electrolyte and analyze the electrochemical stability of SE by energy calculation, as shown in Figure 5.4. When an ion migrates from state A to state B, the average electrochemical potential $\overline{V}_{A\to B}$, for the transition between state A ($Ak_x\prod$) and state B ($Ak_{x+\Delta x}\prod$), with reference to electrochemical potential vs. alkali Ak related to total energies (E_{total}), can be calculated as follows [25, 26]:

$$V_{A-B} = -\frac{1}{z}\{[E_{total}(Ak_{x+\Delta x}\prod) - E_{total}(Ak_x\prod)]/\Delta x - E_{total}(Ak)\} \quad (5.4)$$

where x is the number of alkalis in the formula unit of $Ak_x\prod$, charge value $z = 1$ for alkali ion, Δx is the difference value of alkali atoms, and $\prod$ refers to the collection of other constituents. For SSEs, with the change of lithium content, the corresponding

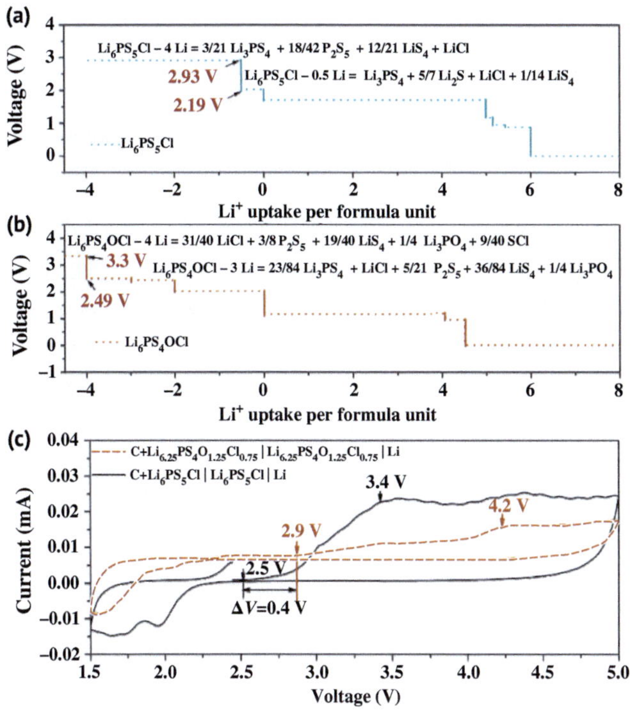

Figure 5.4 The electrochemical windows of electrolytes with different compositions are calculated theoretically, (a) Li_6PS_5Cl and (b) Li_6PS_4OCl against stable constituent phases. (c) Experimental CV tests for the battery. *Source:* [34] / with permission of John Wiley & Sons.

decomposition phase and electrochemical potential are different, and the difference in voltage plateau between two adjacent states is the corresponding electrochemical platform. The corresponding equilibrium phases (EPs) and voltage plateau of Li_6PS_5Cl with different lithium contents during Li^+ delithiation or lithiation are listed in Table 5.1 [34]. This is consistent with the oxidation potential and reduction potential of the sulfide electrolyte obtained by the CV test (Figure 5.4c).

5.3 Intrinsic Properties of Sulfide Electrolytes

From the perspective of large-scale production and application, the stability of sulfide electrolytes in practical use warrants particular attention. The stability of sulfide electrolytes encompasses chemical, electrochemical, thermal, and mechanical stability aspects, with chemical stability requiring resistance to air and polar organic solvents. Specifically, sulfide electrolytes must achieve ionic conductivity > 1 mS cm^{-1} at room temperature to ensure rapid ion transport, attain a densification rate exceeding 90% under 200–300 MPa pressure to guarantee mechanical stability, and maintain a stable crystal structure across a broad temperature range

Table 5.1 The states of Li_6PS_5Cl during the delithiation or lithiation process, with reference to EP, and the corresponding formation energies (E_F) and voltage plateaus.

Delithiation/lithiation	EP	E_F (eV·atom^{-1})	Voltage (V)
$Li_6PS_5Cl + 8\,Li$	$Li_3P + 5Li_2S + LiCl$	0.5929	0.0–0.872
$Li_6PS_5Cl + 6\,Li$	$LiP + 5Li_2S + LiCl$	0.5627	0.0–0.872
$Li_6PS_5Cl + 5.429Li$	$0.143Li_3P_7 + 5Li_2S + LiCl$	0.5507	0.872–0.94
$Li_6PS_5Cl + 5.143Li$	$0.143LiP_7 + 5Li_2S + LiCl$	0.5409	0.94–1.166
$Li_6PS_5Cl + 5Li$	$P + 5Li_2S + LiCl$	0.5354	1.166–1.7
$Li_6PS_5Cl + 0Li$	$Li_3PS_4 + Li_2S + LiCl$	0.0761	1.7–2.02
$Li_6PS_5Cl - 0.5Li$	$Li_3PS_4 + 5/7Li_2S + LiCl + 1/14LiS_4$	0.0744	2.02–2.19
$Li_6PS_5Cl - 4Li$	$3/21Li_3PS_4 + 18/42P_2S_5 + 12/21LiS_4 + LiCl$	0.042	2.19–2.93
$Li_6PS_5Cl - 5.5Li$	$SCl + 0.5P_2S_5 + 0.375LiS_4$	0.084	2.93
$Li_6PS_5Cl - 6Li$	$1/2P_2S_5 + 3/2S + SCl$	1.007	14.42493

from −30 to 200 °C. Furthermore, the electrolyte material should exhibit stability against humid air and resistance to organic solvents to accommodate large-scale production, while possessing a wide electrochemical window to match high-voltage cathode materials.

5.3.1 Synthesis Methods for Sulfide Electrolytes

Methods for synthesizing sulfide electrolytes primarily include solid-state sintering, mechanochemical synthesis, or liquid-phase synthesis. Different synthesis methods directly influence the electrolyte's microstructure, chemical stability, and ion transport properties [35]. Investigating sulfide electrolyte synthesis methods holds significant importance for achieving large-scale industrial production.

Solid-state sintering offers a direct and controllable approach for preparing sulfide electrolytes. In this method, atomic bond rearrangement and bond breaking are driven by thermal energy. Initial mixing ensures uniform dispersion of particles [36]. Subsequently, precursors are loaded into vacuum vessels for high-temperature solid-state reactions. Key reaction parameters include heating temperature, annealing duration, and heating/cooling rates. For thermodynamically stable end products at target temperatures, extended annealing times and slower cooling rates promote grain growth and mechanical stress relief. This enhances the crystallinity of sulfide electrolytes, leading to higher ionic conductivity. Mechanical chemical methods employ high-speed mechanical ball milling followed by thermal treatment. High-energy ball milling initially produces alloys through mechanical energy such as compression, shearing, or friction. During this process, mechanical energy drives material mixing, pulverization, and amorphization. The amorphization process occurs via solid-state diffusion reactions and is governed by point defects and lattice defects introduced by plastic deformation [37]. Thus, the mechanochemical synthesis method offers additional possibilities for preparing novel amorphous glass-phase electrolyte materials. In this method, the speed and duration of mechanical ball milling significantly influence the crystalline structure of the material. An additional annealing process can effectively eliminate grain boundaries and further promote reaction progression [38]. While mechanochemical and solid-state sintering methods yield sulfide SEs with higher ionic conductivity, these approaches are energy-intensive, time-consuming, and challenging to scale for mass production.

Liquid-phase synthesis enables uniform contact between raw materials, thereby shortening reaction times. Additionally, this method allows control over the microstructure of electrolyte particles, forming nanoscale sulfide electrolyte particles that achieve greater ionic contact area with active materials. Based on the state of precursors in the solvent, this method can be categorized into suspension synthesis and solution synthesis [39]. Adding reaction precursors like Li_2S and P_2S_5 to a solvent creates a nonuniform solution composed of intermediate products and supernatant. Solution synthesis refers to dissolving sulfide electrolytes or reaction precursors in specific polar solvents to form uniform, transparent solutions. Both processes require solvent evaporation and thermal treatment. Liang et al. first reported the synthesis of β-Li_3PS_4 using tetrahydrofuran (THF) as the solvent. They first demonstrated the formation of the coordination complex Li_3PS_4 3THF, followed by thermal treatment at 140 °C to form nanoporous β-Li_3PS_4. At 25 °C, this material exhibited a high ionic conductivity of 1.6×10^{-4} S cm^{-1} [40]. Most sulfide electrolytes undergo chemical degradation when exposed to highly polar solvents due to their high reactivity. Currently, only a limited number of electrolyte/solvent combinations can be synthesized via solution methods, such as Li_4SnS_4/methanol [41] and Li_6PS_5Cl/ethanol [42]. Although sulfide electrolytes readily undergo hydrolysis or alcoholysis with protic solvents like water and alcohol, leading to H_2S (hydrogen sulfide) formation, studies indicated that dissolving raw materials like Li_2S in anhydrous ethanol proceeds without side reactions. The homogeneous Li_2S solution reduces the proton activity of anhydrous ethanol, thereby inhibiting its alcoholysis of the PS_4^{3-} unit [43]. Li_4SnS_4 dissolves in methanol without side reactions. According to hard and soft acid–base (HSAB) theory, Sn^{4+} exhibits greater affinity for S^{2-} than for O^{2-}, enabling high ionic conductivity retention in polar organic solvents. The room-temperature ionic conductivity of 0.4LiI–0.6Li_4SnS_4 reaches 4.1×10^{-4} S cm^{-1} [41].

5.3.2 Electrochemical Stability of Sulfide Electrolytes

The "electrochemical window" of an electrolyte is generally determined by quantifying the voltage range over which oxidation and reduction reactions occur, i.e., by determining the voltage difference between the lowest unoccupied molecular orbital (LUMO, reduction potential) and the highest occupied molecular orbital (HOMO, oxidation potential) [44]. For full cells, the electrochemical window of the electrolyte must encompass the redox potential range of both the cathode and anode materials to ensure stable operation of the cell. Therefore, determining the electrochemical window of the electrolyte is of great significance for stabilizing the electrolyte/electrode interface. In the past, inorganic SEs, such as LGPS, were believed to exhibit a wide electrochemical window, covering a voltage range of 0–5 V [16]. However, based on first-principles theoretical calculations, the actual electrochemical window of LGPS electrolyte is much narrower than the experimental value, with a bandgap of 3.6 eV. The LGPS electrolyte begins to decompose at 1.7 V, producing reaction products including Li_4GeS_4, Li_2S, and P. As the potential decreases, the reduction products at 0 V include $Li_{15}Ge_4$, Li_3P, and Li_2S. The LGPS electrolyte starts to oxidize at 2.14 V and undergoes further oxidation at 2.31 V, producing compounds such as P_2S_5, S, and GeS_2 [26]. Han et al. prepared composite cathodes by mixing SEs and conductive carbon in a specific ratio, significantly increasing the contact area between solid particles [45]. The introduction of conductive carbon enhanced the reaction kinetics of electrolyte decomposition. Their results demonstrated reversible redox reactions between 0–1.7 and 2–2.5 V, confirming that the actual thermodynamic electrochemical window of sulfide electrolytes is much narrower than 0–5 V.

Sulfide electrolytes with limited electrochemical windows have been shown to form thermodynamically unstable interfaces with electrode materials, including Li_6PS_5X [46], $Li_7P_3S_{11}$ [47], and LGPS [48]. The electrochemical instability of sulfide electrolytes primarily arises from the reduction of cations (e.g., P^{5+} and Ge^{4+}) and the oxidation of S^{2-}. For instance, some sulfide electrolyte molecules contain high-valent cations, such as P^{5+}, Ge^{4+}, and Ti^{4+}, which readily react with metallic Li, forming highly electron-conductive materials at the interface and exacerbating side reactions. Based on the reaction characteristics at the sulfide/electrode interface, three main types of interfaces can be identified [49]: (i) thermodynamically stable interface. Both components share the same chemical potential, and no spontaneous reactions occur. (ii) Mixed ionic–electronic conducting (MIEC) interface layer, which is thermodynamically and kinetically unstable. Wenzel et al. demonstrated that when LGPS electrolyte comes into contact with Li metal, the decomposition of LGPS leads to the formation of an MIEC layer comprising Li_2S, Li_3P, metallic Ge, and Li–Ge alloys [48]. This interface layer exhibits high electronic conductivity, continuously triggering the decomposition of the sulfide electrolyte and resulting in persistent performance degradation. (iii) Thermodynamically unstable but kinetically stable interface layer. Such an interface exhibits high ionic conductivity and low electronic conductivity, similar to the SE interface (SEI) or cathode electrolyte interface (CEI), and can serve as a kinetically stable passivation layer. Although electron-insulating and ion-conducting buffer layers have been shown to effectively reduce side reactions between sulfide electrolytes and electrodes, they introduce two new interface layers. Comprehensive characterization of the introduced interface layers from physical and chemical perspectives is still required to design highly stable interfacial buffer layers.

5.3.3 Air Stability of Sulfide Electrolytes

The inherent chemical instability of sulfide solid-state electrolytes in humid air poses a critical challenge to their large-scale application. These electrolytes readily undergo vigorous hydrolysis reactions with trace water vapor in the environment, leading to irreversible degradation of their structure and composition. This process is accompanied by the release of toxic, flammable H_2S gas. This hydrolysis process not only causes the electrolyte material itself to powderize, dissolve, or undergo phase transitions, but also drastically degrades its core electrochemical properties. Crucially, the lithium-ion conductivity significantly decreases. The entire process—from synthesis, powder handling, and electrode preparation to battery encapsulation—must be conducted under strictly controlled dry, inert atmospheres (such as argon glove boxes). This stringent operational requirement substantially increases production complexity and cost, severely hindering large-scale synthesis, storage, and application [50].

To characterize sulfide electrolyte stability in humid air, studies typically adopt either macroscopic approaches (e.g., measuring H_2S gas volume or tracking electrolyte weight/morphology changes under different exposure conditions) or microscopic methods (analyzing chemical composition, structure, and electrochemical performance before and after exposure) [51, 52]. The hydrolysis of sulfide electrolytes is accompanied by significant changes in mass and morphology. Using optical imaging, Lu et al. evaluated morphological changes in four sulfide electrolytes after exposure to humid air [51]. Among these, $Li_{9.54}Si_{1.74}P_{1.44}S_{11.7}Cl_{0.3}$ and Li_3PS_4 exhibited a color transition from white to black, accompanied by severe volume changes, indicating significant structural degradation. Muramatsu et al. first investigated structural changes and H_2S gas generation in $75Li_2S{\cdot}25P_2S_5$ glass and glass-ceramic phases exposed to humid atmospheres [52]. They found that both $P_2S_7^{4-}$ and S^{2-} ions tend to decompose by reacting with H_2O molecules in air, forming O—H and S—H bonds, which then hydrolyze further into O—H bonds and H_2S.

HSAB theory explains the chemical stability of sulfide electrolytes in humid air [53]. In this theory, central atoms with dense electron distribution and small volume are termed hard acids, while those with sparse distribution are soft acids. Similarly, highly electronegative, oxidation-resistant coordination atoms are hard bases, and their opposites are soft bases. Hard acids preferentially react with hard bases, while soft acids tend to bond with soft bases. The central atom P^{5+} in sulfide electrolytes belongs to hard acids, while O^{2-} is a strong base. The bond energy of the P—O bond exceeds that of the P—S bond, leading to the spontaneous decomposition of sulfide electrolytes in air.

Researchers have proposed multiple methods to enhance the chemical stability of sulfide electrolytes in humid air: for example, physical additives [54], elemental doping [55–59], and structural modification [60]. Certain physical additives, such as ZnO, Bi_2O_3, and Fe_2O_3, can be directly blended with sulfide electrolytes. Due to their substantial thermodynamic tendency to react with H_2S, these metal oxides can consume H_2S through spontaneous chemical reactions [54]. Certain soft acids (e.g., Sn^{4+}, As^{5+}, Bi^{3+}, and Sb^{5+}) tend to form strong bonds with soft base S^{2-}, thereby blocking attacks from hard base O^{2-} molecules in air [55–58]. Based on HSAB theory, Sahu et al. partially substituted Sn in Li_4SnS_4 with the soft acid element As to prepare $Li_{3.833}Sn_{0.833}As_{0.166}S_4$, demonstrating excellent air stability. Arrhenius plots indicate minimal change in ionic conductivity before and after air exposure for the doped electrolyte [58]. Furthermore, incorporating oxygen into the crystal structure is considered an effective approach to enhance intrinsic air stability, as the P—O bonds formed within the lattice mitigate the instability of P—S bonds in air [59]. Since the hydrolysis of sulfide electrolytes occurs at the surface, methods such as surface coating or designing core–shell structures can also improve their air stability. Xu et al. fabricated a Li^+-conductive film composed of fluorinated polysiloxane and $Li_{1.4}Al_{0.4}Ti_{1.6}(PO_4)_3$ nanoparticles via hydrolysis and condensation reactions, which features a water-stable protective layer with a superhydrophobic surface [60]. By constructing an inert surface coating to resist the chemical attack of O^{2-} in air on the sulfide electrolyte, its air stability was improved.

5.3.4 Solvent Stability of Sulfide Electrolyte

From the perspective of composite electrode fabrication, wet slurry coating is a common method. The wet coating process enables close contact between electrode components and allows precise control over electrode sheet thickness. Additionally, liquid-phase synthesis preparation methods facilitate the formation of nanoscale particles and uniform morphology in sulfide electrolytes. Therefore, the stability of sulfide electrolytes toward organic solvents is crucial for expanding electrolyte synthesis routes and electrode processing technologies.

Key parameters of organic solvents include dielectric constant, polarity index, and vapor pressure. However, studies indicated that most sulfide electrolytes undergo chemical degradation due to high reactivity when exposed to highly polar solvents. Tan et al. investigated the compatibility of the $Li_7P_3S_{11}$ electrolyte with organic solvents via Raman spectroscopy and XRD analysis [40]. They found that the degradation process of sulfide electrolytes is highly correlated with the dielectric constant of organic solvents. The crystal structure of $Li_7P_3S_{11}$ is well preserved in nonpolar or low-polarity solvents (including toluene, xylene, etc.), while structural changes and performance degradation occur in some highly polar solvents (e.g., *N*-methylpyrrolidone and *N,N*-dimethylformamide). Yamamoto et al. [61] similarly validated this conclusion. They observed that the ionic conductivity of sulfide electrolytes remained nearly unchanged when exposed to solvents with a donor number of nine or lower.

The nucleophilic attack process that polar solvents may induce on sulfide electrolytes can also be explained using HSAB theory: highly polar solvents containing electronegative elements like N and O are considered strong Lewis bases. They react with the electrophilic elements P^{5+} and S^{2-} in sulfide electrolytes, forming low-ionic-conductivity phases such as $P_2S_7^{4-}$ and $P_2S_6^{4-}$ [34]. Furthermore, incompatibilities between sulfide electrolytes, organic solvents, and binders significantly reduce the manufacturability of sulfide electrolytes. Traditional binders like polyvinylidene fluoride (PVDF) and polyvinyl alcohol (PVA) can only be dispersed using highly polar solvents. Consequently, the high reactivity of sulfide electrolytes toward polar solvents limits the range of usable binders. Jung et al. proposed a cosolvent strategy where the synergistic effect of dibromomethane and hexyl butyrate simultaneously dissolves nonpolar binders and polar lithium salts while minimizing impact on the ionic conductivity of sulfide electrolytes [61]. Results demonstrated this cosolvent approach is suitable for sulfide electrolytes and maintains excellent electrochemical performance.

5.3.5 Thermal Stability of Sulfide Electrolytes

The synthesis of SEs typically requires elevated temperatures, leading to the assumption that SEs possess high thermal stability. This enables them to withstand high-temperature environments caused by internal short circuits or other factors, thereby preventing further thermal runaway. Recently, Wu et al. summarized the thermal stability of different types of solid and liquid electrolytes, defining the upper limit temperature T_{ts} for thermal stability [62]. For inorganic SEs, T_{ts} is defined as the decomposition temperature or phase transition temperature. If the material is thermally stable before melting, the melting point is recorded as T_{ts}. For polymer-based SEs, polymer softening may lead to internal battery short circuits. Therefore, T_{ts} is generally defined as the melting temperature of the pure polymer (except for amorphous polymers without a fixed melting point). The thermal stability of polymer-based composite electrolytes depends on the combined thermal stability of the polymer matrix and the lithium salt.

Among different types of SEs, oxide and sulfide SEs exhibit superior intrinsic thermal stability. According to relevant reports, the Li_2S–P_2S_5 glass phase can be transformed into the $Li_7P_3S_{11}$ crystalline phase through heat treatment at 240–360 °C. However, upon heating beyond 550 °C, the $Li_7P_3S_{11}$ crystal structure disappears, replaced by thermodynamically stable phases including $Li_4P_2S_6$. This transformation causes the ionic conductivity to decrease by several orders of magnitude [63]. According to the phase diagram mapped by Hori et al., LGPS exhibits thermal stability up to 650 °C, decomposing into Li_4GeS_4 and a liquid phase at 700 °C—a finding consistent with DTA results [64]. Polymer electrolytes decompose between 200 and 400 °C, yet their thermal reaction temperature when combined with cathode materials ranges only from 150 to 350 °C. Although oxides exhibit the highest thermal stability, Kang et al. observed that $Li_{1.5}Al_{0.5}Ge_{1.5}(PO_4)_3$ (LAGP) undergoes violent combustion with Li metal when heated to 200 °C under an inert atmosphere (Ar), releasing substantial heat [65]. They attributed this to the high reactivity between the LAGP electrolyte and Li metal, which releases oxygen during the reaction, thereby reducing the system's thermal stability.

Research on the thermal stability of high ionic conductivity sulfide electrolytes is limited. This is because sulfide electrolytes undergo severe side reactions with the characterization equipment, producing toxic gases (H_2S, SO_2, etc.) and causing significant corrosion and environmental pollution. Wang et al. systematically investigated the thermal stability of a series of sulfide electrolytes, finding the following stability hierarchy: $Li_6PS_5Cl > Li_4SnS_4 > Li_{9.54}Si_{1.74}P_{1.44}S_{11.7}Cl_{0.3} > Li_3PS_4 > Li_7P_3S_{11}$ [66]. Due to the generally high oxidation potential of lithium-ion battery cathode materials, oxidized $Li_{1-x}CoO_2$ exhibits greater oxidizing power compared to pristine $LiCoO_2$. The thermal stability of sulfides is influenced by their inherent stability, the depth of delithiation of the active material $Li_{1-x}CoO_2$, and volatile decomposition products. The thermal reaction between sulfide electrolytes and LCO begins at 400–500 °C, with by-products from the Li_6PS_5Cl–LCO reaction including Co_3S_4, CoS_2, Li_2S, and others.

Compared to conventional electrolytes, sulfide electrolytes exhibit higher thermal stability. However, under varying temperatures and charge–discharge states, electrode materials may undergo side reactions with the electrolyte, accelerating system failure. Currently, research on the thermal stability of sulfide electrolytes remains in its infancy. The thermal safety of all-solid-state batteries requires further validation, and a comprehensive, systematic understanding of thermal stability remains elusive.

5.3.6 Mechanical Stability of Sulfide Electrolyte

Mechanical properties represent crucial physical characteristics of solid materials, with their performance metrics serving as key parameters for ASSLB assembly and stable cycling. During ASSLB operation, volume changes occurring in electrodes during lithium deintercalation/intercalation and side reactions at interfaces lead to accumulated mechanical stresses. This may cause contact failure between electrodes and electrolytes, resulting in elevated interfacial impedance. If stresses exceed the material's strength, sustained mechanical stress may cause phase separation at the electrolyte/electrode interface.

The mechanical property parameters of sulfide electrolytes primarily include elastic modulus and fracture toughness. Elastic modulus parameters encompass Young's modulus (E), shear modulus (G), and bulk modulus (B), representing a solid material's ability to maintain its initial state. According to the Monroe–Newman model, SEs with shear moduli exceeding twice that of metallic shear moduli (4.25 GPa) can effectively suppress lithium dendrite growth [67]. Young's modulus indicates the ease of elastic deformation in materials; a higher Young's modulus implies reduced susceptibility to elastic deformation in electrolytes, correlating with elevated shear moduli to inhibit dendrite growth. Generally, the Young's modulus of sulfide electrolytes (≈10–30 GPa [68]) is significantly lower than that of oxide electrolytes (e.g., $Li_7La_3Zr_2O_{12}$, 149.8 GPa [69]; $Li_{0.33}La_{0.56}TiO_3$, 200 GPa [70]). Among sulfide electrolytes with different structures, the Young's modulus of the glass phase is slightly lower than that of the glass-ceramic and crystalline phases [71]. This difference can be attributed to the absence of long-range ordered structures in the glass phase and lower intermolecular bond energies. Materials with low Young's modulus exhibit excellent strain adaptation properties, ensuring close contact between the electrode and electrolyte and enhancing interfacial stability.

Fracture toughness (K_{Ic}) represents a material's resistance to brittle fracture. Sulfide electrolytes are reported to possess relatively poor fracture resistance, with values of only 0.2–0.4 MPa $m^{1/2}$ [72]. In contrast, "softer" polymer electrolytes, despite their lower Young's moduli (ranging from 1 MPa to 1 GPa), typically exhibit higher fracture toughness (≈0.5 MPa·$m^{1/2}$ [73]). Consequently, sulfide electrolytes are typically cold-pressed into thick sheets (>600 μm) to compensate for their inherent brittleness. However, the resulting longer Li^+ diffusion paths lead to extremely high sheet resistances (>100 Ω cm^2) in the electrolyte layer.

Sulfide electrolytes are considered soft materials with significantly reduced grain boundary and interfacial resistances compared to oxide SEs. Their flexibility enables good contact with electrodes, but high brittleness may pose challenges during the preparation of ultrathin electrolyte films and battery assembly. Therefore, during battery assembly, appropriate pressure parameters are crucial for ensuring tight contact and optimal electrochemical performance in sulfide-based all-solid-state batteries.

5.4 Study on Sulfide-based Composite Electrolyte

Sulfide SE has ultrahigh ionic conductivity (10^{-2}–10^{-3} S cm^{-1}) at room temperature, comparable to liquid electrolytes, and excellent mechanical properties and flame retardancy, making it the most suitable electrolyte material for high-safety lithium solid-state batteries. However, its shortcomings, which cannot be ignored, limit its practical applications, such as poor electrochemical stability, poor wettability with electrodes, and large interfacial impedance-.

Polymer electrolyte is softer than inorganic electrolyte. Moreover, some polymer electrolytes are often sticky and can be in close contact with the electrode surface, effectively reducing interface impedance. In addition, polymer electrolytes are more suitable for large-scale production of solid-state batteries because of their ease of processing and good flexibility. However, compared with inorganic SEs, these electrolytes have difficulty in ensuring high ionic conductivity (10^{-6} S cm^{-1}) due to the lack of transmission channels like fast ion conductors [74]. They often need to be positive at higher temperatures.

It often plays a role, resulting in its application being greatly limited. In addition, these electrolytes are less safe than inorganic SEs because they are easily deformed or burned by heat. In view of the above reasons, people consider combining these two electrolytes to prepare a composite SE, so that it can carry forward the advantages of these two electrolytes as much as possible and avoid their shortcomings.

The main advantages of the composite SE include: (i) rigidity and flexibility. The composite SE can ensure contact with the electrode, has good mechanical strength and thermal stability, and has a good inhibition of lithium dendrites, making it suitable for large-scale processing and production. (ii) Wide voltage window. The electrochemical stability window of a single inorganic SE or polymer electrolyte is not ideal, but the composite polymer electrolyte (CPE) often has better oxidation resistance and reduction characteristics, so it has a wider electrochemical stability window.

The polyethylene oxide (PEO)–lithium salt system has been widely studied for decades. The PEO matrix has a good ability to dissolve lithium salts. It has high flexibility and certain mechanical properties to prevent the growth of lithium dendrites.

However, its drawbacks include low room-temperature ionic conductivity (10^{-5}–10^{-7} S cm^{-1}) and a narrow electrochemical window. At present, the works on sulfide SE and polymer matrix composite electrolyte are relatively limited, and the research on the composite mechanism is not deep enough. Based on this, the main focus of this chapter is to introduce sulfide electrolyte particles into PEO–LiTFSI matrix and study its internal mechanism changes through various characterization and electrochemical test methods. The purpose is to prepare a composite SE that can integrate the advantages of organic and inorganic SEs and can truly use high-safety lithium solid-state batteries. (iii) High ionic conductivity. Compared with pure polymer electrolytes, composite SEs often have higher ionic conductivity and better stability against lithium metals, so that the full battery has better cycle performance.

The composite methods for preparing composite SE can be roughly divided into three categories: (i) inorganic ceramic filler is incorporated into a polymer matrix, (ii) the composite is prepared with an equivalent mass ratio of inorganic ceramic and polymer, and (iii) the polymer filler is incorporated into the inorganic ceramic matrix [75]. However, most sulfide electrolytes have the drawback of sensitivity to air, water, and uncontrolled electrochemical side reactions with lithium metal. They are also difficult to process and show poor contact and electrochemical stability with lithium metal. Solid polymer electrolytes (SPEs) have attracted much attention due to their ability to dissolve a variety of lithium salts by the interaction of specific functional groups with the salt cations. In general, the matrix of SPEs includes PEO, poly(methyl methacrylate) (PMMA), poly(vinylidene fluoride-*co*-hexafluopropylene) (PVDF-HFP), and poly(acrylonitrile) (PAN). PEO-based electrolytes are flexible and easily form an electrochemically stable interface toward lithium metal anode, but they always exhibit a low conductivity of 10^{-6}–10^{-8} S cm^{-1} at room temperature, which limits their applications.

Among all SPEs, CPEs, consisting of the polymer matrix and ceramic particles, have garnered great interest because they combine the advantages of organic electrolytes and inorganic electrolytes. In a PEO-based composite, the inorganic fillers can inhibit local recombination of chains in the PEO matrix, thus it can effectively promote local amorphous regions and improve the ability of the polymer matrix to transport Li ions.

Generally, the CPEs possess higher ionic conductivity and better mechanical strength with the addition of the ceramic fillers. The ceramic fillers in CPEs include inert fillers with no intrinsic Li^+ conductivity, e.g., SiO_2 [76], Al_2O_3 [77, 78], and ZrO_2 [79], and active fillers, such as LATP [80–82], LAGP [83], LLZTO [75, 84, 85], and LLTO [86–88]. However, sulfide CPEs are rarely reported [89, 90]. Recently, some groups have used a theoretical method to design a series of superionic lithium-ion conductors, such as Li_6PS_5X, where X is a halogen [38, 91]. Off-stoichiometric Li_6PS_5X was theoretically reported to possess high ionic conductivity ranging from 10^{-4} to 10^{-2} S cm^{-1} [38, 91, 92]. However, the electrolytes with optimized composition and transport properties have not been fabricated successfully [93]. In this chapter, a designer fast Li conducting $Li_{6.25}PS_{5.25}Cl_{0.75}$ is fabricated and applied as a filler for PEO polymer. A pronounced evolution in the condensed structure of the polymer triggered by the LPSCl is observed.

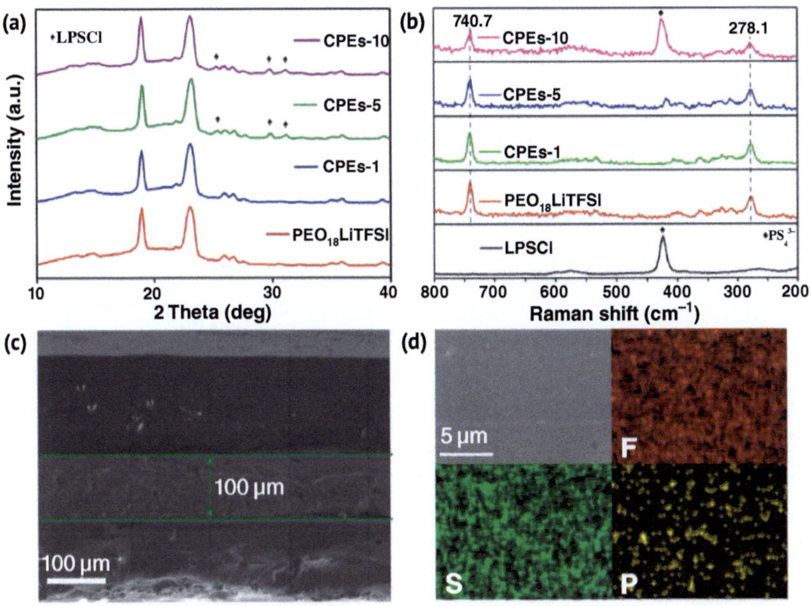

Figure 5.5 Characterization of CPEs with various contents of LPSCL. (a) WAXS patterns of the CPEs with various contents of LPSCL. The dots indicate the phase of LPSCL. (b) Raman spectra of the CPEs with various contents of LPSCL. (c) The cross-sectional SEM images of CPEs-5. (d) EDS maps of F, P, and S elements in the CPEs-5. *Source:* [96] / with permission of Elsevier.

The $Li_{6.25}PS_{5.25}Cl_{0.75}$-based CPEs are fabricated through a simple solution casting method. The WAXD scattering angle vs. scattering intensity patterns in Figure 5.5a are used to investigate the structural evolution of CPEs. Two typical dominant peaks at 19° and 23° can be indexed as the characteristic diffraction peaks of crystallization of PEO. The Raman spectra (Figure 5.5b) are also used to characterize the structure of the PEO matrix with the addition of LPSCl particles. The shift at 278.1 cm^{-1} corresponds to O—C—C and C—O—C bending and C—C rocking in the PEO backbone. The shift at 740.7 cm^{-1} is attributable to $TFSI^-$ group according to the previous report [94]. The Raman peak at 424 cm^{-1} from PS_4^{3-} vibration is also seen in CPEs-5 and CPEs-10. A dual phase structure of organic polymer and ceramic is confirmed in CPEs. The cross-section image of the CPEs in Figure 5.5c shows that the CPEs have a thickness of around 100 μm. The energy-dispersive X-ray spectroscopy (EDX) mapping images of the CPEs-5 are presented in Figure 5.5d. It shows that F and S elements are distributed uniformly in PEO matrix, exhibiting good solubility of Li salt in continuous PEO matrix. The distribution of P element also confirms that the sulfide electrolyte particles are randomly dispersed in PEO matrix [95].

The interfacial stability of CPEs with the lithium electrode is crucial because lithium metal is a promising high-energy-density anode material. The schematic of the symmetric cell setup is shown in Figure 5.6a. The CPEs are sandwiched between two pieces of lithium metal foils and sealed in a coin cell for one week at 60 °C. Figure 5.6b shows the variation of the interfacial resistance (R) of the sandwich symmetric Li / CPEs / Li battery at 60 °C within one week. Figure 5.6c shows the cycle performance

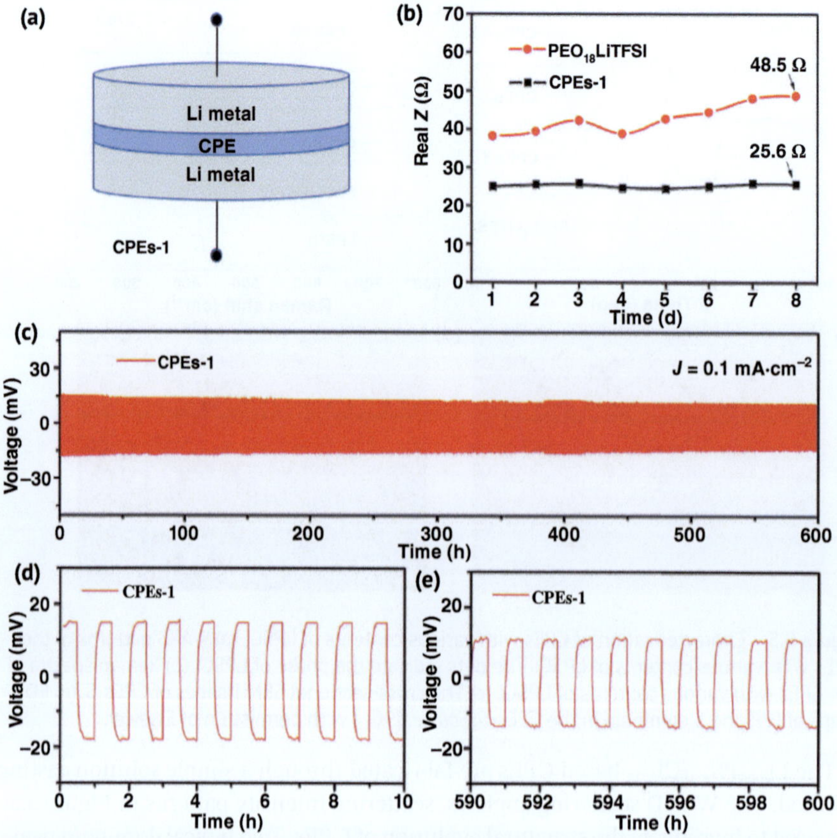

Figure 5.6 The interfacial stability of the CPEs-1 and $PEO_{18}LiTFSI$ electrolyte against lithium metal. (a) Schematic diagram of the symmetrical cycling experiment for lithium plating/stripping. (b) The changes of the interfacial resistance (*R*) for the sandwiched symmetrical Li/CPEs/Li cells at 60 °C over a period of one week. (c) The lithium plating/stripping cycling test of CPE at 60 °C with a current density of 0.1 mA cm^{-2}. (d) The lithium plating/stripping cycling at the first 10 cycles and (e) the last 10 cycles. *Source*: [96] / with permission of Elsevier.

of lithium plating / delithiation at a current density of 0.1 mA cm^{-2} at 60 °C was tested. Figure 5.6d and e tested the lithium deposition / stripping cycle curves for the first 10 weeks and the last 10 weeks, respectively. The performance of the symmetrical battery modified by PEO is still steadily improved.

The LFP/Li cells with CPEs-1 and $PEO_{18}LiTFSI$ electrolytes are assembled to exhibit the electrochemical performance of the CPEs (Figure 5.7). The galvanostatic charge and discharge cycles of the composite electrolytes are tested at 60 °C. As shown in Figure 5.7a, the cell with CPEs-1 displays a smaller semicircle and a shorter Warburg region compared with that of $PEO_{18}LiTFSI$ electrolyte, which indicates a faster lithium diffusion in CPEs-1 cells. We can see from the charge–discharge curve in Figure 5.7c that two typical platforms appear at 3.35 and 3.45 V corresponding to both the charge and discharge processes in LFP/Li cells at 0.2 C,

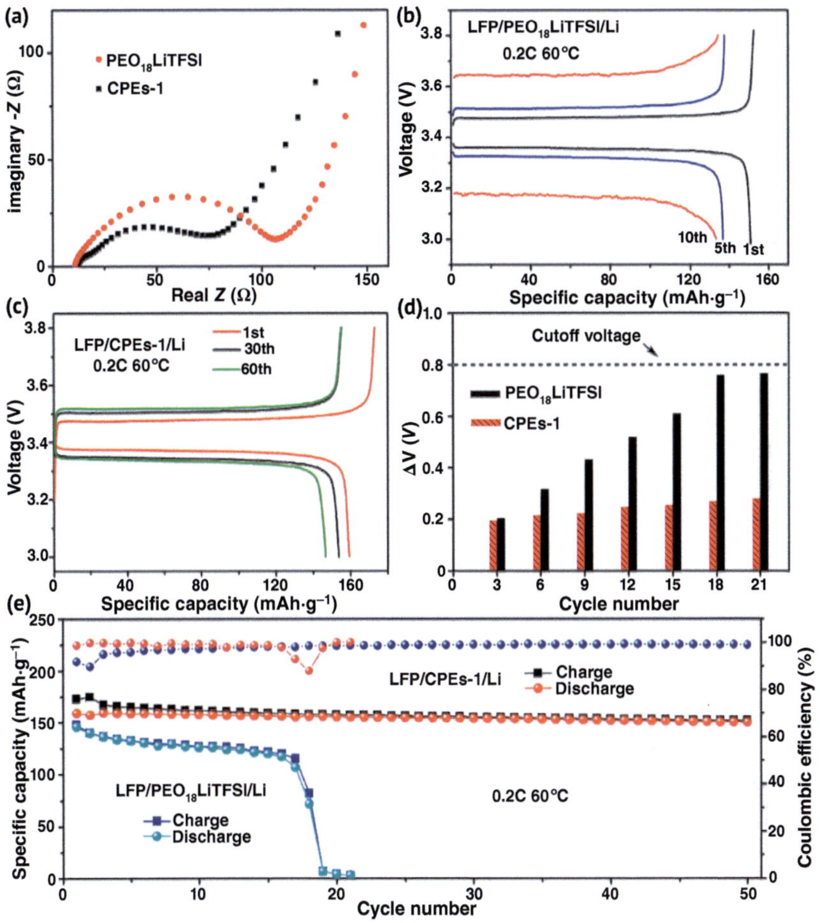

Figure 5.7 Electrochemical performance of the cells with $PEO_{18}LiTFSI$ and CPEs-1. (a) Nyquist plots for the cell using $PEO_{18}LiTFSI$ and CPEs-1 at 60 °C in the frequency range of 10 mHz–1 MHz. (b) The charge and discharge plots of the cells with $PEO_{18}LiTFSI$ at 0.2C (1C = 170 mAh g^{-1}) and 60 °C. (c) The charge and discharge plots of the cells with CPEs-1 at 0.2C and 60 °C. (d) The difference between the charge and discharge voltages during the galvanostatic test at 60 °C and 0.2C. The dotted line indicates the high polarization of $PEO_{18}LiTFSI$ cell approached the cutoff voltage. (e) Cycling performance of the cell using $PEO_{18}LiTFSI$ and CPEs-1 at 60 °C and 0.2C. *Source:* [96] / with permission of Elsevier.

and it also shows a very flat voltage platform and retains low voltage polarization in charge and discharge profiles during 60th cycling. However, the voltage polarization of LFP/$PEO_{18}LiTFSI$/Li cells is more serious after 10 cycles, as shown in Figure 5.7b. Figure 5.7c is the charge–discharge cycle curve of the battery assembled with CPE. The voltage drop (ΔV) between the midpoint voltage of the charge and discharge is regarded as the polarization extent for the battery cycling. Figure 5.7d summarizes the ΔV of the cells running in 21 cycles.

References

1 Kennedy, J. H. (1989) Ionically conductive glasses based on SiS_2. *Materials Chemistry and Physics*, 23, 29.

2 Mercier, R., Malugani, J.-P., Fahys, B. and Robert, G. (1981) Superionic conduction in Li_2S-P_2S_5-LiI-glasses. *Solid State Ionics*, 5, 663.

3 Malugani, J. and Robert, G. (1980) Preparation and electrical properties of the 0,37Li_2S-0,18P_2S_5-0,45LiI glass. *Solid State Ionics*, 1, 519.

4 Aotani, N., Iwamoto, K., Takada, K. and Kondo, S. (1994) Synthesis and electrochemical properties of lithium ion conductive glass, $Li_3PO_4Li_2SSiS_2$. *Solid State Ionics*, 68, 35.

5 Mizuno, F., Hayashi, A., Tadanaga, K. and Tatsumisago, M. (2005) New, highly ion-conductive crystals precipitated from Li_2S–P_2S_5 glasses. *Advanced Materials*, 17, 918.

6 Yamane, H., Shibata, M., Shimane, Y., Junke, T., Seino, Y., Adams, S., Minami, K., Hayashi, A. and Tatsumisago, M. (2007) Crystal structure of a superionic conductor, $Li_7P_3S_{11}$. *Solid State Ionics*, 178, 1163.

7 Seino, Y., Ota, T., Takada, K., Hayashi, A. and Tatsumisago, M. (2014) A sulphide lithium super ion conductor is superior to liquid ion conductors for use in rechargeable batteries. *Energy & Environmental Science*, 7, 627.

8 Bruce, P. G. and West, A. R. (1982) Ionic conductivity of LISICON solid solutions, $Li_{2+2x}Zn_{1-x}GeO_4$. *Journal of Solid State Chemistry*, 44, 354.

9 Murayama, M., Kanno, R., Irie, M., Ito, S., Hata, T., Sonoyama, N. and Kawamoto, Y. (2002) Synthesis of new lithium ionic conductor thio-LISICON—lithium silicon sulfides system. *Journal of Solid State Chemistry*, 168, 140.

10 Winkler, C. (1886) Germanium, Ge, ein neues, nichtmetallisches element. *Berichte der deutschen chemischen Gesellschaft*, 19, 210.

11 Deiseroth, H., Kong, S., Eckert, H., Vannahme, J., Reiner, C., Zaiß, T. and Schlosser, M. (2008) Li_6PS_5X: a class of crystalline Li-rich solids with an unusually high Li^+ mobility. *Angewandte Chemie International Edition*, 47, 755.

12 Rao, R. P. and Adams, S. (2011) Studies of lithium argyrodite solid electrolytes for all-solid-state batteries. *Physica Status Solidi A*, 208, 1804.

13 Adeli, P., Bazak, J. D., Park, K. H., Kochetkov, I., Huq, A., Goward, G. R. and Nazar, L. F. (2019) Boosting solid-state diffusivity and conductivity in lithium superionic argyrodites by halide substitution. *Angewandte Chemie International Edition*, 58, 8681.

14 Kraft, M. A., Culver, S. P., Calderon, M., Böcher, F., Krauskopf, T., Senyshyn, A., Dietrich, C., Zevalkink, A., Janek, J. and Zeier, W. G. (2017) Influence of lattice polarizability on the ionic conductivity in the lithium superionic argyrodites Li_6PS_5X (X = Cl, Br, I). *Journal of the American Chemical Society*, 139, 10909.

15 Kraft, M. A., Ohno, S., Zinkevich, T., Koerver, R., Culver, S. P., Fuchs, T., Senyshyn, A., Indris, S., Morgan, B. J. and Zeier, W. G. (2018) Inducing high ionic conductivity in the lithium superionic argyrodites $Li_{6+x}P_{1-x}Ge_xS_5I$ for all-solid-state batteries. *Journal of the American Chemical Society*, 140, 16330.

16 Kamaya, N., Homma, K., Yamakawa, Y., Hirayama, M., Kanno, R., Yonemura, M., Kamiyama, T., Kato, Y., Hama, S., Kawamoto, K. and Mitsui, A. (2011) A lithium superionic conductor. *Nature Materials*, 10, 682.

17 Ong, S. P., Mo, Y., Richards, W. D., Miara, L., Lee, H. S. and Ceder, G. (2013) Phase stability, electrochemical stability and ionic conductivity of the $Li_{10\pm1}MP_2X_{12}$ (M = Ge, Si, Sn, Al or P, and X = O, S or Se) family of superionic conductors. *Energy & Environmental Science*, 6, 148.

18 Bron, P., Johansson, S., Zick, K., Schmedt auf der Günne, J., Dehnen, S. and Roling, B. (2013) $Li_{10}SnP_2S_{12}$: an affordable lithium superionic conductor. *Journal of the American Chemical Society*, 135, 15694.

19 Whiteley, J. M., Woo, J. H., Hu, E., Nam, K.-W. and Lee, S.-H. (2014) Empowering the lithium metal battery through a silicon-based superionic conductor. *Journal of The Electrochemical Society*, 161, A1812.

20 Kato, Y., Hori, S., Saito, T., Suzuki, K., Hirayama, M., Mitsui, A., Yonemura, M., Iba, H. and Kanno, R. (2016) High-power all-solid-state batteries using sulfide superionic conductors. *Nature Energy*, 1, 16030.

21 Li, Y., Song, S., Kim, H., Nomoto, K., Kim, H., Sun, X., Hori, S., Suzuki, K., Matsui, N., Hirayama, M., Mizoguchi, T., Saito, T., Kamiyama, T. and Kanno, R. (2023) A lithium superionic conductor for millimeter-thick battery electrode. *Science*, 381, 50.

22 Xu, H., Su, Y., Zheng, C., Wang, Y., Tong, Y., Yang, Z., & Hu, J. (2024). Theoretical and experimental design in the study of sulfide-based solid-state battery and interfaces. *Chinese Chemical Letters*, 35(2), 109173.

23 Boukamp, B. A. and Huggins, R. A. (1978) Fast ionic conductivity in lithium nitride. *Materials Research Bulletin*, 13, 23.

24 He, X., Zhu, Y. and Mo, Y. (2017) Origin of fast ion diffusion in super-ionic conductors. *Nature Communications*, 8, 15893.

25 Jalem, R., Yamamoto, Y., Shiiba, H., Nakayama, M., Munakata, H., Kasuga, T. and Kanamura, K. (2013) Concerted migration mechanism in the Li Ion dynamics of garnet-type $Li_7La_3Zr_2O_{12}$. *Chemistry of Materials*, 25, 425.

26 Mo, Y., Ong, S. P. and Ceder, G. (2012) First principles study of the $Li_{10}GeP_2S_{12}$ lithium super ionic conductor material. *Chemistry of Materials*, 24, 15.

27 Lee, J. E., Park, K., Kim, J. C., Wi, T., Ha, A. R., Song, Y. B., Oh, D. Y., Woo, J., Kweon, S. H., Yeom, S. J., Cho, W., Kim, K., Lee, H., Kwak, S. K. and Jung, Y. S. (2022) Universal solution synthesis of sulfide solid electrolytes using alkahest for all-solid-state batteries. *Advanced Materials*, 34, 2200083.

28 Kim, K.-H. and Martin, S. W. (2019) Structures and properties of oxygen-substituted $Li_{10}SiP_2S_{12-x}O_x$ solid-state electrolytes. *Chemistry of Materials*, 31, 3984.

29 Zhu, Y., He, X. and Mo, Y. (2016) First principles study on electrochemical and chemical stability of solid electrolyte–electrode interfaces in all-solid-state Li-ion batteries. *Journal of Materials Chemistry A*, 4, 3253.

30 van de Walle, A., Asta, M. and Ceder, G. (2002) The Alloy Theoretic Automated Toolkit: a user guide. *Calphad*, 26, 539.

31 van de Walle, A. and Ceder, G. (2002) Automating first-principles phase diagram calculations. *Journal of Phase Equilibria*, 23, 348.

32 Richards, W. D., Miara, L. J., Wang, Y., Kim, J. C. and Ceder, G. (2016) Interface stability in solid-state batteries. *Chemistry of Materials*, 28, 266.

33 Banerjee, A., Wang, X., Fang, C., Wu, E. A. and Meng, Y. S. (2020) Interfaces and interphases in all-solid-state batteries with inorganic solid electrolytes. *Chemical Reviews*, 120, 6878.

34 Xu, H., Cao, G., Shen, Y., Yu, Y., Hu, J., Wang, Z. and Shao, G. (2022) Enabling argyrodite sulfides as superb solid-state electrolyte with remarkable interfacial stability against electrodes. *Energy & Environmental Materials*, 5, 852.

35 Yu, C., Zhao, F., Luo, J., Zhang, L. and Sun, X. (2021) Recent development of lithium argyrodite solid-state electrolytes for solid-state batteries: synthesis, structure, stability and dynamics. *Nano Energy*, 83, 105858.

36 Zhou, L., Assoud, A., Zhang, Q., Wu, X. and Nazar, L. F. (2019) New family of argyrodite thioantimonate lithium superionic conductors. *Journal of the American Chemical Society*, 141, 19002.

37 Morimoto, H., Yamashita, H., Tatsumisago, M. and Minami, T. (1999) Mechanochemical synthesis of new amorphous materials of $60Li_2S{\cdot}40SiS_2$ with high lithium ion conductivity. *Journal of the American Ceramic Society*, 82, 1352.

38 Yu, C., Ganapathy, S., Hageman, J., Van Eijck, L., Van Eck, E. R. H., Zhang, L., Schwietert, T., Basak, S., Kelder, E. M. and Wagemaker, M. (2018) Facile synthesis toward the optimal structure-conductivity characteristics of the argyrodite Li_6PS_5Cl solid-state electrolyte. *ACS Applied Materials & Interfaces*, 10, 33296.

39 Park, K. H., Bai, Q., Kim, D. H., Oh, D. Y., Zhu, Y., Mo, Y. and Jung, Y. S. (2018) Design strategies, practical considerations, and new solution processes of sulfide solid electrolytes for all-solid-state batteries. *Advanced Energy Materials*, 8, 1800035.

40 Tan, D. H. S., Banerjee, A., Deng, Z., Wu, E. A., Nguyen, H., Doux, J.-M., Wang, X., Cheng, J., Ong, S. P., Meng, Y. S. and Chen, Z. (2019) Enabling thin and flexible solid-state composite electrolytes by the scalable solution process. *ACS Applied Energy Materials*, 2, 6542.

41 Park, K. H., Oh, D. Y., Choi, Y. E., Nam, Y. J., Han, L., Kim, J., Xin, H., Lin, F., Oh, S. M. and Jung, Y. S. (2016) Solution-processable glass LiI-Li_4SnS_4 superionic conductors for all-solid-state Li-Ion batteries. *Advanced Materials*, 28, 1874.

42 Yubuchi, S., Teragawa, S., Aso, K., Tadanaga, K., Hayashi, A. and Tatsumisago, M. (2015) Preparation of high lithium-ion conducting Li_6PS_5Cl solid electrolyte from ethanol solution for all-solid-state lithium batteries. *Journal of Power Sources*, 293, 941.

43 Wu, F., Lee, J. T., Magasinski, A., Kim, H. and Yushin, G. (2014) Solution-based processing of graphene–Li_2S composite cathodes for lithium-ion and lithium–sulfur batteries. *Particle & Particle Systems Characterization*, 31, 639.

44 Goodenough, J. B. and Park, K.-S. (2013) The Li-ion rechargeable battery: a perspective. *Journal of the American Chemical Society*, 135, 1167.

45 Han, F., Zhu, Y., He, X., Mo, Y. and Wang, C. (2016) Electrochemical stability of $Li_{10}GeP_2S_{12}$ and $Li_7La_3Zr_2O_{12}$ solid electrolytes. *Advanced Energy Materials*, 6, 1501590.

46 Wenzel, S., Sedlmaier, S. J., Dietrich, C., Zeier, W. G. and Janek, J. (2018) Interfacial reactivity and interphase growth of argyrodite solid electrolytes at lithium metal electrodes. *Solid State Ionics*, 318, 102.

47 Wenzel, S., Weber, D. A., Leichtweiss, T., Busche, M. R., Sann, J. and Janek, J. (2016) Interphase formation and degradation of charge transfer kinetics between a lithium metal anode and highly crystalline $Li_7P_3S_{11}$ solid electrolyte. *Solid State Ionics*, 286, 24.

48 Wenzel, S., Randau, S., Leichtweiß, T., Weber, D. A., Sann, J., Zeier, W. G. and Janek, J. (2016) Direct observation of the interfacial instability of the fast ionic conductor $Li_{10}GeP_2S_{12}$ at the lithium metal anode. *Chemistry of Materials*, 28, 2400.

49 Wenzel, S., Leichtweiss, T., Krüger, D., Sann, J. and Janek, J. (2015) Interphase formation on lithium solid electrolytes—an in situ approach to study interfacial reactions by photoelectron spectroscopy. *Solid State Ionics*, 278, 98.

50 Nikodimos, Y., Huang, C.-J., Taklu, B. W., Su, W.-N. and Hwang, B. J. (2022) Chemical stability of sulfide solid-state electrolytes: stability toward humid air and compatibility with solvents and binders. *Energy & Environmental Science*, 15, 991.

51 Lu, P., Liu, L., Wang, S., Xu, J., Peng, J., Yan, W., Wang, Q., Li, H., Chen, L. and Wu, F. (2021) Superior all-solid-state batteries enabled by a gas-phase-synthesized sulfide electrolyte with ultrahigh moisture stability and ionic conductivity. *Advanced Materials*, 33, 2100921.

52 Muramatsu, H., Hayashi, A., Ohtomo, T., Hama, S. and Tatsumisago, M. (2011) Structural change of $Li_2S–P_2S_5$ sulfide solid electrolytes in the atmosphere. *Solid State Ionics*, 182, 116.

53 Pearson, R. G. (1963) Hard and soft acids and bases. *Journal of the American Chemical Society*, 85, 3533.

54 Hayashi, A., Muramatsu, H., Ohtomo, T., Hama, S. and Tatsumisago, M. (2013) Improvement of chemical stability of Li_3PS_4 glass electrolytes by adding M_xO_y (M = Fe, Zn, and Bi) nanoparticles. *Journal of Materials Chemistry A*, 1, 6320.

55 Zhao, F., Liang, J., Yu, C., Sun, Q., Li, X., Adair, K., Wang, C., Zhao, Y., Zhang, S., Li, W., Deng, S., Li, R., Huang, Y., Huang, H., Zhang, L., Zhao, S., Lu, S. and Sun, X. (2020) A versatile Sn-substituted argyrodite sulfide electrolyte for all-solid-state Li metal batteries. *Advanced Energy Materials*, 10, 1903422.

56 Liang, J., Chen, N., Li, X., Li, X., Adair, K. R., Li, J., Wang, C., Yu, C., Norouzi Banis, M., Zhang, L., Zhao, S., Lu, S., Huang, H., Li, R., Huang, Y. and Sun, X. (2020) $Li_{10}Ge(P_{1-x}Sb_x)_2S_{12}$ lithium-ion conductors with enhanced atmospheric stability. *Chemistry of Materials*, 32, 2664.

57 Liu, H., Zhu, Q., Wang, C., Wang, G., Liang, Y., Li, D., Gao, L. and Fan, L. (2022) High air stability and excellent Li metal compatibility of argyrodite-based electrolyte enabling superior all-solid-state Li metal batteries. *Advanced Functional Materials*, 32, 2203858.

58 Sahu, G., Lin, Z., Li, J., Liu, Z., Dudney, N. and Liang, C. (2014) Air-stable, high-conduction solid electrolytes of arsenic-substituted Li_4SnS_4. *Energy & Environmental Science*, 7, 1053.

59 Chen, T., Zhang, L., Zhang, Z., Li, P., Wang, H., Yu, C., Yan, X., Wang, L. and Xu, B. (2019) Argyrodite solid electrolyte with a stable interface and superior dendrite suppression capability realized by ZnO co-doping. *ACS Applied Materials & Interfaces*, 11, 40808.

60 Xu, J., Li, Y., Lu, P., Yan, W., Yang, M., Li, H., Chen, L. and Wu, F. (2022) Water-stable sulfide solid electrolyte membranes directly applicable in all-solid-state batteries enabled by superhydrophobic Li^+ -conducting protection layer. *Advanced Energy Materials*, 12, 2102348.

61 Kim, K. T., Oh, D. Y., Jun, S., Song, Y. B., Kwon, T. Y., Han, Y. and Jung, Y. S. (2021) Tailoring slurries using cosolvents and Li salt targeting practical all-solid-state batteries employing sulfide solid electrolytes. *Advanced Energy Materials*, 11, 2003766.

62 Wu, Y., Wang, S., Li, H., Chen, L. and Wu, F. (2021) Progress in thermal stability of all-solid-state-Li-ion-batteries. *InfoMat*, 3, 827.

63 Mizuno, F., Hayashi, A., Tadanaga, K. and Tatsumisago, M. (2005) New lithium-ion conducting crystal obtained by crystallization of the $Li_2S–P_2S_5$ glasses. *Electrochemical and Solid-State Letters*, 8, A603.

64 Kato, Y., Hori, S. and Kanno, R. (2020) $Li_{10}GeP_2S_{12}$-type superionic conductors: synthesis, structure, and ionic transportation. *Advanced Energy Materials*, 10, 2002153.

65 Chung, H. and Kang, B. (2017) Mechanical and thermal failure induced by contact between a $Li_{1.5}Al_{0.5}Ge_{1.5}(PO_4)_3$ solid electrolyte and Li metal in an all solid-state Li cell. *Chemistry of Materials*, 29, 8611.

66 Wang, S., Wu, Y., Ma, T., Chen, L., Li, H. and Wu, F. (2022) Thermal stability between sulfide solid electrolytes and oxide cathode. *ACS Nano*, 16, 16158.

67 Monroe, C. and Newman, J. (2005) The impact of elastic deformation on deposition kinetics at lithium/polymer interfaces. *Journal of The Electrochemical Society*, 152, A396.

68 Deng, Z., Wang, Z., Chu, I.-H., Luo, J. and Ong, S. P. (2016) Elastic properties of alkali superionic conductor electrolytes from first principles calculations. *Journal of The Electrochemical Society*, 163, A67.

69 Ni, J. E., Case, E. D., Sakamoto, J. S., Rangasamy, E. and Wolfenstine, J. B. (2012) Room temperature elastic moduli and Vickers hardness of hot-pressed LLZO cubic garnet. *Journal of Materials Science*, 47, 7978.

70 Cho, Y.-H., Wolfenstine, J., Rangasamy, E., Kim, H., Choe, H. and Sakamoto, J. (2012) Mechanical properties of the solid Li-ion conducting electrolyte: $Li_{0.33}La_{0.57}TiO_3$. *Journal of Materials Science*, 47, 5970.

71 Kim, K. J., Balaish, M., Wadaguchi, M., Kong, L. Rupp, J. and L. M. (2021) Solid-state Li–metal batteries: challenges and horizons of oxide and sulfide solid electrolytes and their interfaces. *Advanced Energy Materials*, 11, 2002689.

72 McGrogan, F. P., Swamy, T., Bishop, S. R., Eggleton, E., Porz, L., Chen, X., Chiang, Y. and Van Vliet, K. J. (2017) Compliant yet brittle mechanical behavior of $Li_2S–P_2S_5$ lithium-ion-conducting solid electrolyte. *Advanced Energy Materials*, 7, 1602011.

73 Builes, D. H., Hernández-Ortiz, J. P., Corcuera, M. A., Mondragon, I. and Tercjak, A. (2014) Effect of poly(ethylene oxide) homopolymer and two different poly(ethylene oxide-*b*-poly(propylene oxide)-*b*-poly(ethylene oxide) triblock copolymers on morphological, optical, and mechanical properties of nanostructured unsaturated polyester. *ACS Applied Materials & Interfaces*, 6, 1073.

74 Manuel Stephan, A. and Nahm, K. S. (2006) Review on composite polymer electrolytes for lithium batteries. *Polymer*, 47, 5952.

75 Chen, L., Li, Y., Li, S.-P., Fan, L.-Z., Nan, C.-W. and Goodenough, J. B. (2018) PEO/garnet composite electrolytes for solid-state lithium batteries: from "ceramic-in-polymer" to "polymer-in-ceramic." *Nano Energy*, 46, 176.
76 Lin, D., Liu, W., Liu, Y., Lee, H. R., Hsu, P.-C., Liu, K. and Cui, Y. (2016) High ionic conductivity of composite solid polymer electrolyte via in situ synthesis of monodispersed SiO_2 nanospheres in poly(ethylene oxide). *Nano Letters*, 16, 459.
77 Croce, F., Appetecchi, G. B., Persi, L. and Scrosati, B. (1998) Nanocomposite polymer electrolytes for lithium batteries. *Nature*, 394, 456.
78 Zhang, X., Xie, J., Shi, F., Lin, D., Liu, Y., Liu, W., Pei, A., Gong, Y., Wang, H., Liu, K., Xiang, Y. and Cui, Y. (2018) Vertically aligned and continuous nanoscale ceramic–polymer interfaces in composite solid polymer electrolytes for enhanced ionic conductivity. *Nano Letters*, 18, 3829.
79 Hassoun, J. and Scrosati, B. (2010) Moving to a solid-state configuration: a valid approach to making lithium-sulfur batteries viable for practical applications. *Advanced Materials*, 22, 5198.
80 Li, D., Chen, L., Wang, T. and Fan, L.-Z. (2018) 3D fiber-network-reinforced bicontinuous composite solid electrolyte for dendrite-free lithium metal batteries. *ACS Applied Materials & Interfaces*, 10, 7069.
81 Yang, L., Wang, Z., Feng, Y., Tan, R., Zuo, Y., Gao, R., Zhao, Y., Han, L., Wang, Z. and Pan, F. (2017) Flexible composite solid electrolyte facilitating highly stable "soft contacting" Li–electrolyte interface for solid state lithium-ion batteries. *Advanced Energy Materials*, 7, 1701437.
82 Zhai, H., Xu, P., Ning, M., Cheng, Q., Mandal, J. and Yang, Y. (2017) A flexible solid composite electrolyte with vertically aligned and connected ion-conducting nanoparticles for lithium batteries. *Nano Letters*, 17, 3182.
83 Zhao, Y., Huang, Z., Chen, S., Chen, B., Yang, J., Zhang, Q., Ding, F., Chen, Y. and Xu, X. (2016) A promising PEO/LAGP hybrid electrolyte prepared by a simple method for all-solid-state lithium batteries. *Solid State Ionics*, 295, 65.
84 Zhang, X., Liu, T., Zhang, S., Huang, X., Xu, B., Lin, Y., Xu, B., Li, L., Nan, C.-W. and Shen, Y. (2017) Synergistic coupling between $Li_{6.75}La_3Zr_{1.75}Ta_{0.25}O_{12}$ and poly(vinylidene fluoride) induces high ionic conductivity, mechanical strength, and thermal stability of solid composite electrolytes. *Journal of the American Chemical Society*, 139, 13779.
85 Zhang, J., Zang, X., Wen, H., Dong, T., Chai, J., Li, Y., Chen, B., Zhao, J., Dong, S., Ma, J., Yue, L., Liu, Z., Guo, X., Cui, G. and Chen, L. (2017) High-voltage and free-standing poly(propylene carbonate)/$Li_{6.75}La_3Zr_{1.75}Ta_{0.25}O_{12}$ composite solid electrolyte for wide temperature range and flexible solid lithium ion battery. *Journal of Materials Chemistry A*, 5, 4940.
86 Le, H. T. T., Ngo, D. T., Kalubarme, R. S., Cao, G., Park, C.-N. and Park, C.-J. (2016) Composite gel polymer electrolyte based on poly(vinylidene fluoride-hexafluoropropylene) (PVDF-HFP) with modified aluminum-doped lithium lanthanum titanate (A-LLTO) for high-performance lithium rechargeable batteries. *ACS Applied Materials & Interfaces*, 8, 20710.
87 Bae, J., Li, Y., Zhang, J., Zhou, X., Zhao, F., Shi, Y., Goodenough, J. B. and Yu, G. (2018) Cover picture: A 3D nanostructured hydrogel-framework-derived

high-performance composite polymer lithium-ion electrolyte (Angew. Chem. Int. Ed. 8/2018). *Angewandte Chemie International Edition*, 57, 2007.

88 Liu, W., Liu, N., Sun, J., Hsu, P.-C., Li, Y., Lee, H.-W. and Cui, Y. (2015) Ionic conductivity enhancement of polymer electrolytes with ceramic nanowire fillers. *Nano Letters*, 15, 2740.

89 Zhao, Y., Wu, C., Peng, G., Chen, X., Yao, X., Bai, Y., Wu, F., Chen, S. and Xu, X. (2016) A new solid polymer electrolyte incorporating $Li_{10}GeP_2S_{12}$ into a polyethylene oxide matrix for all-solid-state lithium batteries. *Journal of Power Sources*, 301, 47.

90 Zhang, Z., Zhao, Y., Chen, S., Xie, D., Yao, X., Cui, P. and Xu, X. (2017) An advanced construction strategy of all-solid-state lithium batteries with excellent interfacial compatibility and ultralong cycle life. *Journal of Materials Chemistry A*, 5, 16984.

91 Wang, S., Zhang, Y., Zhang, X., Liu, T., Lin, Y.-H., Shen, Y., Li, L. and Nan, C.-W. (2018) High-conductivity argyrodite Li_6PS_5Cl solid electrolytes prepared via optimized sintering processes for all-solid-state lithium–sulfur batteries. *ACS Applied Materials & Interfaces*, 10, 42279.

92 Boulineau, S., Courty, M., Tarascon, J.-M. and Viallet, V. (2012) Mechanochemical synthesis of Li-argyrodite Li_6PS_5X (X=Cl, Br, I) as sulfur-based solid electrolytes for all solid state batteries application. *Solid State Ionics*, 221, 1.

93 Wang, Z. and Shao, G. (2017) Theoretical design of solid electrolytes with superb ionic conductivity: alloying effect on Li^+ transportation in cubic Li_6PA_5X chalcogenides. *Journal of Materials Chemistry A*, 5, 21846.

94 Chaurasia, S. K., Singh, R. K. and Chandra, S. (2013) Ion–polymer complexation and ion-pair formation in a polymer electrolyte PEO:$LiPF_6$ containing an ionic liquid having same anion: a Raman study. *Vibrational Spectroscopy*, 68, 190.

95 Li, D., Cao, L., Liu, C., Cao, G., Hu, J., Chen, J. and Shao, G. (2019) A designer fast Li-ion conductor $Li_{6.25}PS_{5.25}Cl_{0.75}$ and its contribution to the polyethylene oxide based electrolyte. *Applied Surface Science*, 493, 1326.

6

Research Progress of Anode Materials in Solid-state Lithium Batteries

6.1 Research on Anode Materials

Despite significant efforts and progress that have been made in Li metal anode, the uncontrollable interfacial side reactions, continuous lithium dendrite growth in solid electrolytes (SEs), and colossal volume changes in lithium metal still pose great challenges for short-term applications. Other composite anode materials, such as silicon and graphite, thus have been explored in all-solid-state lithium batteries (ASSLBs). Unlike the lithium metal anodes, a homogeneous distribution of sulfide SEs in composite anodes is inevitable for mimicking the uniform ion percolation of liquid electrolytes in lithium-ion batteries (LIBs). This chapter will focus on notable works of graphite and silicon materials in sulfide-based ASSLBs. Additionally, elucidating the electrolyte/electrode interfacial evolution and degradation mechanism may provide meaningful guidance for applying high-performance anode materials in ASSLBs.

6.1.1 Graphite Anodes

Intercalation-based anodes, particularly graphite, are extensively employed in commercial LIBs owing to their favorable electrochemical reversibility, natural abundance, and economic benefits. The graphite structure offers stable insertion sites for lithium ions, ensures consistent physical contact, and presents a promising strategy for reducing the risk of lithium dendrite formation. Early investigations by Takada et al. in 2003 marked the first application of graphite anodes in sulfide-based ASSLBs [1, 2]. Nevertheless, the performance of graphite in sulfide-based ASSLBs falls short of that in liquid electrolytes, primarily due to restricted ion transport pathways. Issues such as high tortuosity, inadequate wetting of the SE, and loss of contact resulting from volumetric changes in graphite (approximately 10%) during cycling are also significant. Using operando microscopy, Otoyama et al. visualized dynamic variations in ionic conduction paths within graphite-based ASSLBs, noting that irreversible expansion leading to voids and cracks reduced ionic connectivity and contributed to capacity fading [3]. As with lithium metal anodes, sustaining efficient ion transport and suppressing the reduction of sulfide SEs at low potentials remain challenging.

Recent studies on graphite in ASSLBs mainly followed two directions. One approach utilized graphite as an intercalation anode material in sulfide-based ASSLBs. For instance, Kim et al. developed a densely compacted graphite anode that completely excluded sulfide SEs, achieving significant gains in volumetric energy density [4]. However, cells incorporating thick or electrolyte-free graphite electrodes showed limited performance, particularly at high C-rates, owing to slow basal-plane diffusion of Li^+. In a subsequent study, the same research group enhanced the anode by incorporating 10 wt.% silicon nanoparticles into graphite, which improved lithium interdiffusion and yielded a high volumetric capacity of 997 mAh cm^{-3} [5]. More recently, Yang et al. fabricated composite electrodes by blending Li_3PS_4 with LiI-coated graphite at a 6:4 weight ratio. The LiI coating effectively inhibited decomposition of the sulfide SE when in contact with lithiated graphite at low potentials, leading to an improved discharge capacity of 348 mAh g^{-1} compared to 248 mAh g^{-1} for uncoated graphite [6].

Another emerging application employed graphite as a lithiophilic layer to improve interfacial stability between lithium metal and sulfide SEs. Duan et al. thoroughly investigated the wetting behavior of lithium on graphite within Li–graphite composites, showing that fully lithiated graphite (forming LiC_6) exhibits intrinsic lithiophilicity [7]. The mechanical properties of graphite also promote intimate contact with sulfide SEs under pressure. For example, Su et al. used a 30 μm graphite layer to coat lithium metal, effectively preventing side reactions with LGPS [8]. The lithiated graphite demonstrated exceptional chemical stability against LGPS, even after annealing at 500 °C for 36 hours. Remarkably, symmetric cells showed no lithium dendrite growth under high current density of 10 mA cm^{-2} and applied pressure of 100–250 MPa. The same team later developed a multilayer cell configuration, Li/graphite/Li_6PS_5Cl (LPSCl)/LGPS/LPSCl/graphite/Li, which sustained low overpotential and avoided short circuits even at 20 mA cm^{-2} [9]. Xing et al. recently proposed a 3D hybrid anode consisting of lithiated graphite and LPS particles, which forms a mixed conductive network. This architecture accommodates volume fluctuations and guides lithium deposition into graphite interstices or voids, resulting in a critical current density (CCD) of 1.4 mA cm^{-2}—three times higher than that of conventional lithium metal anodes [10].

6.1.2 Silicon Anodes

Among the various anode materials, silicon has attracted significant attention as one of the most promising next-generation alternatives due to its natural abundance, high theoretical capacity (e.g., 3,572 mAh g^{-1} for $Li_{15}Si_4$), and moderate operating voltage (~0.4 V vs. Li/Li^+) [11]. Nevertheless, the practical deployment of Si-based anodes remains challenging, primarily because of extreme volume changes (up to 280% for $Li_{15}Si_4$) during lithiation, which induce mechanical fracture and active material pulverization. These issues are compounded by continuous interfacial side reactions between liquid electrolytes and the highly reactive lithiated silicon, along with the inherently low electrical conductivity of silicon ($\approx 3 \times 10^{-5}$ S cm^{-1}) [12]. In ASSLBs, the solid–solid interface between Si and sulfide SEs helps stabilize the interphase by limiting contact area. Moreover, external pressure in sulfide ASSLBs can mitigate mechanical degradation such as cracking and particle disintegration.

Lee and coworkers first demonstrated the use of Si nanoparticles in sulfide-based ASSLBs in 2009, observing improved cycling stability compared to liquid systems [13]. Since then, multiple strategies have been developed to enhance Si anode performance in ASSLBs, including material design (morphology tuning [14], electrolyte modification [15], and binder optimization [16]) and operational management (voltage control [17], applied pressure [18]). Nanostructured silicon generally outperforms micron-sized particles due to shorter ion diffusion paths [14]. Piper et al. showed that high stack pressure suppresses deep lithiation beyond $Li_{15}Si_4$, improving cyclability [18]. Similarly, setting a discharge cutoff voltage above 50 mV also enhances cycling stability [17]. Kato et al. observed that adding LiI to $75Li_2S{\cdot}25P_2S_5$ glass electrolyte reduced its Young's modulus [15]. The softer $70(0.50Li_2S{\cdot}0.50P_2S_5){\cdot}30LiI$ SEs better accommodated volume changes in silicon, leading to improved performance. Dunlap et al. employed a hybrid polyacrylonitrile (PAN)-based binder that, upon low-temperature treatment, formed a conjugated conductive network (cPAN) with strong adhesion and electronic conductivity [17]. This design enabled Si-based electrodes with 70 wt.% active loading to deliver 1,600 mAh g^{-1} after 200 cycles.

To address poor interfacial contact in dry-processed electrodes, Kim et al. developed a solution-infiltration method to incorporate Li_6PS_5Cl into Si electrodes [19]. Using ethanol as solvent, the argyrodite electrolyte retained an ionic conductivity of ~0.1 mS cm^{-1} and formed intimate interfaces with Si, enabling a high reversible capacity of 3,000 mAh g^{-1}. In another study, Tan et al. designed a carbon-free anode comprising 99.9 wt.% micro-Si and 0.1 wt.% PVDF, minimizing SE decomposition at low potentials [20]. The in situ-formed lithiated silicon exhibited high mixed conductivity, allowing high areal capacity (5 mAh cm^{-2}) and > 99.9% average coulombic efficiency over 500 cycles. Conversely, Cao et al. introduced carbon black into nano-Si/SE composites prepared via ball milling, achieving a high capacity of 2,773 mAh g^{-1} and initial coulombic efficiency of 85.6% [21]. The carbon additive improved electronic percolation, facilitating better Si utilization. Notably, the composite anode demonstrated compatibility with thin sulfide electrolyte membranes, and full cells reached an energy density of 285 Wh kg^{-1}, highlighting the potential of Si anodes for practical ASSLBs.

6.1.3 Alloys

Alloy-based anodes (e.g., Li–In, Li–Sn, Li–Au, and Li–Al) are commonly employed in ASSLBs owing to their high lithium diffusion rates and low nucleation overpotentials, which promote uniform lithium deposition and stripping [22, 23]. By serving as a host matrix for lithium, alloy materials help circumvent issues such as lithium dendrite formation and short circuits in sulfide-based ASSLBs. Moreover, the elevated electrochemical potential of such alloys reduces the thermodynamic tendency for reactions between sulfide SEs and lithium anodes [24].

Recently, an Li–Ag intermetallic layer fabricated via a simple roll-pressing process was shown to form a highly lithiophilic interface with LPSCl, suppressing dendrite growth by promoting homogeneous lithium plating/stripping. As a result, cells incorporating this Li–Ag interlayer demonstrated stable discharge performance over 200 cycles without short-circuiting, even at an ultrahigh current density of 12C [25]. Similarly, the Li–In alloy is widely used in sulfide-based ASSLBs due to its dendrite-suppressing characteristics, enhanced interfacial stability, and good deformability, which helps fill voids within the SE [23, 26, 27]. Using in-situ atomic force microscopy, Wan et al. observed the formation of a homogeneous solid electrolyte interface (SEI) between Li–In and sulfide electrolytes [26]. They attributed the uniform lithium deposition to the alloy's high ionic diffusion coefficient, compliant SEI layer, and low ion migration barrier.

Despite these advantages, the relatively high redox potential of Li–In alloys (~0.62 V vs. Li^+/Li) lowers the overall energy density of ASSLBs. Additionally, the high cost of indium restricts its large-scale application [27]. Furthermore, Luo et al. reported that Li–In alloys exhibit kinetic instability under high current densities and prolonged cycling [23]. They observed lateral stripe-like growth of Li–In dendrites that tightly enveloped SE particles, accompanied by the formation of In_2S_3. This configuration maintained close contact between the Li–In anode and LPSCl. In contrast, lithium dendrites were found to grow vertically into the sulfide electrolyte in a sharper manner, owing to high reactivity and localized stress concentration.

6.1.4 Other

In light of the significant challenge of dendrite formation in lithium metal batteries, the anode-free battery concept has recently emerged as a promising strategy to overcome the limitations associated with conventional anode materials [28]. This design employs the current collector directly as the anode, offering benefits such as reduced cost and flexibility in current collector coating design. Crucially, anode-free configurations using sulfide SEs can substantially increase energy density. However, several issues remain unresolved. As lithium is solely supplied from the cathode, capacity fading in anode-free batteries tends to be more pronounced over cycles [29]. One effective approach to promote uniform lithium deposition and stripping is the modification of the current collector with lithiophilic and mechanically robust materials. For instance, Lee et al. developed an anode-free ASSLB using a 5 μm Ag–C composite layer as the anode [28]. The Li–Ag alloying process significantly lowered the nucleation barrier for lithium, enabling highly reversible lithium plating on the current collector. This design achieved an energy density exceeding 900 Wh L^{-1} in a 0.6 Ah pouch cell and retained 89% of its initial capacity after 1,000 cycles. More recently, Lee et al. constructed an anode-free interface through the in-situ reaction of metal fluoride (MF_x) with lithium, forming a composite of lithium metal alloy (Li–M) and lithium fluoride (LiF) [30]. Here, the electronically insulating LiF suppressed dendrite growth, while the Li–M alloy served as a nucleation promoter for uniform plating. Among the various metal fluorides (e.g., M = Ag, Sn[II], In, Zn, Al, and Sn[IV]), AgF exhibited the lowest nucleation overpotential (0.92 mV) during alloy formation. Remarkably, the ASSLB with an AgF-modified interface and a high-loading composite cathode (50 mg cm^{-2}) delivered an areal capacity of 9.7 mAh cm^{-2} and maintained 85.4% of its initial capacity after 50 cycles, underscoring the importance of interface engineering in anode-free systems.

An alternative strategy to enhance the performance of anode-free cells focuses on the structural and material design of the current collector. Given its critical role in initial lithium nucleation and growth, optimizing the substrate is essential for improved lithium deposition. Yan et al. investigated the nucleation overpotential of lithium on various substrates (e.g., Au, Ag, Cu, and Zn) and found that those with the largest crystallographic mismatch with lithium generally induced higher overpotentials [31]. Interestingly, despite the significant difference between the crystal structures of lithium (BCC) and gold (FCC), lithium still deposited favorably on these lithiophilic materials due to partial solubility or alloying, which effectively eliminates nucleation barriers. In contrast, copper—a widely used current collector—showed no solubility in lithium and led to high deposition overpotentials [29].

Three-dimensional (3D) current collectors have been demonstrated as effective hosts for reducing local current density and accommodating volume changes in liquid-electrolyte cells [32, 33]. However, their application in sulfide-based ASSLBs remains limited, primarily due to poor Li^+ transport within 3D architectures. Thus, engineering lithiophilic properties and a stable SEI on the current collector is highly desirable for anode-free configurations. It should be noted, however, that the volumetric energy density advantage of anode-free cells may be compromised when using bulky 3D current collectors. Further investigation into the morphological evolution and nucleation mechanisms of lithium within 3D frameworks is necessary to advance this approach.

6.2 Dendrite Growth Mechanism

Owing to their exceptional lithium-ion conductivity and moderate mechanical rigidity, sulfide SEs are considered promising physical barriers to inhibit the growth of detrimental lithium dendrites in metal anode batteries. According to the well-established Monroe–Newman linear elasticity model, the shear modulus of SEs—approximately twice that of lithium metal—should theoretically prevent dendrite propagation [34]. Indeed, the mechanical properties of many sulfide SEs are sufficient to resist the initiation of Li dendrites. However, experimental observations with inorganic SEs often contradict these predictions.

In reality, multiple factors beyond shear modulus influence lithium filament penetration, including interfacial chemistry, interfacial resistance, and microstructural features of the SEs (e.g., grain boundaries [GBs], pores, and cracks). Furthermore, poor interfacial contact, such as point-contact scenarios between sulfide SEs and lithium metal, leads to nonuniform ion flux distribution and increased interfacial resistance. The presence of numerous active sites on irregular surfaces generates localized electric field enhancement, facilitating continued lithium dendrite growth [26, 35]. Concurrently, substantial volume expansion at dendritic sites induces mechanical stress concentration and crack propagation within the SEs, ultimately resulting in cell failure [36]. Additionally, most sulfide SEs are thermodynamically unstable against lithium metal, leading to the formation of interphases with low ionic conductivity. In this section, we systematically examine the degradation mechanisms at the Li/sulfide SE interface, which is inherently multiscale and involves interplay among chemical stability, electrochemical behavior, and mechanical properties.

6.2.1 Chemical Failures

Here, we systematically discuss the chemical stability of sulfide SEs with Li metal under open-circuit conditions (a resting interface). Based on the theoretical calculations and in-situ characterizations, the well-known sulfide SEs have been demonstrated to be intrinsically thermodynamically unstable against Li metal anode, including Li_6PS_5X [37], Li_3PS_4 [22], $Li_7P_3S_{11}$ [38], $Li_{10}GeP_2S_{12}$ [39, 40], etc. These SEs contain high-valent cations such as Sn^{4+}, Ge^{4+}, and Ti^{4+} in their matrices, which can easily react with the metal Li to form high-electronic-conductivity materials at the Li/SEs interface. The formation of the interphase could be attributed to the potential "window" of SEs, i.e., the bottom of the conduction band (CB) and the top of the valence band (VB) [41]. For sulfide SE materials, the bottom of the CB at the interface is lower than the Li^+/Li^0 potential energy. It spontaneously leads to the electrons transfer from the lithium anode to the electrolyte GBs, which reduces the lithium ion to lithium metal at the GBs, eventually forming dendrites [42].

Three main types of interfaces can be identified based on the interfacial behavior: (i) thermodynamically stable interface with Li metal, in which no side reactions occur, such as the binary compounds Li_3N, LiF, and Li_2S [43]; (ii) thermodynamically unstable for the mixed ionic–electronic conducting (MIEC) interphase, which triggers the continuous electrochemical reduction in the SEs by its higher electronic conductivity; and (iii) thermodynamically unstable but kinetically stable with SEI. Such an interface endows both high ion conductivity and negligible electron conductivity to impede the chemical degradation of SEs. Through in-situ observation and theoretical predictions, Wenzel et al. demonstrated that the decomposition of LGPS led to the MIEC interphase formation of Li_2S, Li_3P, and Li–Ge alloy/Ge metal when in contact with Li metal, which accelerated the flow of electrons and led to a rapid deterioration of the SEs [40]. Through ab-initio molecular dynamics (AIMD) simulations, Cheng et al. studied the Li/LPSCl interface, where the decomposition products contain Li_2S, Li_3P, LiCl, and LiP due to weak interaction between P and S bonds [44]. Despite that these decomposition products alleviate the interfacial degradation between SEs and Li, the ionic transport kinetics of LPSCl will be diminished due to the formation of poor ionic conduction materials.

6.2.2 Electric Failures

Ideally, uniform and reversible lithium-ion nucleation occurs at the interface between lithium metal and SEs during cycling. However, the occurrence of lithium filament growth remains widespread in sulfide-based ASSLBs employing lithium anodes. This subsection concentrates on the dynamic morphological and structural evolution at the Li/SEs interface throughout charge and discharge processes.

Lithium ions tend to deposit at defects such as preexisting voids and cracks within sulfide SEs, as these sites provide active nucleation points with minimal overpotential. Using synchrotron X-ray tomography, Shen et al. observed microstructure-driven failure of SEs, noting that interconnected pores served as pathways for lithium growth and were associated with lower CCD compared to structures with limited pore connectivity [45]. Tian et al. reported that imperfect surface features (e.g., pores and cracks) exhibited a reduced bandgap compared to the bulk SE material, promoting electron trapping at internal defects [46]. Their results also indicated that SEs with finer grain sizes were more prone to isolated lithium nucleation due to a greater density of extended internal defects. Cheng et al. were among the first to note that lithium deposition preferentially follows GBs in polycrystalline ceramics, forming web-like structures that ultimately cause cell failure [47]. Although polycrystallinity is typical in sulfide SEs, even single-crystal SEs are susceptible to cracking from lithium filament expansion due to surface defects [48]. The reason for preferential Li penetration along GBs has remained unclear. Yu and Seigel proposed a "softening" mechanism wherein the shear modulus of GBs is approximately 50% lower than that of the bulk material, facilitating lithium dendrite propagation into otherwise mechanically rigid SEs [49]. However, this mechanism does not fully account for the low CCD values commonly observed in sulfide SEs.

Conventional models assumed that lithium dendrites grow unidirectionally from one electrode to the other. By employing operando neutron depth profiling (NDP), Han et al. investigated lithium dendrite growth mechanisms in three representative SEs: Li_3PS_4, LLZO, and LiPON [50]. Their findings suggested that electronic conductivity exceeding ~10^{-9} S cm^{-1} may enable lithium nucleation and growth within the SEs. High electronic conductivity along GBs can interlink isolated lithium clusters inside the electrolyte, correlating strongly with reduced CCD. Despite these macroscopic observations, the precise mechanisms of Li dendrite formation at GBs remain incompletely understood. More recently, Liu et al. demonstrated that the narrow bandgap of GBs plays a critical role in modulating lithium-ion flux [51]. Unlike the bulk material, GBs with reduced bandgap cannot block electron leakage once the local potential exceeds the bandgap energy, thereby acting as electron sources that reduce Li^+ ions and ultimately induce short circuits. Thus, identifying the origin of high electronic conductivity—whether from impurities, dopants, electrochemical by-products, or GBs—is essential for guiding future development of SEs.

6.2.3 Mechanical Failures

Mechanical degradation often results from the combined effects of chemical and electrical failure mechanisms. The continuous formation of additional compounds and localized volume expansion at the interface contribute to the mechanical deterioration of MIEC interphases in sulfide-based SEs. Furthermore, microstructural defects at the Li/SEs interface can cause heterogeneous electrodeposition, which in turn accelerates unstable morphological changes. This section emphasizes mechanical failure mechanisms at the Li/SEs interface, covering evolution behavior, stress distribution, and the mechanical properties of lithium metal.

Kasemchainan et al. investigated lithium plating and stripping processes at the Li/LPSCl interface [52]. They introduced the concept of a critical current for stripping (CCS), defined as the current density beyond which lithium removal outpaces replenishment. Current densities exceeding CCS result in void formation and accumulation at the lithium interface. The consequent reduction in contact area raises local current density, promoting dendrite formation and eventual cell failure. Applied external stack pressure was shown to mitigate void formation and suppress lithium dendrite growth during cycling. Alternatively, Porz et al. proposed a chemomechanical model in which crack-tip stress facilitates lithium infiltration through various SEs [48]. Their work highlighted the significant influence of surface defects in SEs on lithium dendrite penetration: once lithiation overpotential and mechanical stress exceed a threshold, lithium deposition drives crack propagation within the SEs.

Sulfide SEs possess specific mechanical properties. Unlike in liquid electrolytes, the volume changes during lithium plating and stripping generate substantial stresses at the Li/SEs interface that cannot be alleviated in ASSLBs due to the rigid nature of SEs. Thus, understanding the mechanical characteristics of lithium metal—including its elastic, plastic, and creep behaviors—is essential, as these properties strongly influence its dynamic evolution during cycling. Progress has been made in elucidating the mechanical role of lithium metal. Using acoustic measurements, Masias et al. determined that lithium metal has a Young's modulus of 7.82 GPa and a yield strength between 0.4 and 0.9 MPa [53]. The high ductility of lithium helps maintain interfacial contact with sulfide SEs through plastic deformation. However, pronounced deformation—or creep—also poses a risk by facilitating lithium filament propagation into SE defects. Creep refers to the continuous deformation of a material under constant stress. In contrast, Zhang et al. combined atomic force microscopy and transmission electron microscopy to show that growing Li whiskers at the submicrometer scale could sustain stresses up to 130 MPa, with yield strengths reaching 244 MPa—significantly higher than those of bulk lithium [54]. In a related study, Doux et al. correlated stack pressure with mechanical failure and short-circuit behavior, noting that a moderate stack pressure of 25 MPa improved interfacial contact, whereas higher pressure (75 MPa) induced mechanical short circuits as ductile lithium crept into SE pores, forming electronic percolation pathways [55].

6.3 Modification Strategy of Lithium Metal Anodes

Lithium metal anodes are regarded as the ultimate anode choice due to their exceptional gravimetric and volumetric energy densities. However, sulfide-based ASSLBs incorporating lithium metal anodes remain prone to short-circuiting caused by lithium dendrite growth. The origins and mechanisms of lithium dendrite formation have been extensively studied. Key requirements for a stable Li/SEs interface include: (i) low and consistent interfacial resistance, (ii) a protective SEI that prevents electrochemical and mechanical degradation, and (iii) high coulombic efficiency during lithium stripping and plating. This section presents state-of-the-art strategies for suppressing lithium dendrite growth, such as improving the intrinsic properties of SEs, constructing artificial protective interlayers, and engineering lithiophilic lithium-rich alloy interfaces. In conventional graphite-based LIBs, an SEI is formed on the anode surface, enabling reversible lithium-ion intercalation. An ideal interphase layer should exhibit high thermodynamic and electrochemical stability against lithium metal, a wide bandgap to block electron leakage, and sufficient ionic conductivity to facilitate efficient ion transport across the Li/SEs interface [56]. To alleviate parasitic reactions between lithium metal and sulfide SEs, artificial protective layers have been widely employed to maintain interfacial stability. These layers can be fabricated using inorganic compounds, organic solvents, lithium salts, polymers, alloys, and other materials. Advanced coating techniques such as spin-coating, physical/chemical vapor deposition, and in-situ solution reactions are commonly utilized to apply these protective layers [43, 57, 58].

6.3.1 Ex-situ SEI

To understand the influence of interface structures on failure mechanisms at the Li/SEs interface, Mo and coworkers employed atomistic modeling via molecular dynamics (MD) simulations to study lithium plating and stripping processes [59]. Beyond facilitating rapid Li^+ transport, establishing a strongly adhesive and coherent interface between SEs and lithium metal is crucial for suppressing dendrite formation. Density functional theory (DFT) calculations have further indicated that high-interfacial-energy components in the SEI layer can effectively impede lithium dendrite penetration [60, 61]. Recently, using high-throughput DFT screening, Li et al. identified 28 potential protective-layer materials from the Inorganic Crystal Structure Database suitable for use between LPSCl and lithium metal [61]. Most candidates were halides and polyanionic oxides with appropriate band alignment, including LiF, LiCl, LiI, and Li_2SiO_3.

Using a chemical iodine-vapor deposition technique, Duan et al. fabricated a densely structured LiI layer approximately 1.5 μm thick on lithium metal [57]. This nanostructured LiI interlayer improved Li^+ transport kinetics at the Li/LGPS interface, and its high Young's modulus (6.62 GPa) effectively resisted lithium dendrite propagation. Symmetric cells with this modified interface achieved a CCD of 2.3 mA cm^{-2}. In a related approach, Park et al. applied chemical vapor deposition to prepare nanostructured lithium selenide (Li_2Se) layers with tailored morphologies [62]. The strong chemical bonding between Li_2Se and lithium metal ensured intimate interfacial contact, mitigating side reactions. More recently, Su et al. introduced a thin LiPON interlayer via radio-frequency sputtering between argyrodite LPSCl and lithium metal [63]. The LiPON layer formed a self-limiting reaction interface, producing an ion-conductive SEI composed of Li_3PO_4, Li_3P, Li_3N, and Li_2O, which effectively blocked electron leakage. Symmetric cells with this protection layer demonstrated an ultrahigh CCD of 4.1 mA cm^{-2}.

Although LiF possesses the highest interfacial energy among SEI components, it exhibits poor wettability [43]. Single-component interlayers often fail to simultaneously meet the requirements of high interfacial energy, ionic conductivity, and chemical stability. Addressing this, Ji et al. incorporated lithiophilic and highly ion-conductive Li_3N into a LiF-based interface. After activation, LPS electrolytes with the Li_3N–LiF hybrid layer achieved a remarkable room-temperature CCD exceeding 6 mA cm^{-2} [58].

6.3.2 In-situ SEI

The use of liquid electrolytes or organic solvents to improve interfacial wettability represents a compromise strategy for practical ASSLBs. In-situ-formed SEI can produce ultrathin layers that achieve intimate contact with lithium anodes. Inspired by successful SEI designs in liquid systems, Fan et al. [60] fabricated a thin LiF-rich interlayer through the reduction of 6 M LiFSI in dimethoxyethane (DME) (~20 μL) at the Li_3PS_4/Li interface. The resulting symmetric cells exhibited a high CCD exceeding 2 mA cm^{-2}. Similarly, Gao et al. employed constant-current electrochemical decomposition to generate a nanocomposite SEI—comprising organic species ($LiO{-}(CH_2O)_n{-}Li$) and inorganic compounds (LiF, $-NSO_2-Li$, and Li_2O)—that prevented reduction of LGPS by lithium metal [64]. Cells with an SEI derived from 1 M LiTFSI in DOL/DME (1:1, v/v) showed stable electrodeposition for over 3,000 hours. In another study, $Li_{10}GeP_2S_{12}$ was stabilized using a bifunctional interphase composed of a lithiophilic Li_xMg alloy and a lithiophobic LiF-rich layer, which reduced interfacial resistance and suppressed dendrite growth. This modification increased the CCD of $Li_{10}GeP_2S_{12}$ from 0.6 to 1.3 mA cm^{-2} [65].

DFT calculations have been used to estimate interfacial energies of SEI components. For instance, Li_3PO_4 exhibits high interfacial energy (0.603 J m^{-2}) and Young's modulus (78 GPa), but its low ionic conductivity (~10^{-8} S cm^{-1}) limits its utility. To address this, Su et al. incorporated highly conductive and adhesive Li_3N. A Li_3N–Li_3PO_4 layer synthesized via two-step reactions involving H_3PO_4 and $LiNO_3$ with lithium metal enabled stable cycling for 300 hours at 1 mA cm^{-2} [43].

Ionic liquids (ILs) are also promising modifiers due to their nonflammability, tunable viscosity, and thermal stability [66]. Zheng et al. improved the Li/$Li_{10}SnP_2S_{12}$ (LSPS) interface using 1.5 M LiTFSI in *N*-propyl-*N*-methylpyrrolidinium bis(trifluoromethanesulfonyl)imide (Pyr13TFSI), which enhanced wettability and compatibility [67]. Although sulfide electrolytes are susceptible to nucleophilic attack by polar solvents [68], solvate ionic liquids (SILs) such as Li(G3)TFSI (G3: triethylene glycol dimethyl ether) exhibit good compatibility because Li^+–glycol interactions mitigate nucleophilicity [69]. Gao and coworkers further showed that Li(G3)TFSI reduced interfacial resistance and promoted uniform lithium deposition at Li/LGPS interfaces [70].

6.3.3 Buffer Layer

Unlike ultrathin SEI layers, buffer layers made from polymer electrolytes or inorganic SEs can regulate Li^+ flux and enhance interfacial stability. The Li^+ diffusion coefficient within such interlayers critically influences dendrite suppression.

Soft polymers are often used to address interfacial contact loss and side reactions due to their flexibility and stress-buffering capability. Polymer electrolytes—typically comprising a polymer matrix and lithium salt—offer advantages in flexibility and adhesion. Poly(ethylene oxide) (PEO)-based electrolytes, for example, exhibit ionic conductivities of ~10^{-5}–10^{-4} S cm^{-1} at 60 °C and good compatibility with lithium metal [71]. However, their low intrinsic conductivity and transference number limit their application. To improve ion transport, Wang et al. developed a plastic crystal electrolyte (PCE) interlayer from succinonitrile (SN) with 5 mol.% LiTFSI, achieving a high room-temperature conductivity of 1.47×10^{-3} S cm^{-1} [72]. This SN-based layer enhanced stability between LGPS and lithium and enabled high capacity retention (0.14% decay per cycle over 100 cycles) in Li–S ASSLBs. Chen et al. designed a gel-like artificial layer from poly(propylene carbonate) (PPC) and LiTFSI (6:1 wt.% ratio), which prevented reactions between Li and LPSCl [73]. Degradation products from PPC (e.g., propylene carbonate) reduced interfacial resistance and promoted LiF-rich SEI formation, enabling stable symmetric cell operation for 300 hours at 0.5 mA cm^{-2}.

References

1 Takada, K., Inada, T., Kajiyama, A., Sasaki, H., Kondo, S., Watanabe, M., Murayama, M. and Kanno, R. (2003) Solid-state lithium battery with graphite anode. *Solid State Ionics*, 158(3), 269–274.

2 Takada, K., Nakano, S., Inada, T., Kajiyama, A., Sasaki, H., Kondo, S. and Watanabe, M. (2003) Compatibility of lithium ion conductive sulfide glass with carbon-lithium electrode. *Journal of The Electrochemical Society*, 150(3), A274.

3 Otoyama, M., Kowada, H., Sakuda, A., Tatsumisago, M. and Hayashi, A. (2020) Operando confocal microscopy for dynamic changes of Li^+ ion conduction path in graphite electrode layers of all-solid-state batteries. *The Journal of Physical Chemistry Letters*, 11(3), 900–904.

4 Kim, J. Y., Park, J., Lee, M. J., Kang, S. H., Shin, D. O., Oh, J., Kim, J., Kim, K. M., Lee, Y.-G. and Lee, Y. M. (2020) Diffusion-dependent graphite electrode for all-solid-state batteries with extremely high energy density. *ACS Energy Letters*, 5(9), 2995–3004.

5 Kim, J. Y., Jung, S., Kang, S. H., Park, J., Lee, M. J., Jin, D., Shin, D. O., Lee, Y. G. and Lee, Y. M. (2021) Graphite-silicon diffusion-dependent electrode with short effective diffusion length for high-performance all-solid-state batteries. *Advanced Energy Materials*, 12(3), 2103108.

6 Yang, S., Yamamoto, K., Mei, X., Sakuda, A., Uchiyama, T., Watanabe, T., Takami, T., Hayashi, A., Tatsumisago, M. and Uchimoto, Y. (2021) High rate capability from a graphite anode through surface modification with lithium iodide for all-solid-state batteries. *ACS Applied Energy Materials*, 5(1), 667–673.

7 Duan, J., Zheng, Y., Luo, W., Wu, W., Wang, T., Xie, Y., Li, S., Li, J. and Huang, Y. (2020) Is graphite lithiophobic or lithiophilic? *National Science Review*, 7(7), 1208–1217.

8 Su, Y., Ye, L., Fitzhugh, W., Wang, Y., Gil-González, E., Kim, I. and Li, X. (2020) A more stable lithium anode by mechanical constriction for solid state batteries. *Energy & Environmental Science*, 13(3), 908–916.

9 Ye, L. and Li, X. (2021) A dynamic stability design strategy for lithium metal solid state batteries. *Nature*, 593(7858), 218–222.

10 Xing, X., Li, Y., Wang, S., Liu, H., Wu, Z., Yu, S., Holoubek, J., Zhou, H. and Liu, P. (2021) Graphite-based lithium-free 3D hybrid anodes for high energy density all-solid-state batteries. *ACS Energy Letters*, 6(5), 1831–1838.

11 Wu, H. and Cui, Y. (2012) Designing nanostructured Si anodes for high energy lithium ion batteries. *Nano Today*, 7(5), 414–429.

12 Chen, Z., Soltani, A., Chen, Y., Zhang, Q., Davoodi, A., Hosseinpour, S., Peukert, W. and Liu, W. (2022) Emerging organic surface chemistry for Si anodes in lithium-ion batteries: advances, prospects, and beyond. *Advanced Energy Materials*, 12(32), 2200924.

13 Trevey, J., Jang, J. S., Jung, Y. S., Stoldt, C. R. and Lee, S.-H. (2009) Glass–ceramic $Li_2S–P_2S_5$ electrolytes prepared by a single step ball billing process and their application for all-solid-state lithium–ion batteries. *Electrochemistry Communications*, 11(9), 1830–1833.

14 Dunlap, N. A., Kim, S., Jeong, J. J., Oh, K. H. and Lee, S.-H. (2018) Simple and inexpensive coal-tar-pitch derived Si-C anode composite for all-solid-state Li-ion batteries. *Solid State Ionics*, 324, 207–217.

15 Kato, A., Yamamoto, M., Sakuda, A., Hayashi, A. and Tatsumisago, M. (2018) Mechanical properties of $Li_2S-P_2S_5$ glasses with lithium halides and application in all-solid-state batteries. *ACS Applied Energy Materials*, 1(3), 1002–1007.

16 Trevey, J. E., Rason, K. W., Stoldt, C. R. and Lee, S.-H. (2010) Improved performance of all-solid-state lithium-ion batteries using nanosilicon active material with multiwalled-carbon-nanotubes as a conductive additive. *Electrochemical and Solid-State Letters*, 13(11), A154.

17 Dunlap, N. A., Kim, J., Guthery, H., Jiang, C.-S., Morrissey, I., Stoldt, C. R., Oh, K. H., Al-Jassim, M. and Lee, S.-H. (2020) Towards the commercialization of the all-solid-state Li-ion battery: local bonding structure and the reversibility of sheet-style Si-PAN anodes. *Journal of The Electrochemical Society*, 167(6), 060522.

18 Piper, D. M., Yersak, T. A. and Lee, S.-H. (2012) Effect of compressive stress on electrochemical performance of silicon anodes. *Journal of The Electrochemical Society*, 160(1), A77.

19 Kim, D. H., Lee, H. A., Song, Y. B., Park, J. W., Lee, S.-M. and Jung, Y. S. (2019) Sheet-type Li_6PS_5Cl-infiltrated Si anodes fabricated by solution process for all-solid-state lithium-ion batteries. *Journal of Power Sources*, 426, 143–150.

20 Tan, D. H. S., Chen, Y.-T., Yang, H., Bao, W., Sreenarayanan, B., Doux, J.-M., Li, W., Lu, B., Ham, S.-Y., Sayahpour, B., Scharf, J., Wu, E. A., Deysher, G., Han, H. E., Hah, H. J., Jeong, H., Lee, J. B., Chen, Z. and Meng, Y. S. (2021) Carbon-free high-loading silicon anodes enabled by sulfide solid electrolytes. *Science*, 373(6562), 1494–1499.

21 Cao, D., Sun, X., Li, Y., Anderson, A., Lu, W. and Zhu, H. (2022) Long-cycling sulfide-based all-solid-state batteries enabled by electrochemo-mechanically stable electrodes. *Advanced Materials*, 34(24), 2200401.

22 Kato, A., Kowada, H., Deguchi, M., Hotehama, C., Hayashi, A. and Tatsumisago, M. (2018) XPS and SEM analysis between Li/Li_3PS_4 interface with Au thin film for all-solid-state lithium batteries. *Solid State Ionics*, 322, 1–4.

23 Luo, S., Wang, Z., Li, X., Liu, X., Wang, H., Ma, W., Zhang, L., Zhu, L. and Zhang, X. (2021) Growth of lithium-indium dendrites in all-solid-state lithium-based batteries with sulfide electrolytes. *Nature Communications*, 12(1), 6968.

24 Lewis, J. A., Cavallaro, K. A., Liu, Y. and McDowell, M. T. (2022) The promise of alloy anodes for solid-state batteries. *Joule*, 6, 1418–1430.

25 Choi, H. J., Kang, D. W., Park, J. W., Park, J. H., Lee, Y. J., Ha, Y. C., Lee, S. M., Yoon, S. Y. and Kim, B. G. (2022) In situ formed Ag-Li intermetallic layer for stable cycling of all-solid-state lithium batteries. *Advanced Science*, 9(1), 2103826.

26 Wan, J., Song, Y. X., Chen, W. P., Guo, H. J., Shi, Y., Guo, Y. J., Shi, J. L., Guo, Y. G., Jia, F. F., Wang, F. Y., Wen, R. and Wan, L. J. (2021) Micromechanism in all-solid-state alloy-metal batteries: regulating homogeneous lithium precipitation and flexible solid electrolyte interface evolution. *Journal of the American Chemical Society*, 143(2), 839–848.

27 Santhosha, A. L., Medenbach, L., Buchheim, J. R. and Adelhelm, P. (2019) The indium-lithium electrode in solid-state lithium-ion batteries: phase formation, redox potentials, and interface stability. *Batteries & Supercaps*, 2(6), 524–529.

28 Lee, Y.-G., Fujiki, S., Jung, C., Suzuki, N., Yashiro, N., Omoda, R., Ko, D.-S., Shiratsuchi, T., Sugimoto, T., Ryu, S., Ku, J. H., Watanabe, T., Park, Y., Aihara, Y., Im, D. and Han, I. T. (2020) High-energy long-cycling all-solid-state lithium metal batteries enabled by silver–carbon composite anodes. *Nature Energy*, 5(4), 299–308.

29 Nanda, S., Gupta, A. and Manthiram, A. (2020) Anode-free full cells: a pathway to high-energy density lithium-metal batteries. *Advanced Energy Materials*, 11(2), 2000804.

30 Lee, J., Choi, S. H., Im, G., Lee, K. J., Lee, T., Oh, J., Lee, N., Kim, H., Kim, Y., Lee, S. and Choi, J. W. (2022) Room-temperature anode-less all-solid-state batteries via the conversion reaction of metal fluorides. *Advanced Materials*, 34(40), e2203580.

31 Yan, K., Lu, Z., Lee, H.-W., Xiong, F., Hsu, P.-C., Li, Y., Zhao, J., Chu, S. and Cui, Y. (2016) Selective deposition and stable encapsulation of lithium through heterogeneous seeded growth. *Nature Energy*, 1(3), 16010.

32 Shi, Y., Wang, Z., Gao, H., Niu, J., Ma, W., Qin, J., Peng, Z. and Zhang, Z. (2019) A self-supported, three-dimensional porous copper film as a current collector for advanced lithium metal batteries. *Journal of Materials Chemistry A*, 7(3), 1092–1098.

33 Jin, L., Zhang, H., Li, S., Lei, S., Liu, M., Zeng, Z., Yu, C., Cheng, S. and Xie, J. (2021) Exchange of Li and $AgNO_3$ enabling stable 3D lithium metal anodes with embedded lithophilic nanoparticles and a solid electrolyte interface inducer. *ACS Applied Materials & Interfaces*, 13(32), 38425–38431.

34 Monroe, C. and Newman, J. (2005) The impact of elastic deformation on deposition kinetics at lithium/polymer interfaces. *Journal of The Electrochemical Society*, 152(2), A396.

35 Kim, S., Jung, C., Kim, H., Thomas-Alyea, K. E., Yoon, G., Kim, B., Badding, M. E., Song, Z., Chang, J., Kim, J., Im, D. and Kang, K. (2020) The role of interlayer chemistry in Li-metal growth through a garnet-type solid electrolyte. *Advanced Energy Materials*, 10(12), 1903993.

36 LePage, W. S., Chen, Y., Kazyak, E., Chen, K.-H., Sanchez, A. J., Poli, A., Arruda, E. M., Thouless, M. D. and Dasgupta, N. P. (2019) Lithium mechanics: roles of strain rate and temperature and implications for lithium metal batteries. *Journal of The Electrochemical Society*, 166(2), A89.

37 Wenzel, S., Sedlmaier, S. J., Dietrich, C., Zeier, W. G. and Janek, J. (2018) Interfacial reactivity and interphase growth of argyrodite solid electrolytes at lithium metal electrodes. *Solid State Ionics*, 318, 102–112.

38 Wenzel, S., Weber, D. A., Leichtweiss, T., Busche, M. R., Sann, J. and Janek, J. (2016) Interphase formation and degradation of charge transfer kinetics between a lithium metal anode and highly crystalline $Li_7P_3S_{11}$ solid electrolyte. *Solid State Ionics*, 286, 24–33.

39 Bron, P., Roling, B. and Dehnen, S. (2017) Impedance characterization reveals mixed conducting interphases between sulfidic superionic conductors and lithium metal electrodes. *Journal of Power Sources*, 352, 127–134.

40 Wenzel, S., Randau, S., Leichtweiß, T., Weber, D. A., Sann, J., Zeier, W. G. and Janek, J. (2016) Direct observation of the interfacial instability of the fast ionic conductor $Li_{10}GeP_2S_{12}$ at the lithium metal anode. *Chemistry of Materials*, 28(7), 2400–2407.

41 Goodenough, J. B. and Park, K. S. (2013) The Li-ion rechargeable battery: a perspective. *Journal of the American Chemical Society*, 135(4), 1167–1176.

42 Han, F., Zhu, Y., He, X., Mo, Y. and Wang, C. (2016) Electrochemical stability of $Li_{10}GeP_2S_{12}$ and $Li_7La_3Zr_2O_{12}$ solid electrolytes. *Advanced Energy Materials*, 6(8), 1501590.

43 Su, H., Liu, Y., Zhong, Y., Li, J., Wang, X., Xia, X., Gu, C. and Tu, J. (2022) Stabilizing the interphase between Li and argyrodite electrolyte through synergistic phosphating process for all-solid-state lithium batteries. *Nano Energy*, 96, 107104.

44 Cheng, T., Merinov, B. V., Morozov, S. and Goddard, W. A. (2017) Quantum mechanics reactive dynamics study of solid Li-electrode/Li_6PS_5Cl-electrolyte interface. *ACS Energy Letters*, 2(6), 1454–1459.

45 Shen, F., Dixit, M. B., Xiao, X. and Hatzell, K. B. (2018) Effect of pore connectivity on Li dendrite propagation within LLZO electrolytes observed with synchrotron X-ray tomography. *ACS Energy Letters*, 3(4), 1056–1061.

46 Tian, H.-K., Liu, Z., Ji, Y., Chen, L.-Q. and Qi, Y. (2019) Interfacial electronic properties dictate Li dendrite growth in solid electrolytes. *Chemistry of Materials*, 31(18), 7351–7359.

47 Cheng, E. J., Sharafi, A. and Sakamoto, J. (2017) Intergranular Li metal propagation through polycrystalline $Li_{6.25}Al_{0.25}La_3Zr_2O_{12}$ ceramic electrolyte. *Electrochimica Acta*, 223, 85–91.

48 Porz, L., Swamy, T., Sheldon, B. W., Rettenwander, D., Frömling, T., Thaman, H. L., Berendts, S., Uecker, R., Carter, W. C. and Chiang, Y. M. (2017) Mechanism of lithium metal penetration through inorganic solid electrolytes. *Advanced Energy Materials*, 7(20), 1701003.

49 Yu, S. and Siegel, D. J. (2018) Grain boundary softening: a potential mechanism for lithium metal penetration through stiff solid electrolytes. *ACS Applied Materials & Interfaces*, 10(44), 38151–38158.

50 Han, F., Westover, A. S., Yue, J., Fan, X., Wang, F., Chi, M., Leonard, D. N., Dudney, N. J., Wang, H. and Wang, C. (2019) High electronic conductivity as the origin of lithium dendrite formation within solid electrolytes. *Nature Energy*, 4(3), 187–196.

51 Liu, X., Garcia-Mendez, R., Lupini, A. R., Cheng, Y., Hood, Z. D., Han, F., Sharafi, A., Idrobo, J. C., Dudney, N. J., Wang, C., Ma, C., Sakamoto, J. and Chi, M. (2021) Local electronic structure variation resulting in Li 'filament' formation within solid electrolytes. *Nature Materials*, 20(11), 1485–1490.

52 Kasemchainan, J., Zekoll, S., Jolly, D. S., Ning, Z., Hartley, G. O., Marrow, J. and Bruce, P. G. (2019) Critical stripping current leads to dendrite formation on plating in lithium anode solid electrolyte cells. *Nature Materials*, 18(10), 1105–1111.

53 Masias, A., Felten, N., Garcia-Mendez, R., Wolfenstine, J. and Sakamoto, J. (2018) Elastic, plastic, and creep mechanical properties of lithium metal. *Journal of Materials Science*, 54(3), 2585–2600.

54 Zhang, L., Yang, T., Du, C., Liu, Q., Tang, Y., Zhao, J., Wang, B., Chen, T., Sun, Y., Jia, P., Li, H., Geng, L., Chen, J., Ye, H., Wang, Z., Li, Y., Sun, H., Li, X., Dai, Q., Tang, Y., Peng, Q., Shen, T., Zhang, S., Zhu, T. and Huang, J. (2020) Lithium whisker growth and stress generation in an in situ atomic force microscope-environmental transmission electron microscope set-up. *Nature Nanotechnology*, 15(2), 94–98.

55 Doux, J. M., Nguyen, H., Tan, D. H. S., Banerjee, A., Wang, X., Wu, E. A., Jo, C., Yang, H. and Meng, Y. S. (2019) Stack pressure considerations for room-temperature all-solid-state lithium metal batteries. *Advanced Energy Materials*, 10(1), 1903253.

56 Wang, T., Duan, J., Zhang, B., Luo, W., Ji, X., Xu, H., Huang, Y., Huang, L., Song, Z., Wen, J., Wang, C., Huang, Y. and Goodenough, J. B. (2022) A self-regulated gradient interphase for dendrite-free solid-state Li batteries. *Energy & Environmental Science*, 15(3), 1325–1333.

57 Duan, C., Cheng, Z., Li, W., Li, F., Liu, H., Yang, J., Hou, G., He, P. and Zhou, H. (2022) Realizing compatibility of Li metal anode in all-solid-state Li-S battery by chemical iodine-vapor deposition. *Energy & Environmental Science*, 15, 3236–3245.

58 Ji, X., Hou, S., Wang, P., He, X., Piao, N., Chen, J., Fan, X. and Wang, C. (2020) Solid-state electrolyte design for lithium dendrite suppression. *Advanced Materials*, 32(46), 2002741.

59 Yang, M., Liu, Y., Nolan, A. M. and Mo, Y. (2021) Interfacial atomistic mechanisms of lithium metal stripping and plating in solid-state batteries. *Advanced Materials*, 33(11), e2008081.

60 Fan, X., Ji, X., Han, F., Yue, J., Chen, J., Chen, L., Deng, T., Jiang, J. and Wang, C. (2018) Fluorinated solid electrolyte interface enables highly reversible solid-state Li metal battery. *Science Advances*, 4(12), eaau9245.

61 Li, S., Chen, Z., Zhang, W., Li, S. and Pan, F. (2022) High-throughput screening of protective layers to stabilize the electrolyte-anode interface in solid-state Li-metal batteries. *Nano Energy*, 102, 107640.

62 Park, H., Kim, J., Lee, D., Park, J., Jo, S., Kim, J., Song, T. and Paik, U. (2021) Epitaxial growth of nanostructured Li_2Se on lithium metal for all solid-state batteries. *Advanced Science*, 8(11), e2004204.

63 Su, J., Pasta, M., Ning, Z., Gao, X., Bruce, P. G. and Grovenor, C. R. M. (2022) Interfacial modification between argyrodite-type solid-state electrolytes and Li metal anodes using LiPON interlayers. *Energy & Environmental Science*, 15, 3805–3814.

64 Gao, Y., Wang, D., Li, Y. C., Yu, Z., Mallouk, T. E. and Wang, D. (2018) Salt-based organic-inorganic nanocomposites: towards a stable lithium metal/$Li_{10}GeP_2S_{12}$ solid electrolyte interface. *Angewandte Chemie International Edition*, 57(41), 13608–13612.

65 Wan, H., Liu, S., Deng, T., Xu, J., Zhang, J., He, X., Ji, X., Yao, X. and Wang, C. (2021) Bifunctional interphase-enabled $Li_{10}GeP_2S_{12}$ electrolytes for lithium-sulfur battery. *ACS Energy Letters*, 6(3), 862–868.

66 Hwang, J., Matsumoto, K., Chen, C.-Y. and Hagiwara, R. (2021) Pseudo-solid-state electrolytes utilizing the ionic liquid family for rechargeable batteries. *Energy & Environmental Science*, 14(11), 5834–5863.

67 Zheng, B., Zhu, J., Wang, H., Feng, M., Umeshbabu, E., Li, Y., Wu, Q. H. and Yang, Y. (2018) Stabilizing $Li_{10}SnP_2S_{12}$/Li interface via an in situ formed solid electrolyte interface layer. *ACS Applied Materials & Interfaces*, 10(30), 25473–25482.

68 Yamamoto, M., Terauchi, Y., Sakuda, A. and Takahashi, M. (2018) Binder-free sheet-type all-solid-state batteries with enhanced rate capabilities and high energy densities. *Scientific Reports*, 8(1), 1212.

69 Oh, D. Y., Nam, Y. J., Park, K. H., Jung, S. H., Cho, S.-J., Kim, Y. K., Lee, Y.-G., Lee, S.-Y. and Jung, Y. S. (2015) Excellent compatibility of solvate ionic liquids with sulfide solid electrolytes: toward favorable ionic contacts in bulk-type all-solid-state lithium-ion batteries. *Advanced Energy Materials*, 5(22), 1500865.

70 Cao, Y., Lou, S., Sun, Z., Tang, W., Ma, Y., Zuo, P., Wang, J., Du, C., Gao, Y. and Yin, G. (2020) Solvate ionic liquid boosting favorable interfaces kinetics to achieve the excellent performance of $Li_4Ti_5O_{12}$ anodes in $Li_{10}GeP_2S_{12}$ based solid-state batteries. *Chemical Engineering Journal*, 382, 123046.

71 Fenton, D. E., Parker, J. M. and Wright, P. V. (1973) Complexes of alkali metal ions with poly(ethylene oxide). *Polymer*, 14(11), 589.

72 Wang, C., Adair, K. R., Liang, J., Li, X., Sun, Y., Li, X., Wang, J., Sun, Q., Zhao, F., Lin, X., Li, R., Huang, H., Zhang, L., Yang, R., Lu, S. and Sun, X. (2019) Solid-state plastic crystal electrolytes: effective protection interlayers for sulfide-based all-solid-state lithium metal batteries. *Advanced Functional Materials*, 29(26), 1900392.

73 Chen, Y., Li, W., Sun, C., Jin, J., Wang, Q., Chen, X., Zha, W. and Wen, Z. (2020) Sustained release-driven formation of ultrastable SEI between Li_6PS_5Cl and lithium anode for sulfide-based solid-state batteries. *Advanced Energy Materials*, 11(4), 2002545.

7

Magnesium-ion Batteries: Recent Progress and Challenges

7.1 Background and Introduction

Li-ion batteries have been widely adopted owing to their high energy density and portability for energy storage [1]. However, Li resources in China are limited, accounting for only approximately 3% of the known world reserves. Therefore, the prospects of Li resource development are uncertain. As global Li resources continue to diminish, the steady price rise of Li can seriously affect and potentially hinder the development of two pillar industries: Li-based energy storage and electric cars [2–4]. Furthermore, the safety and reliability of Li-ion batteries in large-scale energy storage applications are contentious, being associated with high safety risks, environmental concerns, and low utilization efficiency [5, 6]. Under these constraints, attention has gradually shifted to alternative rechargeable battery systems, such as Na-, Mg-, and Al-based batteries (Figure 7.1).

Mg metal is an attractive anode material because it is chemically similar to Li but offers several advantages over Li such as abundant natural reserves, lower cost, high volumetric capacity, and improved safety. Mg exhibits a volumetric capacity of 3,833 mAh cm^{-3} and a relatively low redox potential (−2.37 V vs. SHE) [7, 8]. Because Mg^{2+} is divalent, by enabling the transfer of two electrons during oxidation, it can increase its energy density when intercalated into cathode materials. Moreover, as is widely recognized, passivation may occur during Mg deposition and stripping, but the formation of dendrites—which compromise the interfacial stability—is largely suppressed. Therefore, Mg-ion batteries are safer than Li-ion batteries over prolonged charge–discharge cycles. Owing to these advantages, Mg-ion batteries are considered a promising candidate for next-generation energy storage devices.

Therefore, Li-based systems can potentially be replaced with rechargeable Mg batteries offering a high theoretical capacity, low cost, and favorable chemical stability. Within the typical cell configuration, current collectors such as stainless-steel foil, copper foil, or carbon paper are coated with an active cathode material mixed with a binder. The anode comprises Mg metal or Mg alloy. The working mechanism is illustrated in Figure 7.2. Mg^{2+} ions are transported through the electrolyte, while the separator—a material permeable to ions but not electrons—prevents

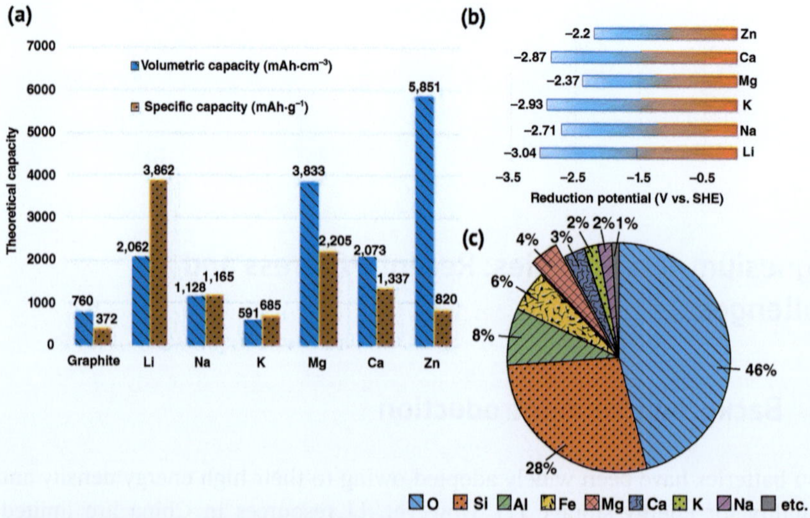

Figure 7.1 Comparison of battery anode characteristics: (a) theoretical capacities of graphite and various metal anodes, (b) reduction potential of various metal anodes, and (c) abundances of battery anode elements in the crust.

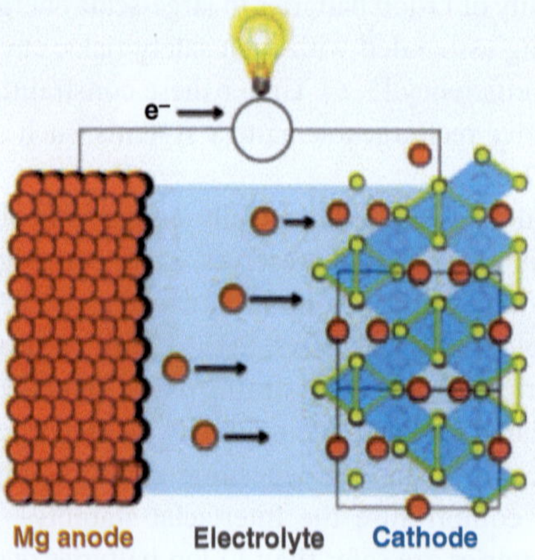

Figure 7.2 Schematic diagram of the discharge process of a rechargeable Mg battery. *Source*: [9] / with permission of the American Chemical Society.

direct contact between the electrodes, thereby avoiding short circuits and potential hazardous events. During the discharge process of a rechargeable Mg battery, the anode is oxidized ($Mg \rightarrow Mg^{2+} + 2e^-$), and the Mg^{2+} ions then deintercalate from the anode, migrate through the electrolyte and separator, and intercalate into the cathode. During charging, this process is reversed: Mg ions are driven back by the electric field, deintercalated from the cathode, and deposited on the anode (Mg metal or Mg alloy), thus restoring the cell to its initial state.

In 1990, Gregory et al. first introduced rechargeable Mg batteries [10] by constructing a cell with a metallic Mg anode, a Co_3O_4 cathode, and a magnesium dibutyl diphenyl borate ($Mg[BBu_2Ph_2]_2$) solution in tetrahydrofuran (THF)/dimethoxyethane (DME) as the electrolyte. In 2000, Aurbach et al. [11] developed the first highly reversible Mg battery with the configuration Mg//0.25 M $Mg(AlCl_2BuEt)_2$/THF//Mo_6S_8. This system demonstrated stable performance over 2,000 cycles, marking a major milestone in rechargeable Mg battery research.

However, to successfully use Mg anodes in rechargeable batteries, one must overcome several challenges, the most critical being the formation of passivation layers [12]. As Mg is a highly reactive metal, an Mg anode rapidly establishes a surface passivation layer when exposed to water, oxygen, or strongly reducing organic compounds. Although Li-ion batteries may develop an ion-conductive solid electrolyte interface (SEI) during cycling, the passivating layer formed on Mg completely blocks Mg^{2+} ion transport [8]. Therefore, the solvent and salt must remain stable at the Mg interface, imposing stringent requirements on the electrolyte [13]. Another major challenge is the development of cathode materials that allow reversible electrochemical insertion and extraction of Mg^{2+}. High-charge-density Mg^{2+} ions undergo strong electrostatic interactions with ions in the host cathode material, boosting the polarity and considerably slowing the insertion kinetics [14]. Screening of various cathode materials has shown that Mg^{2+} intercalation is hindered in most metal oxides and sulfides [15].

In summary, the most crucial steps in Mg-ion battery development are the design and optimization of suitable cathode materials with appropriate structural and electrochemical properties, along with electrolytes compatible with the Mg-metal anode interface. These efforts represent the primary pathway toward rechargeable Mg batteries with low cost, high safety, and long-term cycling stability.

7.2 Research on Cathode Materials

Mg ions are divalent, meaning that they carry two positive charges. Despite their relatively small ionic radius, Mg ions also exhibit high charge density, giving rise to strong electrostatic interactions and sluggish diffusion kinetics that hinder the intercalation of Mg^{2+} into cathode materials. Nevertheless, as Li-ion batteries are limited by stagnant energy density and safety concerns—especially thermal runaway at temperatures below 150 °C—enhancing the performance of Mg batteries has become a research focus. To this end, researchers must identify or develop cathode materials fulfilling the following criteria: (i) high operating voltage and specific capacity, (ii) good compatibility with the electrolyte interface, (iii) tailored structures enabling fast and reversible Mg^{2+} migration kinetics, and (iv) excellent structural stability and cycling longevity. Cathode materials for rechargeable Mg batteries, including transition-metal oxides, chalcogenides, polyanionic compounds, Prussian blue analogs (PBAs), and organic compounds, are primarily categorized into intercalation-based or conversion-based materials.

7.2.1 Transition-metal Oxides

With their straightforward synthesis, robust structural stability, strong charge compensation compatibility, and high anodic oxidation potential, transition-metal oxides have become one of the earliest cathode systems explored for Mg batteries and have been extensively utilized in Li-ion battery literature. The V-, Mo-, and Mn-based oxide systems, as well as AB_2O_4 spinel-structured compounds, have been thoroughly investigated and are now considered well established.

V_2O_5 is a layered metal oxide; its crystal structure is shown in Figure 7.3a. Novák and Desilvestro in 1993 [16] first recognized its potential as a cathode material in rechargeable Mg batteries. Their work first sparked its widespread interest. Levi et al. [17] fabricated a V_2O_5-based cathode material and tested it in a Mg battery with a Mg perchlorate electrolyte dissolved in acetonitrile (AN). The initial discharge delivered a specific capacity of 170 mAh g^{-1}. However, the structural instability of V_2O_5 cathodes reduces the battery cycle life. Yu and Zhang [18] constructed a Mg-ion battery with a V_2O_5 cathode and 0.1 M $Mg(ClO_4)_2/H_2O$/propylene carbonate (PC) as the electrolyte. The specific capacity rapidly decreased to 61.6 mAh g^{-1} after one cycle. Recent studies have demonstrated promising reactivity between Mg^{2+} and various morphological forms of V_2O_5, such as V_2O_5 nanotubes, V_2O_5 gels, and V_2O_5/C composites. The specific capacity of V_2O_5 has been improved by the establishment of a three-dimensional electronic conductive network. Le et al. [19] synthesized V_2O_5 aerogels and investigated the effects of their intercalation with different cations. These aerogels exhibited a high specific surface area and short diffusion distances for ion transport. An et al. [20] reported a graphene-doped V_2O_5 nanocomposite cathode with a specific capacity of 330 mAh g^{-1} for rechargeable Mg batteries. The nanocomposite retained 81% of its initial capacity after 200 cycles. Jang et al. [21] investigated a unique Mg^{2+} storage mechanism in a hierarchically structured, accordion-like vanadium oxide/carbon heterointerface (V_2O_3@C). After the initial cycle, the V_2O_3 crystal structure was reconfigured into a $MgV_3O_7{\cdot}H_2O$ phase through a cathodic hydration reaction, which improved the kinetics and enhanced the cycling performance.

Molybdenum oxides are widely used in various fields owing to their structural diversity and low cost. Among these compounds, MoO_3 has been investigated as a cathode material for Mg batteries. MoO_3 possesses a characteristic-layered structure (Figure 7.3b) of bilayers formed by edge- and corner-sharing MoO_6 octahedra. These layers are connected by weak van der Waals forces, creating two-dimensional ion diffusion channels, whereas the layered configuration facilitates the intercalation

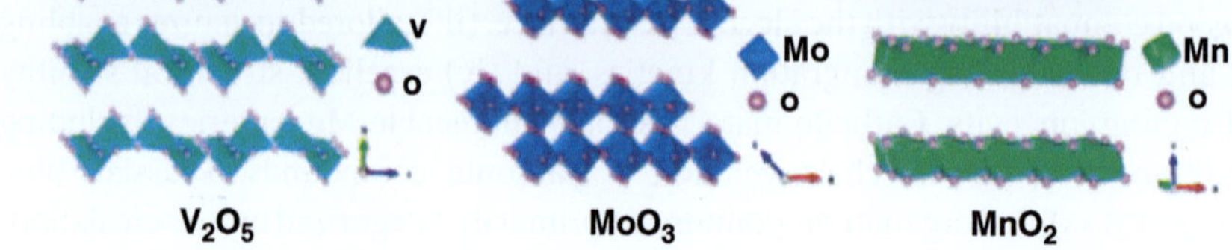

Figure 7.3 Crystal structures of transition-metal oxides: (a) V_2O_5, (b) MoO_3, and (c) MnO_2.

and deintercalation of Mg^{2+} ions. Gregory et al. [11] reported the electrochemical intercalation of 0.5 mol of Mg^{2+} into 1 mol of MoO_3 in an *n*-butylmagnesium electrolyte. Recent studies have shown that F dopant atoms partially substitute O atoms, forming layered $MoO_{2.8}F_{0.2}$ compounds that substantially reduce the activation energy of Mg-ion migration [22]. Layered MnO_2 (commonly referred to as δ-MnO_2), with the structure shown in Figure 7.3c, is also widely used in environmentally friendly energy storage applications because Mn resources are abundantly available. Such oxides have been modified via a strategy involving composite formation with carbon materials. When used as a cathode in Mg-ion batteries, δ-MnO_2/acetylene black composite exhibits a low initial coulombic efficiency but high cycling stability. The dissociation of trace water generates protons that induce the pre-intercalation of Mg, thereby facilitating interfacial transfer and diffusion of Mg^{2+} ions [23].

Oxides with spinel structures (general formula: AB_2O_4) establish a three-dimensional lattice framework in which Mg-ion intercalation and deintercalation are more efficient than two-dimensional-layered materials. Batteries incorporating these materials can operate at high voltages, showing broad application prospects. The three-dimensional ion diffusion channels and multimetal redox characteristics offer distinct advantages of AB_2O_4 cathode materials in Mg batteries. Performance variations among the typical spinel materials—such as $MgCo_2O_4$, $MgMn_2O_4$, $MgFe_2O_4$, and $MgCr_2O_4$—primarily originate from the electronic structure and crystal field effects of the B-site metal (Figure 7.4) [24]. A Co_3O_4//Mg battery system developed by Gregory's research team exhibited unique energy storage

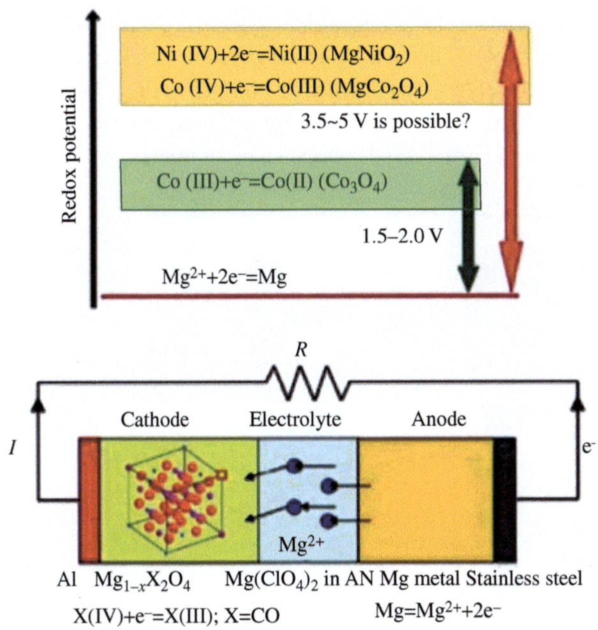

Figure 7.4 Method for increasing the open-circuit voltage in Mg-ion batteries (upper) and schematic of an Mg cell with a positive $MgCo_2O_4$ electrode and an Mg anode (lower). *Source*: [24] / with permission of the Royal Society of Chemistry.

behavior in a 0.25 M $Mg[B(Bu_2Ph_2)]_2$/THF–DME electrolyte, delivering a reversible capacity of 186 mAh g^{-1} with a high coulombic efficiency (~100%). However, the capacity of such batteries is limited by substantial polarization and the low voltage plateau [1, 25].

7.2.2 Transition-metal Sulfides/Selenides

Transition-metal sulfides are dominated by Chevrel phases and two-dimensional layered materials. The most representative Chevrel phase compound is Mo_6S_8 with a crystal structure composed of stacked Mo_6S_8 units (Figure 7.5, upper right). Chevrel phase Mo_6S_8 was among the earliest intercalation cathode materials used in Mg batteries. Mo_6S_8, synthesized via $CuMo_3S_4$ at high temperatures, enables relatively fast and reversible Mg^{2+} de/intercalation in various nonaqueous electrolytes used in rechargeable Mg batteries. Mg^{2+} intercalation into Chevrel phase Mo_6S_8 occurs via the following electrochemical reaction: $Mo_6S_8 + 2Mg^{2+} + 4e^- \rightarrow Mg_2Mo_6S_8$ [17, 26]. With a general structural formula of M_6T_8 or MMo_6T_8 (where M is a metal element and T denotes a chalcogen such as S [sulfur], Se [selenium], or Te), Chevrel phases exhibit similar structures and reaction mechanisms. Aurbach et al. [11] fabricated an $Mg_xMo_3S_4$ cathode and investigated its Mg storage behavior in $Mg(AlCl_2BuEt)_2$/THF electrolyte. The cathode delivered an initial discharge specific capacity of approximately 100 mAh g^{-1} and excellent cycling stability, with a 94% capacity retention after 100 cycles. Saha et al. [27] synthesized Chevrel phase Mo_6S_8 through a facile liquid phase chemical method, followed by heat treatment under a mixed H_2/Ar atmosphere. Their method synthesized cubic-shaped $Cu_2Mo_6S_8$ particles (sized 1–1.5 μm) and converted them to Mo_6S_8 via thermal treatment. In a Mg battery system employing an all-phenyl complex (APC)-type electrolyte—specifically, 2(PhMgCl)–$AlCl_3$/THF—the material delivered a capacity of 76 mAh g^{-1} after 50 charge/discharge cycles at a low current density of 20 mA g^{-1}. Murgia et al. [28] developed a rapid microwave-assisted

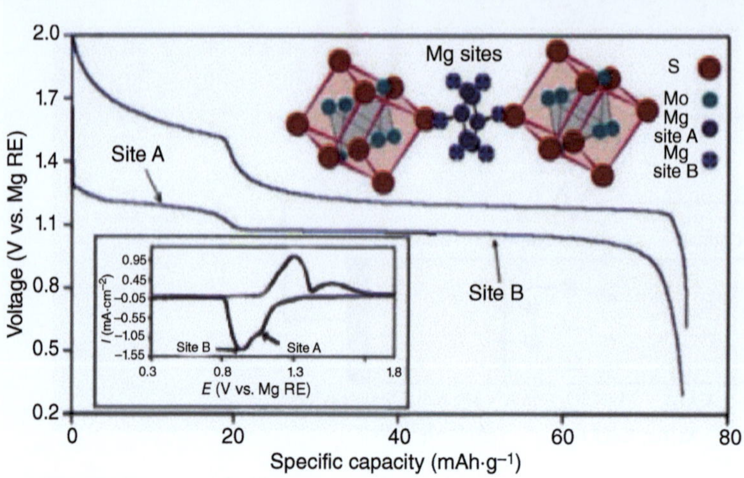

Figure 7.5 Crystal structure, cyclic voltammetry, and charge/discharge curves of Mo_6S_8. *Source:* [11] / with permission of Springer Nature.

synthesis route for the large-scale production of $Cu_2Mo_6S_8$, followed by acid leaching to obtain Chevrel-phase Mo_6S_8. In an EtMgCl–Et_2AlCl–THF electrolyte, the electrochemical performance of the resulting Mo_6S_8 cathode material (with a reversible capacity exceeding 80 mAh g^{-1}) was comparable to that of materials synthesized via conventional high-temperature solid-state methods.

Cobalt sulfide (CoS) is a promising conversion-type cathode material for rechargeable Mg-ion batteries. The overall charge–discharge reaction is $CoS + Mg^{2+} + 2e^- \leftrightarrow MgS + Co$. Mg batteries with a CoS as the cathode exhibit good cycling stability, maintaining a specific capacity of approximately 60 mAh g^{-1} after 65 cycles [29]. Hollow nanospheres of heterostructured cobalt sulfide (Co_3S_4/CoS_2) have enhanced the efficiency of Mg^{2+} migration. Co_3S_4 adsorbs Mg polysulfide intermediates, inhibiting their dissolution from CoS_2 and thereby improving the performance of Mg batteries [30].

CuS is a copper-based conversion-type cathode material with a high theoretical capacity of 560 mAh g^{-1}. The reversible Mg^{2+} insertion and extraction process in CuS can be described as a two-step conversion reaction: $2CuS + Mg^{2+} + 2e^- \leftrightarrow Cu_2S + MgS$ and $Cu_2S + Mg^{2+} + 2e^- \leftrightarrow 2Cu + MgS$ [31]. Using $Mg(ClO_4)_2$ in AN as the electrolyte, a Mg-ion battery assembled with a flower-like CuS cathode achieved a reversible Mg storage capacity exceeding 300 mAh g^{-1}. Furthermore, after intercalation with the cationic surfactant cetyltrimethylammonium bromide, the interlayer spacing of the flower-like CuS expanded, and the battery stably operated through 1,000 cycles with an ultralow capacity decay rate of 0.055% per cycle [32]. Similarly, a nanoporous NiS_x thin-film electrode prepared via electrochemical deposition, etching, and anodization exhibited excellent reversible Mg-ion insertion/extraction performance and long-term cycling stability. The micron-scale pores reduce the electrode mass, whereas the nanoscale pores increase the surface area, broadening the pathways for ion diffusion and reducing the diffusion energy barrier [33].

As a typical layered material with its crystal structure shown in Figure 7.6a [34, 35], TiS_2 serves as a model system for studying the kinetics and thermodynamics

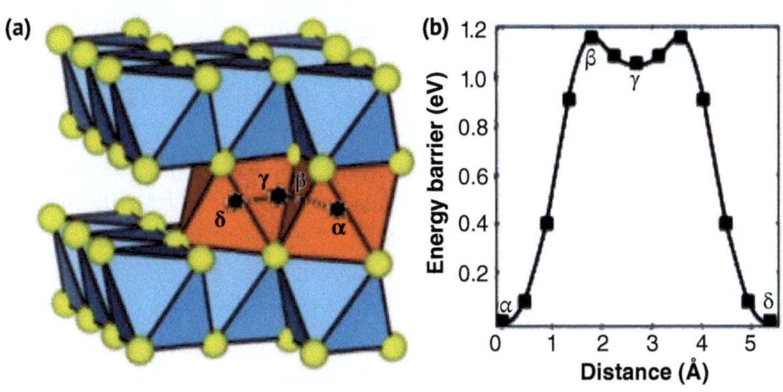

Figure 7.6 (a) Crystal structure of layered TiS_2 and migration paths. *Source*: [36] / with permission of the American Chemical Society. (b) Migration barriers in layered TiS_2. *Source*: [37] / with permission of John Wiley & Sons.

associated with Mg intercalation into layered structures. Layered TiS_2 can deliver a reversible capacity of 115 mAh g^{-1} [34]. Each monolayer consists of a close-packed triangular lattice of S. The octahedral coordination sites within the layers host the Ti atoms and provide the sites on which the Mg^{2+} ions diffuse via a hopping mechanism. First-principles calculations indicated a substantial diffusion energy barrier to this process, with an activation energy of 0.35 eV for a Mg^{2+} ion migrating from an initial octahedral site (energy minimum) to an adjacent site. Further first-principles studies revealed sloped voltage profiles and steps at stoichiometric compositions caused by Mg vacancy ordering in layered and spinel Mg_xTiS_2. The migration barrier under dilute Mg concentration conditions, when Mg^{2+} passes through the triangular face formed by three sulfur atoms of an MgS_6 octahedron (denoted as point β in Figure 7.6a and b), was calculated as 1,160 meV. The size of the diffusion channel is determined by the Ti – S bond length. In a related study, Arsentev et al. [36] investigated the structural properties of TiS_3 using first-principles calculations. They proposed the material as a promising cathode material for rechargeable Mg batteries.

MoS_2 and VS_2 have been extensively studied as layered cathode materials for Mg batteries. Liang et al. [37] successfully synthesized MoS_2 nanomaterials with a typical graphene-like layered stacking structure via a solvothermal method. In an electrolyte system based on alkylaluminum chloride ($Mg(AlCl_3Bu)_2$/THF), this material delivered an initial discharge capacity of 170 mAh g^{-1}, but the capacity degraded rapidly after 50 charge/discharge cycles. Sun et al. [38] designed a hybrid system enabling the rapid insertion of both Li^+ and Mg^{2+}. They pioneered the strategy of encapsulating VS_2 with graphene oxide, achieving a discharge capacity as high as 235 mAh g^{-1} and maintaining a reversible capacity of 129 mAh g^{-1} even at a high rate of 80C. However, Li^+ intercalation/deintercalation dominates the electrochemical process within the VS_2 layers. Both experimental and computational studies have demonstrated that expanding the interlayer spacing of layered materials or constructing three-dimensional architectures can accelerate Mg^{2+} diffusion, thereby enhancing the stability and capacity of the battery. For instance, hydrothermal synthesis followed by heat treatment yields MoS_2/C microspheres with a sandwich-like layered structure that provides rapid migration pathways for Mg ions. The maximum discharge capacity reached 118.8 mAh g^{-1} at a current density of 50 mA g^{-1} [39]. Recent studies have focused on polymetallic chalcogenides, such as $CoNi_2S_4$. The introduction of additional metallic elements increases the number of reaction sites, endowing these materials with strong potential in energy storage applications.

Selenium and sulfur, belonging to the same group (chalcogens) in the periodic table, exhibit similar chemical properties. When Se bonds with transition-metal ions as an S substitute, it forms longer bonds than S, weakening the interaction with Mg^{2+} ions in the cathode material and substantially reducing the diffusion energy barrier of Mg^{2+} [40]. Transition-metal chalcogenides demonstrate excellent Mg storage capacity during cycling in Mg batteries. Electrochemically induced structural reconstruction throughout cycling markedly optimizes the charge-transport properties of the cathode material. In situ Raman characterization confirms that Mg^{2+} insertion into WSe_2 nanowires triggers a semiconductor-to-metal phase transition that enhances the electrical conductivity of the electrode material [41].

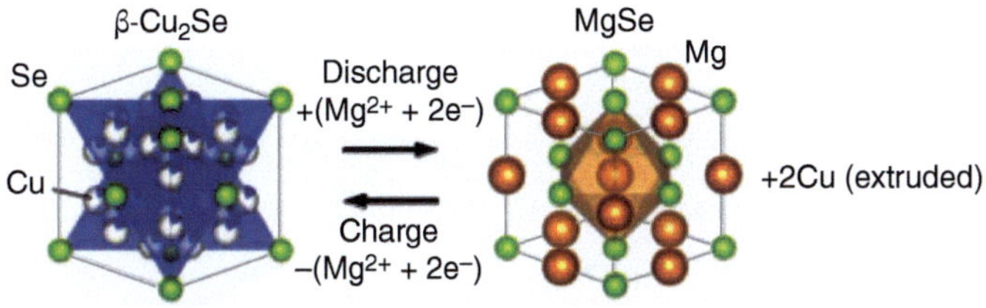

Figure 7.7 Schematic showing the generation of MgSe in β-Cu_2Se. *Source*: [43] / with permission of Elsevier.

Mao et al. [42] investigated the electrochemical performances of VSe_2 and $TiSe_2$ cathodes in rechargeable Mg batteries at a current density of 5 mA g^{-1}. Both materials delivered initial discharge capacities of approximately 127 mAh g^{-1} and retained high capacities after 40 cycles. Copper selenide is another widely studied cathode material for Mg batteries. The high mobility of Cu^+ ions within the crystal structure of β-Cu_2Se enables rapid ionic conduction [43]. The Mg storage mechanism (Figure 7.7) involves the displacement of Cu^+ by Mg^{2+} to form MgSe during discharge; the reverse process occurs during charging. At a low current density (5 mA g^{-1}), the reversible capacity of β-Cu_2Se reaches up to 120 mAh g^{-1}. After nanostructuring, the reversible capacity of β-Cu_2Se is further enhanced to approximately 230 mAh g^{-1}, 88.5% of its theoretical capacity.

7.2.3 Polyanionic Compounds

Polyanionic compounds are a class of three-dimensional open-framework materials composed of transition-metal centers and strongly covalent $[XO_4]^{n-}$ polyanionic units (X = S, P, and Si). Such a structure effectively promotes the migration kinetics of magnesium ions. Among the polyanionic compounds, silicates with an olivine structure and phosphates with a sodium superionic conductor (NASICON) structure have attracted substantial interest for their unique three-dimensional ion-channel design. These materials exhibit remarkable structural stability and a high operating voltage.

Silicates with the olivine structure (general formula $MgMSiO_4$, where M = Fe, Mn, or Co) crystallize in an orthorhombic structure in which silicon atoms occupy the tetrahedral sites, while magnesium and transition-metal ions reside in two distinct octahedral sites within an oxygen-atom framework in a distorted hexagonal close packing arrangement (4a and 4c). The 4a sites are occupied by Mg^{2+} ions, whereas the 4c sites are filled by transition-metal ions (M = Fe, Mn, and Co), as illustrated in Figure 7.8a [44]. Each 4a site is connected to its nearest neighboring 4c site, forming a one-dimensional zigzag chain (Figure 7.8b) [44]. Various polyanionic compounds such as $MgFeSiO_4$ [45], $MgMnSiO_4$ [46], $MgCoSiO_4$ [47], and $Mg_{1.03}Mn_{0.97}SiO_4$ [48] have been investigated as cathode materials for Mg batteries. After partial substitution of Fe with Co or Mn in $MgFeSiO_4$, the operating voltage increases from 2.3 V (vs. Mg/Mg^{2+}) to 2.8–3 V [45].

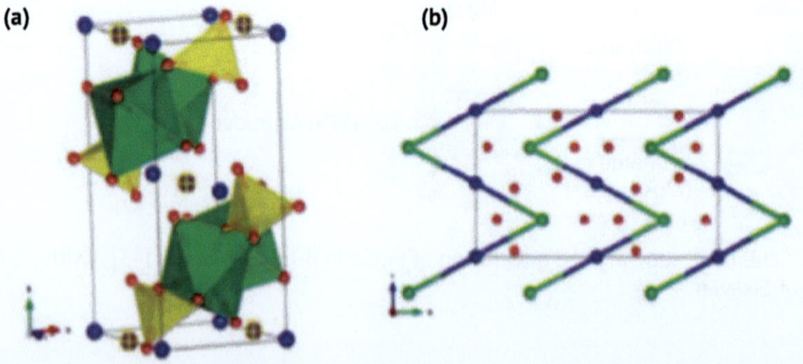

Figure 7.8 (a) Crystal structure of an olivine-type compound (yellow spheres represent Si, green spheres represent transition metals at site 4e, and blue spheres represent Mg at site 4a); (b) two-dimensional view of the olive structure in the (110) plane, which is extended. *Source*: [46] / with permission of the Royal Society of Chemistry.

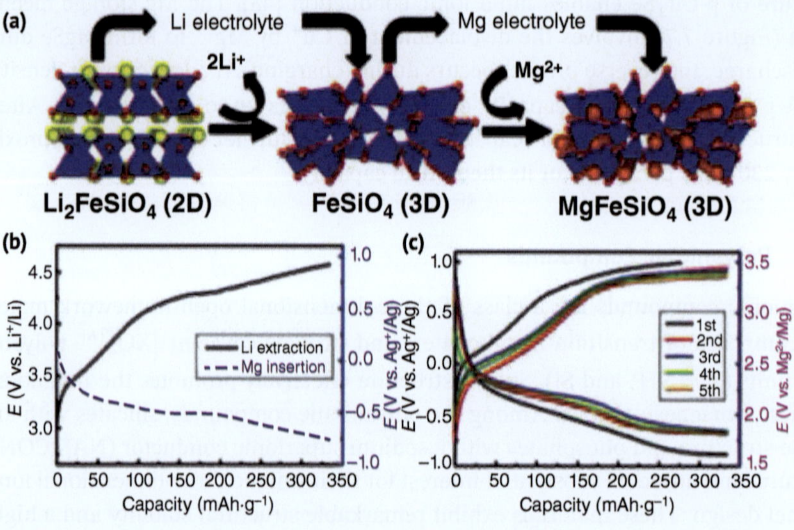

Figure 7.9 (a) Schematic of the sub-stable $MgFeSiO_4$ phase; charge–discharge curves of (b) Li_2FeSiO_4 vs. $MgFeSiO_4$ and (c) $MgFeSiO_4$ ion exchange. All measurements were performed at 55 °C. *Source*: [49] / Springer Nature / CC BY 4.0.

Feng et al. [48] fabricated mesoporous $Mg_{1.03}Mn_{0.97}SiO_4$ cathode materials using two distinct carbon-coating methods. The resulting electrodes delivered Mg storage capacities of 64.5 mAh g^{-1} at 3.1 mA g^{-1} and 92.9 mAh g^{-1} at a rate of 0.02C. Using the less stable tetrahedrally coordinated Mg sites within the silicate structure can enhance the Mg^{2+} diffusion and improve the insertion/extraction kinetics. Orikasa et al. [49] demonstrated that an electrochemically driven ion exchange process can convert Li_2FeSiO_4 into a metastable $MgFeSiO_4$ phase (Figure 7.9), which exhibited a reversible capacity of up to 300 mAh g^{-1}. However, the sluggish diffusion of Mg^{2+} ions at room temperature continues to limit the development of silicate-based cathodes in Mg batteries [50].

NASICON-type fast-ion conductors, with the general formula $A_xM_2(XO_4)_3$ (where A = Li or Na, M = transition metal, and X = S, P, or As), are promising cathode materials for rechargeable Mg batteries owing to their large interstitial spaces and high structural stability [49]. Makino et al. [51, 52] synthesized a series of phosphate-based compounds (with PO_4^{3-} anions) via the sol–gel method, yielding $Mg_{0.5}Ti_2(PO_4)_3$ and $Mg_{0.5+y}\left(M_yTi_{1-y}\right)_2(PO_4)_3$ (with M = Cr, Fe). They achieved reversible Mg^{2+} (de)intercalation in $Mg_{0.5}Ti_2(PO_4)_3$ in PC-based electrolytes, albeit with limited migration rates. The latter compounds, in which Ti is partially substituted by Cr or Fe, exhibited even lower Mg^{2+} insertion capacities. Using a solvothermal method, Xu et al. [53] prepared carbon-coated $LiTi_2(PO_4)_3$ with a micro-flower morphology, which retained 80% of its initial capacity after 3,000 cycles at 10 °C. The observed specific capacities of these phosphate-based materials are often considerably lower than their theoretical values. This phenomenon has been attributed to surface amorphization owing to the formation of thermodynamically metastable intercalation products, which hinders the penetration of electrochemical reactions into the bulk material [54].

7.2.4 Prussian Blue Analogs

PBAs with the general formula $A_xM_y\left[M'(CN)_6\right]\cdot nH_2O$ (where A = alkali metal, M, M′ = transition metals, and CN = ligand) form a three-dimensional framework of transition metals linked via cyanide bridges, creating abundant interstitial sites (Figure 7.10). The triple-bonded CN ligands increase the distance between the M and M′ ions, while the cations A and water molecules occupy the interstitial A sites.

Reversible Mg^{2+} intercalation has been demonstrated in nickel hexacyanoferrate, which delivers an initial capacity of ~40 mAh g^{-1} in $Mg(TFSI)_2$/PC electrolyte (TFSI = bis(trifluoromethylsulfonyl)imide). The capacity increased to 80 mAh g^{-1} by the 50th cycle, accompanied by a relatively high operating voltage and a low overpotential vs. Mg/Mg^{2+} [55]. Chae et al. [56] assembled a Mg battery with a $K_{0.86}Ni[Fe(CN)_6]_{0.954}(H_2O)_{0.766}$ cathode and 0.5 M $Mg(ClO_4)_2$ in AN as the electrolyte.

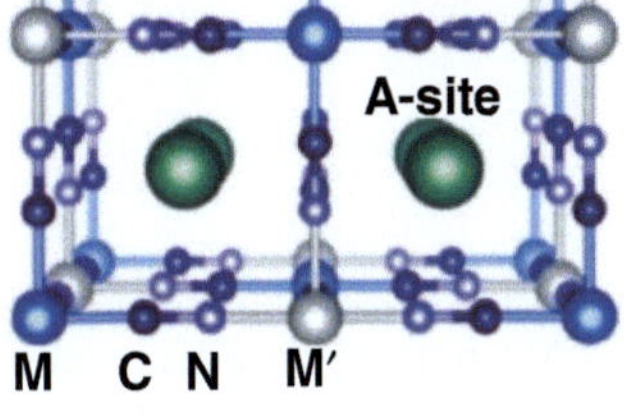

Figure 7.10 Structures of PBAs. *Source*: [49] / Springer Nature / CC BY 4.0.

They reported a discharge capacity and discharge voltage of 48.3 mAh g^{-1} (at 0.2C) and 2.99 V, respectively. Unlike many conventional cathode materials that degrade through irreversible side reactions and active material loss in aqueous electrolytes, PBAs are compatible to some extent with aqueous media. Mizuno et al. [57] demonstrated reversible Mg^{2+} insertion and extraction in $K_{0.1}Cu[Fe(CN)_6]_{0.7}{\cdot}3.6H_2O$ in an aqueous electrolyte system.

7.2.5 Organic Compounds

Redox-active organic materials function as conversion-type cathodes for Mg storage, accommodating Mg^{2+} ions through redox reactions of their electroactive groups or through changes in partial charge states. These materials are advantaged by low cost, highly tunable syntheses, and environmental friendliness. However, the intermolecular interactions are weaker than those in inorganic compounds, reducing the polarization effects on Mg^{2+} ions [50].

A typical class of organic electrode materials is quinones, among which 2,5-dimethoxy-1,4-benzoquinone (DMBQ) exhibits a high theoretical capacity (319 mAh g^{-1}) and good stability in contact with electrolytes. DMBQ also cycles stably with slow capacity decay in Li-ion batteries [58]. However, when applied in magnesium batteries, DMBQ and similar organic materials often exhibit low capacity retention and low discharge plateaus. Pan et al. [59] synthesized two novel quinone-based polymers, 2,6-polyanthraquinone (26PAQ) and 1,4-polyanthraquinone (14PAQ), with rigid and rotationally flexible polymer chains, respectively. The flexibility of 14PAQ enhances the structural stability because the redox-active quinone groups attach to one side of the main polymer chain. In Mg batteries with 14PAQ as the cathode and $Mg(HMDS)_2{\cdot}4MgCl_2$/THF as the electrolyte, the capacity loss was notable during the initial cycles but subsequently stabilized over 1,000 cycles with a low-capacity decay rate.

In 2014, Senoh et al. [58] dissolved $Mg(TFSI)_2$ in a sulfone-based solvent, forming the electrolyte for performance evaluation of a DMBQ-based cathode in an Mg-ion battery. They observed limited cycling stability and a low discharge potential, which they attributed to the poor mobility of Mg^{2+} ions in the sulfone solvent. Rodriguez-Perez et al. [60] investigated the reversible Mg^{2+} (de)intercalation behavior of a 3,4,9,10-perylenetetracarboxylic dianhydride cathode in both aqueous and organic electrolytes. The reversible capacity of the cathode was 125 mAh g^{-1} at 20 mA g^{-1} in aqueous $Mg(NO_3)_2$ electrolyte and was 126 mAh g^{-1} at 200 mA g^{-1} when cycled in a 0.25 M APC organic electrolyte with a 0.2 M LiCl additive.

7.3 Research on Anode Materials

The anode material is another essential component of Mg battery systems, necessitating a low redox potential to enable reversible Mg^{2+} deposition and dissolution.

7.3.1 Metallic Mg Anode

Metallic Mg has long been considered an ideal anode material owing to its high theoretical specific capacity, low reduction potential, abundance of crustal reserves, metallic electrical conductivity, low chemical reactivity, high melting point, and superior safety characteristics. Nevertheless, the practical application and development of Mg-metal anodes faces several challenges: (i) passivation of the Mg anode in electrolytes, which reduces the Mg^{2+} conductivity, slows the ion migration, increases the battery impedance, and enlarges the overpotential [61, 62]; (ii) formation of Mg dendrites (unlike earlier studies suggesting that Mg-metal anodes are free from dendritic growth [63, 64], more recent research has identified fibrous Mg dendrites that can penetrate the separators and cause internal short circuits [65]; and (iii) substantial volume fluctuations during cycling, further exacerbated by nonuniform Mg deposition and dendritic formation, leading to increased mechanical strain and electrode instability [66].

To overcome the challenges associated with Mg-metal anodes, recent research has focused on various modification strategies, such as electrolyte additives [67, 68], artificial SEI design [69, 70], interfacial alloy layers [71], three-dimensional scaffold structures [72], solid-state electrolytes, and advanced characterization techniques. Theoretical investigations have also been undertaken. Among these developments, engineering of artificial SEI layers has emerged as a prominent approach for improving the Mg anode performance. For instance, Zhang et al. [73] through sequential metallic Mg with H_3PO_4 and $SiCl_4$ in situ chemical reactions developed an amorphous (a-)$MgCl_2$@polymer as an artificial SEI. This constructed interface effectively suppresses electrolyte decomposition and facilitates Mg^{2+} transport, serving as a functional solid SEI on the Mg anode. When incorporated into Mg–S and Mg–Mo_6S_8 batteries, the modified anode substantially reduced the voltage hysteresis in both batteries.

7.3.2 Alloy-based Anodes

Intermetallic compounds formed by alloying metals, such as Sn, Sb, and Bi with Mg, exhibit high volumetric capacity [74] (Figure 7.11). In fact, they are a promising class of anode materials for high-energy-density rechargeable Mg batteries.

In 2021, Arthur et al. [70] demonstrated that electrodeposited $Bi_{0.88}Sb_{0.12}$ and Bi anodes deliver stable cycling performance over 100 cycles at 1C. Subsequently, Bi nanotubes (BiNTs) [75] and nanowires [76] have also exhibited promising performance in Mg batteries. Shao et al. [75] employed bismuth nanotubes as an intercalation-type anode for Mg^{2+}, reporting a demagnetization potential around 0.23 V (Figure 7.12). The material showed a high reversible specific capacity (350 mAh g^{-1} and 3,430 mAh cm^{-3}) and excellent cycling stability, retaining 92.3% of its capacity after 200 cycles. A SnSb alloy anode—which converts to Mg_2Sn and Mg_3Sb_2 during magnetization—exhibits a high theoretical capacity (768 mAh g^{-1}) than

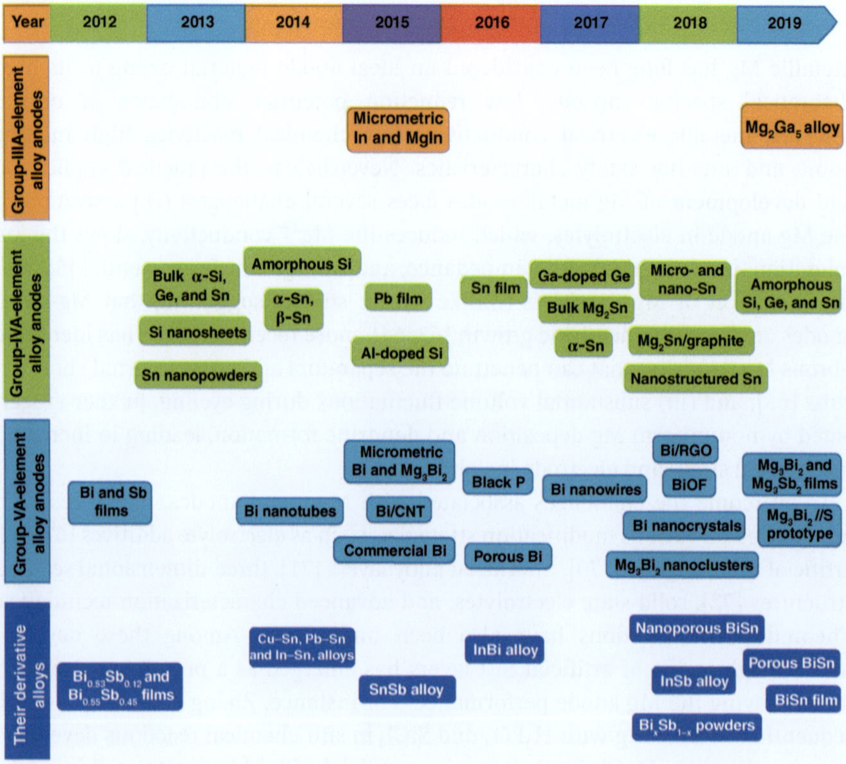

Figure 7.11 Development of rechargeable Mg batteries using alloy anodes based on Group-IIIA, -IVA, and -VA elements and multimetal alloys. *Source*: [74] / with permission of John Wiley & Sons.

Bi (385 mAh g^{-1}). During experiments, the SnSb anode delivered a reversible capacity of approximately 420 mAh g^{-1}, along with good rate capability (~300 mAh g^{-1} at 1 A g^{-1}) and cycling stability (retaining ~270 mAh g^{-1} after 200 cycles at 500 mA g^{-1}) [77]. Combining an experimental study with density functional theory computations, Parent et al. [78] later investigated the Mg^{2+} de/intercalation and conversion mechanisms of SnSb nanoparticles (<40 nm). Mg can react with Ga, a relatively soft metal, to form various intermetallic compounds such as Mg_2Ga_5, $MgGa_2$, MgGa, Mg_2Ga, and Mg_5Ga_2. A Mg_5Ga_2 anode provided exceptional cycling stability, maintaining a specific capacity of ~213 mAh g^{-1} after 1,000 cycles at a current density of 922.5 mA g^{-1} [79].

Despite their attractive advantages and promising performance, alloy-based anodes face critical challenges and limitations. In conventional electrolytes, alloy anodes are compatible with high-voltage and high-capacity transition-metal oxide cathodes. However, this advantage largely stems from their compact structure and higher redox potential than that of metallic Mg. Meanwhile, the cycle life of batteries using alloy anodes is substantially limited by inevitable side reactions.

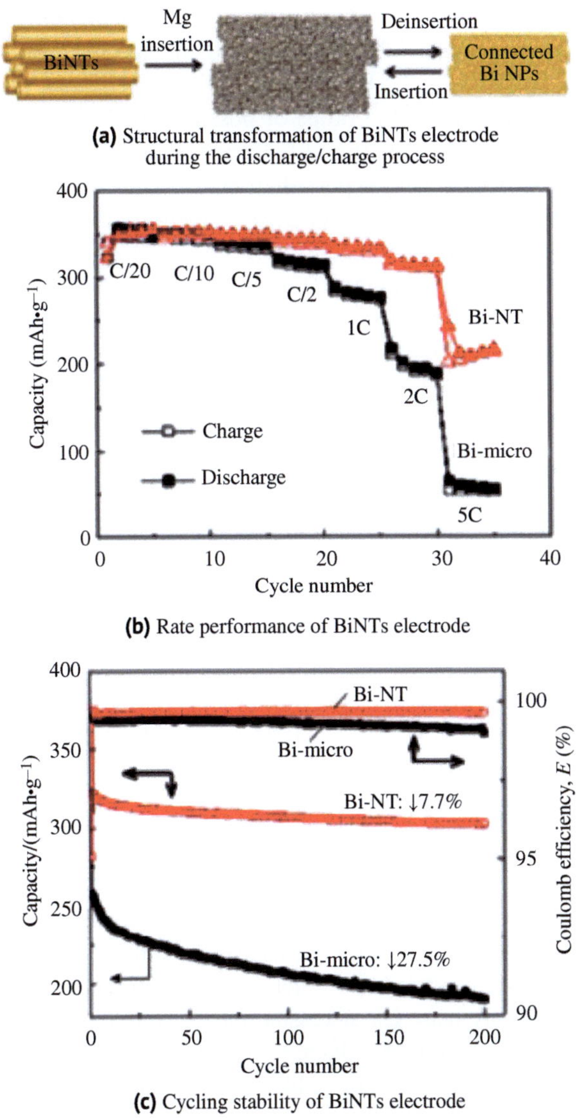

Figure 7.12 Structural transformations and electrochemical performances of electrodes based on BiNTs. *Source*: [75] / with permission of the American Chemical Society.

7.3.3 Intercalation-type Anodes

Inorganic intercalation-type materials are another category of potential anode materials for Mg batteries. Compounds such as $Li_4Ti_5O_{12}$ [80], $LiVO_4$ [81], TiO_2–B [82], and layered Na_2TiO_3 [83] are among the primary candidates under current investigation. However, the doubly charged Mg^{2+} ions undergo strong electrostatic interactions with the host lattice, slowing the diffusion kinetics within these materials.

Furthermore, owing to their relatively high redox potential and low theoretical specific capacity, these intercalation anodes often yield lower-than-expected overall battery performance.

Carbon-based materials have expanded the range of possible anode choices. Er et al. [84] theoretically predicted the adsorption behavior of Mg^{2+} on defective graphene and graphene allotropes. They reported an Mg storage capacity of 1,042 mAh g^{-1} in graphene with 25% double vacancies. The open-circuit voltage is 1.6 V, and it decreases with increasing Mg concentration. In a subsequent computational study, Kim et al. [85] demonstrated that graphite can host Mg ions, enabling their reversible co-intercalation with linear solvents. Phosphorene [86] and borophene [87] have also been proposed as promising anode materials for Mg-ion storage. Nevertheless, the practical application of these materials in Mg battery systems remains unexplored, and their feasibility requires further validation.

7.4 Solid-state Electrolytes

Recently, solid-state electrolytes have gained attention for their excellent safety performance, outstanding mechanical properties, wide voltage window, and high energy density. However, research on Mg solid-state electrolytes remains in its early stages. Solid-state Mg electrolytes used in battery applications should meet the following performance criteria:

1. High ionic conductivity: to function as an ion transport medium, an electrolyte must deliver excellent ionic conductivity. Typically, the room-temperature ionic conductivity of solid-state electrolytes should reach the order of 10^{-4} S cm^{-1}.
2. Wide electrochemical stability window: the electrochemical window refers to the potential difference between the oxidation reaction at the cathode and the reduction reaction at the anode. High-voltage systems are continuously being explored for high-energy-density batteries, necessitating the development of electrolyte materials compatible with high-voltage cathodes.
3. Reversible deposition/dissolution of Mg ions: unlike Li-metal batteries, a feasible Mg-metal battery requires reversible deposition/dissolution of Mg ions on the Mg anode surface. Therefore, passivation reactions between the solid-state electrolyte and Mg-metal anode must be avoided to preserve charge–discharge cycling.
4. High ion transference number: a low ion transference number reduces the effective ionic conductivity and causes substantial concentration polarization during charge–discharge processes, resulting in uneven ion deposition and degrading the cycle rate performance of the battery. Thus, the ion transference number of polymer electrolytes should be maximized.
5. Low electronic conductivity: the electronic conductivity of electrolytes should generally be less than 10^{-6} S cm^{-1} to prevent internal short circuits in the battery.

6. Excellent mechanical properties: as solid-state electrolytes are in direct contact with the cathode and anode, they must exhibit strong toughness to withstand stress changes during battery assembly, storage, and use without fracturing. In addition, they must function as a separator to prevent short circuits between the cathode and anode.
7. Good chemical and thermal stability: solid-state electrolytes should be chemically compatible with all battery components and must not undergo intense chemical reactions with the cathode, anode, and current collectors. Moreover, they must be excellently thermally stable to ensure safe and normal operation at elevated temperatures.
8. Simple preparation process: a straightforward and environmentally friendly preparation process is essential for advancing solid-state electrolytes.
9. Good compatibility with electrode materials: the thermal expansion or compression coefficient of the electrolyte should be similar to that of the electrode materials. This criterion prevents physical separation between the electrode layer and the solid electrolyte layer during frequent changes in battery temperature or pressure.
10. Environmental friendliness: the raw materials and intermediate products should be nontoxic to operators and the environment.

This section categorizes Mg solid electrolytes into three main types: inorganic solid electrolytes, organic solid electrolytes, and organic–inorganic composite solid electrolytes. These three categories are analyzed and discussed in the following subsections.

7.4.1 Inorganic Solid-state Electrolytes

Figure 7.13 summarizes the development of inorganic Mg solid-state electrolytes and their conductivities over the past few decades. This development has been slowed by the low ionic conductivity of solid-state Mg. Inorganic Mg solid-state electrolytes

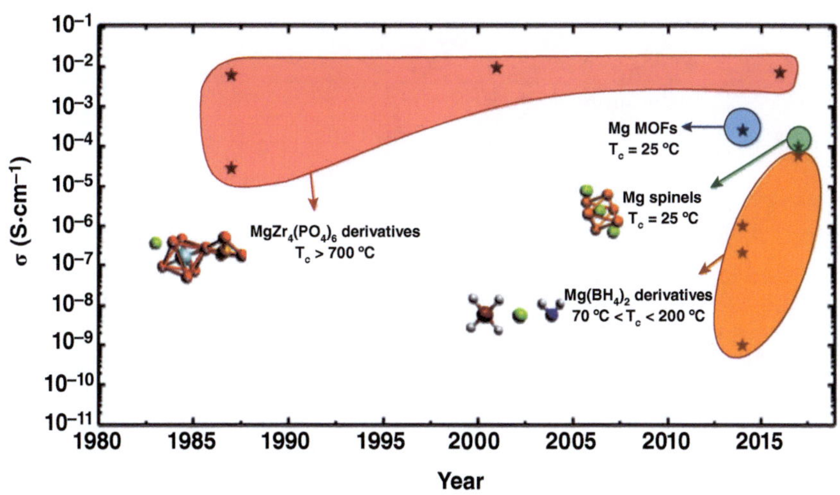

Figure 7.13 Development of ionic conductivity in inorganic Mg solid electrolytes.

were reported as early as 1987, when Mg-ion conduction was identified in Mg zirconium phosphate ($MgZr_4(PO_4)_6$) [88]. Ikeda et al. reported Mg^{2+} conduction in $MgZr_4(PO_4)_6$ at high temperatures, with a conductivity of 2.9×10^{-5} S cm^{-1} at 400 °C and of 6.1×10^{-3} S cm^{-1} at 800 °C. The authors successfully extended this technology to other multivalent ions such as Mn^{2+} and Zn^{2+}, demonstrating the versatility of the structural framework and phase transitions of this zirconium phosphate derivative [89, 90]. Employing the sol–gel method, Adamu and Kale synthesized the inorganic solid-state electrolyte $MgZr_4(PO_4)_6$ with a conductivity of 7.23×10^{-3} S cm^{-1} at 725 °C [91].In 2014, Higashi et al. from Toyota reported Mg^{2+} conductivity in magnesium borohydride ($Mg(BH_4)_2$) and its derivative $Mg(BH_4)(NH_2)$ [92]. The $Mg(BH_4)(NH_2)$ compound delivered an ionic conductivity of 10^{-6} S cm^{-1} at a relatively low temperature (150 °C), was stable against Mg electrodes, and possessed an electrochemical stability window of 3 V. More recently, an ethylenediamine derivative of $Mg(BH_4)_2$ with a high ionic conductivity (6×10^{-5} S cm^{-1}) at 70 °C was reported [93]. Separately, Yamanaka et al. developed an Mg sulfide-based solid electrolyte analogous to Li-ion conducting sulfides ($Li_2S–P_2S_5$). Their $MgS–P_2S_5–MgI_2$ glass-ceramic electrolyte delivered an ionic conductivity of 2.1×10^{-7} S cm^{-1} at 200 °C [94].

Recently, Canepa et al. [95] discovered fast Mg-ion conduction in closely packed MgX_2Z_4 compounds with spinel structures, where X = Y or Sc and Z = S or Se. $MgSc_2Se_4$ exhibited a particularly high ionic conductivity, reaching 1×10^{-4} S cm^{-1} at room temperature. $MgSc_2Se_4$ crystallizes in the $Fd\bar{3}m$ space group as illustrated in Figure 7.14.

However, the high electronic conductivity has limited applications of this inorganic electrolyte. Wang et al. [96] proposed that the electronic conductivity of $MgSc_2Se_4$ could be reduced by (i) synthesizing selenium-rich phases and (ii) doping

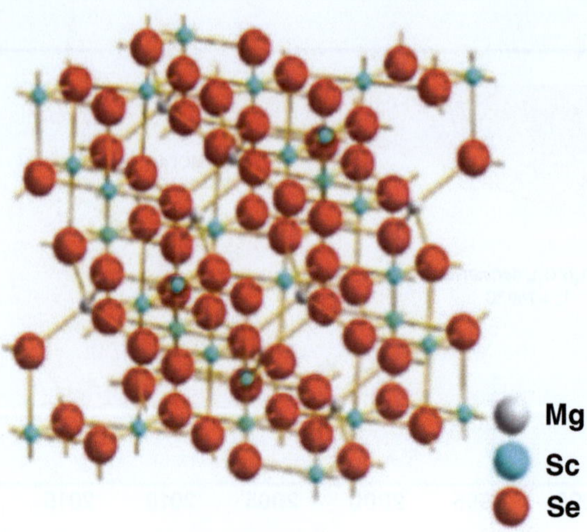

Figure 7.14 Model of the crystal structure of $MgSc_2Se_4$. *Source*: [75] / with permission of the American Chemical Society.

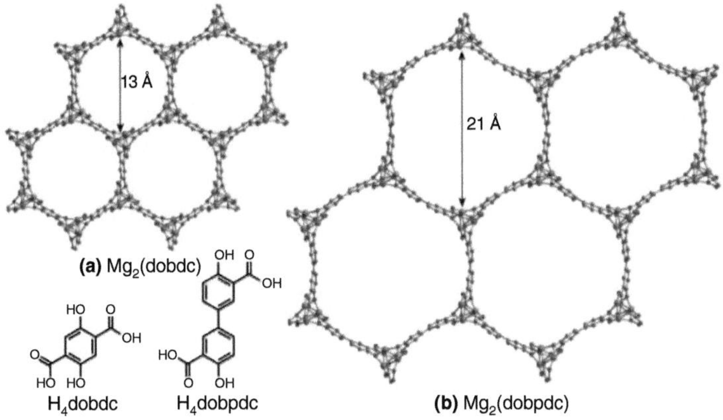

Figure 7.15 Structural arrangement of the MOFs: (a) Mg_2(dobdc) and (b) Mg_2(dobpdc), viewed along the *c*-axis direction. *Source*: [97] / with permission of the Royal Society of Chemistry.

Sc^{3+} sites with Ti^{4+} or Ce^{4+}. Unfortunately, neither approach yielded fully satisfactory results. The authors further suggested that the mixed ionic–electronic conduction properties of $MgSc_2Se_4$ could be leveraged in cathode materials.

In contrast to conventional inorganic solid-state conductors, a metal–organic framework (MOF)-based Mg solid electrolyte with a room-temperature Mg^{2+} conductivity of up to 2.5×10^{-4} S cm^{-1} was reported by Aubrey et al. [97] in 2014. Figure 7.15 shows the structures of Mg_2(2,5-dioxidobenzene-1,4-dicarboxylate) (Mg_2(dobdc)) and Mg_2(4,4′-dioxidobiphenyl-3,3′-dicarboxylate) (Mg_2(dobpdc)) along the *c*-axis. Unlike polymers, MOFs exhibit no anode correlation between the mechanical properties and ionic conductivity. Thus, MOF-based electrolytes offer a promising pathway for designing solid electrolytes with both high ionic conductivity and robust mechanical performance.

7.4.2 Organic Solid-state Electrolytes

Organic solid electrolytes, also referred to as polymer solid electrolytes, are composed of organic polymers combined with Mg salts. Commonly used polymer matrices include poly(ethylene oxide) (PEO) [98, 99], poly(vinylidene fluoride-*co*-hexafluoropropylene) (P(VDF–HFP)) [100–102], poly(vinylidene fluoride) (PVDF) [103], and poly(vinyl alcohol) (PVA) [104].

To enhance their ionic conductivity, electrolyte systems are commonly dosed with plasticizers. Plasticizer additives such as ethylene carbonate (EC), PC, and diethyl carbonate (DEC) can considerably improve the ionic conductivity of batteries. Kumar and Munichandraiah [105, 106] dissolved Mg triflate ($Mg(CF_3SO_3)_2$) in a polyacrylonitrile (PAN) suspension with EC and PC as solvents, forming PAN–$Mg(CF_3SO_3)_2$–EC/PC as a gel polymer electrolyte (GPE) with an approximate room-temperature conductivity of 10^{-3} S cm^{-1}. Employing a cross-linking strategy, Yoshimoto et al. [107] synthesized a Mg polymer electrolyte using cross-linked poly(methacrylate) as the host matrix, $Mg(TFSI)_2$ as the Mg salt, and poly(ethylene glycol) dimethyl ether (PEGDME) as the plasticizer. However, the room-temperature conductivity was below 10^{-3} S cm^{-1}. When PEGDME was replaced with organic carbonate solvents (EC and DMC), the ionic conductivity increased to 2.8×10^{-3} S cm^{-1} [108]. Although carbonate solvents improve the room-temperature conductivity of polymer electrolytes, they are volatile and highly flammable, posing safety risks in battery applications.

Owing to their nonvolatility and low flammability, ionic liquids have attracted much interest as plasticizers for GPEs. Morita and coworkers [109, 110] developed a Mg polymer electrolyte composed of $Mg(TFSI)_2$ salt, PEO-modified poly(methacrylate) (PEO–PMA) as the polymer matrix, and 1-ethyl-3-methylimidazolium TFSI (EMITFSI) as the plasticizer. At 60 °C, the optimal composition (20 wt.% $Mg(TFSI)_2$, 80 wt.% EMITFSI) yielded an ionic conductivity of 3.5 mS cm^{-1}. Subsequently, Pandey et al. synthesized polymer electrolytes from P(VDF–HFP) [102] and PEO [111] as the polymer matrices, with $Mg(Tf)_2$ as the Mg salt, and 1-ethyl-3-methylimidazolium trifluoromethanesulfonate (EMITf) as the ionic liquid plasticizer. The ion transference numbers of the PEO-based electrolyte and P(VDF–HFP)-based system were 0.45 and 0.26, respectively.

Chusid et al. [103] blended PEO and PVDF polymer matrices with a solution of $Mg(AlCl_2EtBu)_2$ in ether, using tetraethylene glycol dimethyl ether as a plasticizer. The resulting Mg GPE exhibited a high ionic conductivity (3.7 mS cm^{-1}) and reversibly cycled over several weeks when paired with a Chevrel-phase Mo_6S_8 cathode (Figure 7.16). More recently, Cui's group [112] developed a GPE system (denoted as PTB@GF–GPE) through an in situ cross-linking reaction between hydroxyl-terminated poly(tetrahydrofuran) (PTHF) and $Mg(BH_4)_2$. The synthetic process is illustrated in Figure 7.17. This polymer electrolyte exhibited a high Mg ionic conductivity (4.76×10^{-4} S cm^{-1}) and a low overpotential (0.1 V at 0.05 mA cm^{-2}). When assembled into a Mo_6S_8/Mg battery, the cell demonstrated stable charge–discharge performance across a broad temperature range from −20 to 60 °C.

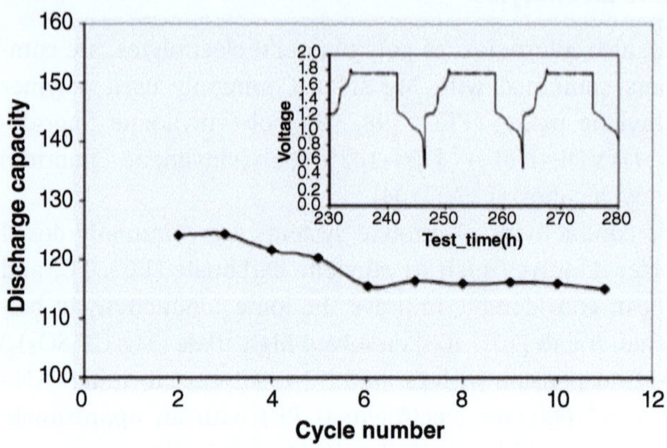

Figure 7.16 Cyclic voltammetry curves of the Mo_6S_8//PVDF/$Mg(AlCl_2EtBu)_2$/tetraglyme//AZ-31 battery. *Source*: [105] / with permission of Elsevier.

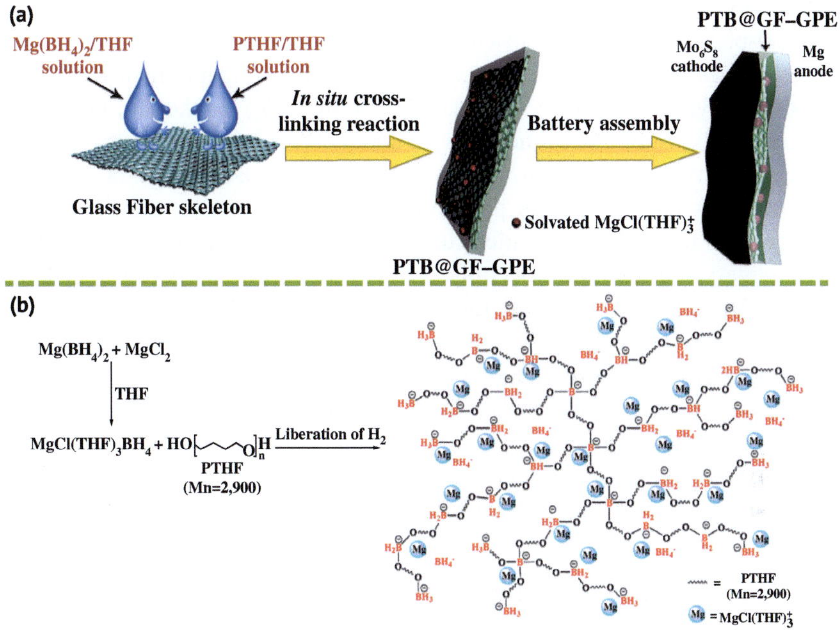

Figure 7.17 (a) Schematic showing the in situ preparation and battery assembly process of PTB@GF–GPE and (b) reaction synthesis route of PTB–GPE. *Source*: [114] / with permission of IOP Publishing.

7.4.3 Organic–Inorganic Composite Solid-state Electrolytes

Compared with Li composite solid electrolytes, Mg composite solid electrolytes with organic fillers have not yet been reported. Currently, organic–inorganic Mg composite solid electrolytes based on polymers are reinforced with inorganic fillers such as MgO [102], Al_2O_3 [99], SiO_2 [100], and TiO_2 [114].

Dissanayake et al. [99] incorporated nano-Al_2O_3 particles into a PEO–$Mg(ClO_4)_2$ polymer electrolyte. The O/OH groups on the Al_2O_3 nanoparticle surfaces interact with migrating ions, creating additional hopping sites (consequently, extra pathways) for ion transport. Although adding nano-Al_2O_3 boosted the conductivity of the PEO–$Mg(ClO_4)_2$ electrolyte, whether Mg^{2+} or ClO_4^- ions played the dominant role in the conduction process remained unclear. Pandey et al. developed $Mg(ClO_4)_2$/P(VDF–HFP) polymer electrolytes with nano-MgO [102, 113] and SiO_2 [101] as fillers. At the optimal filler composition (10 wt.% nano-MgO), the ion transference number reached 0.44, the ionic conductivity was 10^{-3} S cm^{-1}, the stable voltage exceeded 3.5 V, and reversible Mg^{2+} deposition/dissolution was observed. Shao et al. [98] fabricated a PEO-based nanocomposite polymer electrolyte using $Mg(BH_4)$ as the Mg salt and MgO nanoparticles as the filler. Solid-state ^{25}Mg NMR results confirm Mg^{2+} migration within the composite electrolyte. The characteristic second-order peak of $Mg(BH_4)_2$ disappeared, suggesting strong interactions between the Mg^{2+} cations and oxygen atoms in the PEO chains. Cyclic voltammetry tests demonstrated reversible Mg deposition and dissolution. Although the exact conductivity value was not reported, the nanocomposite polymer electrolyte—with a Chevrel-phase Mo_6S_8 cathode in a full cell—maintained a capacity of 100 mAh g^{-1} over 150 cycles. All tests were conducted at an elevated temperature of 100 °C. The composite polymer electrolyte, formulated with $Mg(ClO_4)_2$ as the salt, PEO as the polymer matrix, EMITFSI as the ionic liquid, and SiO_2 as the filler, exhibited an ionic conductivity of 0.5 mS cm^{-1} and a voltage stability of up to 4 V vs. Mg [115]. However, the charge–discharge performance of this electrolyte system was not investigated in full-cell configurations using actual cathode and anode electrodes.

In summary, Mg solid-state electrolytes have been incompletely researched, and their development is hindered by the formation of a passivation layer at the Mg anode interface and the poor mobility of Mg^{2+} ions in solid matrices, limiting their ionic conductivity. Although the room-temperature ionic conductivity of some Mg polymer electrolytes already approaches that of commercial Li polymer electrolytes, whether Mg^{2+} migration dominates the conduction process has not been clarified. Furthermore, few studies have investigated the cycling performance of full solid-state Mg batteries, indicating substantial room for further advancement.

7.5 Conclusion and Outlook

Although research on Mg-ion batteries has markedly progressed, numerous challenges spanning the materials, electrolytes, interfaces, and the entire battery system have hindered the transition of Mg-ion batteries from laboratory to industrialization. Key barriers are sluggish Mg^{2+} diffusion kinetics; the lack of cathode materials that simultaneously offer high capacity, high operating voltage, and excellent cycling stability; and the difficulty of achieving interfacial compatibility between electrolytes and electrodes. Furthermore, current research remains largely focused on half-cell configurations (using Mg counter-electrodes); studies on full cells are relatively limited.

Future research priorities and development trends will likely concentrate on the following areas:

1. **Advanced screening and prediction:** high-throughput computational methods and machine learning can screen for novel cathode materials, electrolyte salts, and solvent molecules on a large scale. This approach can predict critical properties such as structure, voltage profiles, diffusion pathways, and energy barriers, thereby accelerating material discovery and reducing the cost of experimental trial and error.
2. **Electrolyte engineering:** this research avenue will realize non-nucleophilic electrolytes with wide electrochemical windows (> 3.5 V), high Mg^{2+} conductivity, and good compatibility with Mg anodes.
3. **Interdisciplinary integration:** strengthening collaboration across materials science, electrochemistry, theoretical computation, and advanced characterization techniques—especially in situ and operando methods—will deepen our understanding of Mg^{2+} storage and transport mechanisms and interfacial reaction processes, providing a theoretical foundation for rational material design and battery optimization.

References

1 Foot, P. J. S. (2015) Principles and prospects of high-energy magnesium-ion batteries. *Science Progress*, 98, 264–275.

2 Grey, C. P. and Tarascon, J. M. (2016) Sustainability and in situ monitoring in battery development. *Nature Materials*, 16, 45–56.

3 Yoshio, M., Wang, H. and Fukuda, K. (2003) Spherical carbon-coated natural graphite as a lithium-ion battery-anode material. *Angewandte Chemie International Edition*, 42, 4203–4206.

4 Xu, W., Wang, J., Ding, F., Chen, X., Nasybulin, E., Zhang, Y. and Zhang, J.-G. (2014) Lithium metal anodes for rechargeable batteries. *Energy & Environmental Science*, 7, 513–537.

5 Scrosati, B. (2000) Recent advances in lithium ion battery materials. *Electrochimica Acta*, 45, 2461–2466.

6 Xiang, H., Shi, P., Bhattacharya, P., Chen, X., Mei, D., Bowden, M. E., Zheng, J., Zhang, J.-G. and Xu, W. (2016) Enhanced charging capability of lithium metal batteries based on lithium bis(trifluoromethanesulfonyl)imide-lithium bis(oxalato)borate dual-salt electrolytes. *Journal of Power Sources*, 318, 170–177.

7 Liu, F., Wang, T., Liu, X. and Fan, L. Z. (2020) Challenges and recent progress on key materials for rechargeable magnesium batteries. *Advanced Energy Materials*, 11, 2000787.

8 Roy, A., Sotoudeh, M., Dinda, S., Tang, Y., Kübel, C., Groß, A., Zhao-Karger, Z., Fichtner, M. and Li, Z. (2024) Improving rechargeable magnesium batteries through dual cation co-intercalation strategy. *Nature Communications*, 15, 492.

9 Cheng, X. B., Zhang, R., Zhao, C. Z. and Zhang, Q. (2017) Toward safe lithium metal anode in rechargeable batteries: a review. *Chemical Reviews*, 117, 10403–10473.

10 Gregory, T. D., Hoffman, R. J. and Winterton, R. C. (1990) Nonaqueous electrochemistry of magnesium applications to energy storage. *Journal of The Electrochemical Society*, 137, 775.

11 Aurbach, D., Lu, Z., Schechter, A., Gofer, Y., Gizbar, H., Turgeman, R., Cohen, Y., Moshkovich, M. and Levi, E. (2000) Prototype systems for rechargeable magnesium batteries. *Nature*, 407, 724–727.

12 Shi, H., Wang, G., Wang, Z., Yang, L., Zhang, S., Dong, S., Qu, B., Du, A., Li, Z., Zhou, X. and Cui, G. (2024) Understanding the cathode-electrolyte interfacial chemistry in rechargeable magnesium batteries. *Advanced Science*, 11, 2401536.

13 Zhang, J., Zhang, Z., Zhou, H., Liu, J., Du, A., Dong, S., Zhou, X., Wang, Q., Li, G. and Cui, G. (2025) Critical ingredients revitalize magnesium-metal batteries: rationality and challenges. *Advanced Materials*, 37, 2417652.

14 Chen, Y., Atwi, R., Nguyen, D. T., Bazak, J. D., Hahn, N. T., Ryu, J., Sears, J. A., Han, K. S., Song, M., Li, Z., Karkamkar, A. J., Hu, J. Z., Zavadil, K. R., Rajput, N. N., Mueller, K. T. and Murugesan, V. (2024) From bulk to interface: solvent exchange dynamics and their role in ion transport and the interfacial model of rechargeable magnesium batteries. *Journal of the American Chemical Society*, 146, 12984–12999.

15 Shterenberg, I., Salama, M., Gofer, Y., Levi, E. and Aurbach, D. (2014) The challenge of developing rechargeable magnesium batteries. *MRS Bulletin*, 39, 453–460.

16 Novák, P. and Desilvestro, J. (1993) Electrochemical insertion of magnesium in metal oxides and sulfides from aprotic electrolytes. *Journal of The Electrochemical Society*, 140, 140.

17 Levi, E., Gofer, Y. and Aurbach, D. (2009) On the way to rechargeable Mg batteries: the challenge of new cathode materials. *Chemistry of Materials*, 22, 860–868.

18 Yu, L. and Zhang, X. (2004) Electrochemical insertion of magnesium ions into V_2O_5 from aprotic electrolytes with varied water content. *Journal of Colloid and Interface Science*, 278, 160–165.

19 Le, D. B., Passerini, S., Coustier, F., Guo, J., Soderstrom, T., Owens, B. B. and Smyrl, W. H. (1998) Intercalation of polyvalent cations into V_2O_5 aerogels. *Chemistry of Materials*, 10, 682–684.

20 An, Q., Li, Y., Deog Yoo, H., Chen, S., Ru, Q., Mai, L. and Yao, Y. (2015) Graphene decorated vanadium oxide nanowire aerogel for long-cycle-life magnesium battery cathodes. *Nano Energy*, 18, 265–272.

21 Jang, G., Joe, Y. S., Lee, S. J., Cho, H. G., Baek, S. H., Xiong, P., Shin, K. H., Yeon, J. S., Kang, M. S., Oh, S. H. and Park, H. S. (2024) Crystal reconstruction of V_2O_3/carbon heterointerfaces via anodic hydration for ultrafast and reversible Mg-ion battery cathodes. *InfoMat*, 6, e12517.

22 Incorvati, J. T., Wan, L. F., Key, B., Zhou, D., Liao, C., Fuoco, L., Holland, M., Wang, H., Prendergast, D., Poeppelmeier, K. R. and Vaughey, J. T. (2015) Reversible magnesium intercalation into a layered oxyfluoride cathode. *Chemistry of Materials*, 28, 17–20.

23 Xu, M., Lei, S., Qi, J., Dou, Q., Liu, L., Lu, Y., Huang, Q., Shi, S. and Yan, X. (2018) Opening magnesium storage capability of two-dimensional MXene by intercalation of cationic surfactant. *ACS Nano*, 12, 3733–3740.

24 Ichitsubo, T., Adachi, T., Yagi, S. and Doi, T. (2011) Potential positive electrodes for high-voltage magnesium-ion batteries. *Journal of Materials Chemistry*, 21, 11764–11772.

25 Sanchez, L. and Pereira-Ramos, J. P. (1997) Electrochemical insertion of magnesium in a mixed manganese–cobalt oxide. *Journal of Materials Chemistry*, 7, 471–473.

26 Yoo, H. D., Shterenberg, I., Gofer, Y., Gershinsky, G., Pour, N. and Aurbach, D. (2013) Mg rechargeable batteries: an on-going challenge. *Energy & Environmental Science*, 6, 2265–2279.

27 Saha, P., Jampani, P. H., Datta, M. K., Okoli, C. U., Manivannan, A. and Kumta, P. N. (2014) A convenient approach to Mo_6S_8 Chevrel phase cathode for rechargeable magnesium battery. *Journal of The Electrochemical Society*, 161(4), A593.

28 Murgia, F., Antitomaso, P., Stievano, L., Monconduit, L. and Berthelot, R. (2016) Express and low-cost microwave synthesis of the ternary Chevrel phase $Cu_2Mo_6S_8$ for application in rechargeable magnesium batteries. *Journal of Solid State Chemistry*, 242, 151–154.

29 He, D., Wu, D., Gao, J., Wu, X., Zeng, X. and Ding, W. (2015) Flower-like CoS with nanostructures as a new cathode-active material for rechargeable magnesium batteries. *Journal of Power Sources*, 294, 643–649.

30 Wang, J., Ghosh, T., Ju, Z., Ng, M.-F., Wu, G., Yang, G., Zhang, X., Zhang, L., Handoko, A. D., Kumar, S., Busayaporn, W., Kaewsuwan, D., Jiang, C., Wei, M., Yu, G. and Seh, Z. W. (2024) Heterojunction structure of cobalt sulfide cathodes for high-performance magnesium-ion batteries. *Matter*, 7, 1833–1847.

31 Xiong, F., Fan, Y., Tan, S., Zhou, L., Xu, Y., Pei, C., An, Q. and Mai, L. (2018) Magnesium storage performance and mechanism of CuS cathode. *Nano Energy*, 47, 210–216.

32 Shen, Y., Wang, Y., Miao, Y., Yang, M., Zhao, X. and Shen, X. (2019) High-energy interlayer-expanded copper sulfide cathode material in non-corrosive electrolyte for rechargeable magnesium batteries. *Advanced Materials*, 32, e1905524.

33 Liang, K., Marcus, K., Guo, L., Li, Z., Zhou, L., Li, Y., De Oliveira, S. T., Orlovskaya, N., Sohn, Y. H. and Yang, Y. (2017) A freestanding NiS_x porous film as a binder-free electrode for Mg-ion batteries. *Chemical Communications*, 53, 7608–7611.

34 Sun, X., Bonnick, P. and Nazar, L. F. (2016) Layered TiS_2 positive electrode for Mg batteries. *ACS Energy Letters*, 1, 297–301.

35 Emly, A. and Van der Ven, A. (2015) Mg intercalation in layered and spinel host crystal structures for Mg batteries. *Inorganic Chemistry*, 54, 4394–4402.

36 Arsentev, M., Missyul, A., Petrov, A. V. and Hammouri, M. (2017) TiS_3 magnesium battery material: atomic-scale study of maximum capacity and structural behavior. *The Journal of Physical Chemistry C*, 121, 15509–15515.

37 Liang, Y., Feng, R., Yang, S., Ma, H., Liang, J. and Chen, J. (2011) Rechargeable Mg batteries with graphene-like MoS_2 cathode and ultrasmall Mg nanoparticle anode. *Advanced Materials*, 23, 640–643.

38 Sun, R., Pei, C., Sheng, J., Wang, D., Wu, L., Liu, S., An, Q. and Mai, L. (2018) High-rate and long-life VS_2 cathodes for hybrid magnesium-based battery. *Energy Storage Materials*, 12, 61–68.

39 Liu, Y., Jiao, L., Wu, Q., Du, J., Zhao, Y., Si, Y., Wang, Y. and Yuan, H. (2013) Sandwich-structured graphene-like MoS_2/C microspheres for rechargeable Mg batteries. *Journal of Materials Chemistry A*, 1, 5822–5826.

40 Du, C., He, S., Yang, L., Liu, X., Jiang, R., Ma, X., Zhu, Y., Zou, M. and Cao, C. (2024) Initiating cationic-anionic chemistry with stepwise surface-to-inner conversion in copper selenide superstructures for high-energy rechargeable magnesium batteries. *Energy Storage Materials*, 70, 103539.

41 Liu, B., Luo, T., Mu, G., Wang, X., Chen, D. and Shen, G. (2013) Rechargeable Mg-ion batteries based on WSe_2 nanowire cathodes. *ACS Nano*, 7, 8051–8058.

42 Mao, M., Ji, X., Hou, S., Gao, T., Wang, F., Chen, L., Fan, X., Chen, J., Ma, J. and Wang, C. (2019) Tuning anionic chemistry to improve kinetics of Mg intercalation. *Chemistry of Materials*, 31, 3183–3191.

43 Li, S., Zhang, H., Cao, Y., Zhang, S., Liu, Z., Tang, S., Yang, C. and Wen, D. (2022) Micro-nano Cu_2Se as a stable and ultralong cycle life anode material for sodium-ion batteries. *International Journal of Hydrogen Energy*, 47, 28280–28288.

44 Massé, R. C., Uchaker, E. and Cao, G. (2015) Beyond Li-ion: electrode materials for sodium- and magnesium-ion batteries. *Science China Materials*, 58, 715–766.

45 Heath, J., Chen, H. and Islam, M. S. (2017) $MgFeSiO_4$ as a potential cathode material for magnesium batteries: ion diffusion rates and voltage trends. *Journal of Materials Chemistry A*, 5, 13161–13167.

46 Mori, T., Masese, T., Orikasa, Y., Huang, Z. D., Okado, T., Kim, J. and Uchimoto, Y. (2016) Anti-site mixing governs the electrochemical performances of olivine-type $MgMnSiO_4$ cathodes for rechargeable magnesium batteries. *Physical Chemistry Chemical Physics*, 18, 13524–13529.

47 Zheng, Y., NuLi, Y., Chen, Q., Wang, Y., Yang, J. and Wang, J. (2012) Magnesium cobalt silicate materials for reversible magnesium ion storage. *Electrochimica Acta*, 66, 75–81.

48 Feng, Z., Yang, J., NuLi, Y. and Wang, J. (2008) Sol–gel synthesis of $Mg_{1.03}Mn_{0.97}SiO_4$ and its electrochemical intercalation behavior. *Journal of Power Sources*, 184, 604–609.

49 Orikasa, Y., Masese, T., Koyama, Y., Mori, T., Hattori, M., Yamamoto, K., Okado, T., Huang, Z. D., Minato, T., Tassel, C., Kim, J., Kobayashi, Y., Abe, T., Kageyama, H. and Uchimoto, Y. (2014) High energy density rechargeable magnesium battery using earth-abundant and non-toxic elements. *Scientific Reports*, 4, 5622.

50 Mao, M., Gao, T., Hou, S. and Wang, C. (2018) A critical review of cathodes for rechargeable Mg batteries. *Chemical Society Reviews*, 47, 8804–8841.

51 Makino, K., Katayama, Y., Miura, T. and Kishi, T. (2002) Preparation and electrochemical magnesium insertion behaviors of $Mg_{0.5+y}(Me_yTi_{1-y})_2(PO_4)_3$ (Me = Cr, Fe). *Journal of Power Sources*, 112, 85–89.

52 Makino, K., Katayama, Y., Miura, T. and Kishi, T. (2001) Electrochemical insertion of magnesium to $Mg_{0.5}Ti_2(PO_4)_3$. *Journal of Power Sources*, 99, 66–69.

53 Xu, Y., Xu, C., An, Q., Wei, Q., Sheng, J., Xiong, F., Pei, C. and Mai, L. (2017) Robust $LiTi_2(PO_4)_3$ microflowers as high-rate and long-life cathodes for Mg-based hybrid-ion batteries. *Journal of Materials Chemistry A*, 5, 13950–13956.

54 Zhang, R. and Ling, C. (2016) Unveil the chemistry of olivine $FePO_4$ as magnesium battery cathode. *ACS Applied Materials & Interfaces*, 8, 18018–18026.

55 Yagi, S., Fukuda, M., Ichitsubo, T., Nitta, K., Mizumaki, M. and Matsubara, E. (2015) EQCM analysis of redox behavior of CuFe Prussian blue analog in Mg battery electrolytes. *Journal of The Electrochemical Society*, 162, A2356.

56 Chae, M. S., Hyoung, J., Jang, M., Lee, H. and Hong, S.-T. (2017) Potassium nickel hexacyanoferrate as a high-voltage cathode material for nonaqueous magnesium-ion batteries. *Journal of Power Sources*, 363, 269–276.

57 Mizuno, Y., Okubo, M., Hosono, E., Kudo, T., Oh-ishi, K., Okazawa, A., Kojima, N., Kurono, R., Nishimura, S.-i. and Yamada, A. (2013) Electrochemical Mg^{2+} intercalation into a bimetallic CuFe Prussian blue analog in aqueous electrolytes. *Journal of Materials Chemistry A*, 1, 13055–13059.

58 Senoh, H., Sakaebe, H., Sano, H., Yao, M., Kuratani, K., Takeichi, N. and Kiyobayashi, T. (2014) Sulfone-based electrolyte solutions for rechargeable magnesium batteries using 2,5-dimethoxy-1,4-benzoquinone positive electrode. *Journal of The Electrochemical Society*, 161, A1315.

59 Pan, B., Huang, J., Feng, Z., Zeng, L., He, M., Zhang, L., Vaughey, J. T., Bedzyk, M. J., Fenter, P., Zhang, Z., Burrell, A. K. and Liao, C. (2016) Polyanthraquinone-based organic cathode for high-performance rechargeable magnesium-ion batteries. *Advanced Energy Materials*, 6, 1600140.

60 Rodriguez-Perez, I. A., Yuan, Y., Bommier, C., Wang, X., Ma, L., Leonard, D. P., Lerner, M. M., Carter, R. G., Wu, T., Greaney, P. A., Lu, J. and Ji, X. (2017) Mg-ion battery electrode: an organic solid's herringbone structure squeezed upon Mg-ion insertion. *Journal of the American Chemical Society*, 139, 13031–13037.

61 Sun, Y., Ai, F. and Lu, Y. C. (2022) Electrolyte and interphase design for magnesium anode: major challenges and perspectives. *Small*, e2200009.

62 Muldoon, J., Bucur, C. B. and Gregory, T. (2017) Fervent hype behind magnesium batteries: an open call to synthetic chemists—electrolytes and cathodes needed. *Angewandte Chemie International Edition*, 56, 12064–12084.

63 Matsui, M. (2011) Study on electrochemically deposited Mg metal. *Journal of Power Sources*, 196, 7048–7055.

64 Ling, C., Banerjee, D. and Matsui, M. (2012) Study of the electrochemical deposition of Mg in the atomic level: why it prefers the non-dendritic morphology. *Electrochimica Acta*, 76, 270–274.

65 Davidson, R., Verma, A., Santos, D., Hao, F., Fincher, C. D., Zhao, D., Attari, V., Schofield, P., Van Buskirk, J., Fraticelli-Cartagena, A., Alivio, T. E. G., Arroyave, R., Xie, K., Pharr, M., Mukherjee, P. P. and Banerjee, S. (2020) Mapping mechanisms and growth regimes of magnesium electrodeposition at high current densities. *Materials Horizons*, 7, 843–854.

66 Zhang, X., Lv, R., Tang, W., Li, G., Wang, A., Dong, A., Liu, X. and Luo, J. (2020) Challenges and opportunities for multivalent metal anodes in rechargeable batteries. *Advanced Functional Materials*, 30, 2004187.

67 Zhang, J., Guan, X., Lv, R., Wang, D., Liu, P. and Luo, J. (2020) Rechargeable Mg metal batteries enabled by a protection layer formed in vivo. *Energy Storage Materials*, 26, 408–413.

68 Attias, R., Salama, M., Hirsch, B., Goffer, Y. and Aurbach, D. (2019) Anode-electrolyte interfaces in secondary magnesium batteries. *Joule*, 3, 27–52.

69 Son, S. B., Gao, T., Harvey, S. P., Steirer, K. X., Stokes, A., Norman, A., Wang, C., Cresce, A., Xu, K. and Ban, C. (2018) An artificial interphase enables reversible magnesium chemistry in carbonate electrolytes. *Nature Chemistry*, 10, 532–539.

70 Arthur, T.S., Singh, N., and Matsui, M. Electrodeposited Bi, Sb and Bi1–xSbx alloys as anodes for Mg-ion batteries. *Electrochemistry Communications*, 2012, 16: 103–106.

71 Wei, C., Tan, L., Tao, Y., An, Y., Tian, Y., Jiang, H., Feng, J. and Qian, Y. (2021) Interfacial passivation by room-temperature liquid metal enabling stable 5 V-class lithium-metal batteries in commercial carbonate-based electrolyte. *Energy Storage Materials*, 34, 12–21.

72 Li, D., Yuan, Y., Liu, J., Fichtner, M. and Pan, F. (2020) A review on current anode materials for rechargeable Mg batteries. *Journal of Magnesium and Alloys*, 8, 963–979.

73 Zhang, R., Cui, C., Li, R., Li, Y., Du, C., Gao, Y., Huo, H., Ma, Y., Zuo, P. and Yin, G. (2021) An artificial interphase enables the use of $Mg(TFSI)_2$-based electrolytes in magnesium metal batteries. *Chemical Engineering Journal*, 426, 130751.

74 Niu, J., Zhang, Z. and Aurbach, D. (2020) Alloy anode materials for rechargeable Mg ion batteries. *Advanced Energy Materials*, 10, 2000697.

75 Shao, Y., Gu, M., Li, X., Nie, Z., Zuo, P., Li, G., Liu, T., Xiao, J., Cheng, Y., Wang, C., Zhang, J. G. and Liu, J. (2014) Highly reversible Mg insertion in nanostructured Bi for Mg ion batteries. *Nano Letters*, 14, 255–260.

76 Liu, Z., Lee, J., Xiang, G., Glass, H. F. J., Keyzer, E. N., Dutton, S. E. and Grey, C. P. (2017) Insights into the electrochemical performances of Bi anodes for Mg ion batteries using ^{25}Mg solid state NMR spectroscopy. *Chemical Communications*, 53, 743–746.

77 Cheng, Y., Shao, Y., Parent, L. R., Sushko, M. L., Li, G., Sushko, P. V., Browning, N. D., Wang, C. and Liu, J. (2015) Interface promoted reversible Mg insertion in nanostructured tin-antimony alloys. *Advanced Materials*, 27, 6598–6605.

78 Parent, L. R., Cheng, Y., Sushko, P. V., Shao, Y., Liu, J., Wang, C.-M. and Browning, N. D. (2015) Realizing the full potential of insertion anodes for Mg-ion batteries through the nanostructuring of Sn. *Nano Letters*, 15, 1177–1182.

79 Wang, L., Welborn, S. S., Kumar, H., Li, M., Wang, Z., Shenoy, V. B. and Detsi, E. (2019) High-rate and long cycle-life alloy-type magnesium-ion battery anode enabled through (de)magnesiation-induced near-room-temperature solid–liquid phase transformation. *Advanced Energy Materials*, 9, 1902086.

80 Wu, N., Lyu, Y.-C., Xiao, R.-J., Yu, X., Yin, Y.-X., Yang, X.-Q., Li, H., Gu, L. and Guo, Y.-G. (2014) A highly reversible, low-strain Mg-ion insertion anode material for rechargeable Mg-ion batteries. *NPG Asia Materials*, 6, e120.

81 Zeng, J., Yang, Y., Li, C., Li, J., Huang, J., Wang, J. and Zhao, J. (2017) Li_3VO_4: an insertion anode material for magnesium ion batteries with high specific capacity. *Electrochimica Acta*, 247, 265–270.

82 Meng, Y., Wang, D., Zhao, Y., Lian, R., Wei, Y., Bian, X., Gao, Y., Du, F., Liu, B. and Chen, G. (2017) Ultrathin TiO_2-$_B$ nanowires as an anode material for Mg-ion batteries based on a surface Mg storage mechanism. *Nanoscale*, 9, 12934–12940.

83 Chen, C., Wang, J., Zhao, Q., Wang, Y. and Chen, J. (2016) Layered $Na_2Ti_3O_7$/$MgNaTi_3O_7$/$Mg_{0.5}NaTi_3O_7$ nanoribbons as high-performance anode of rechargeable Mg-ion batteries. *ACS Energy Letters*, 1, 1165–1172.

84 Er, D., Detsi, E., Kumar, H. and Shenoy, V. B. (2016) Defective graphene and graphene allotropes as high-capacity anode materials for Mg ion batteries. *ACS Energy Letters*, 1, 638–645.

85 Kim, D.-M., Jung, S. C., Ha, S., Kim, Y., Park, Y., Ryu, J. H., Han, Y.-K. and Lee, K. T. (2018) Cointercalation of Mg^{2+} ions into graphite for magnesium-ion batteries. *Chemistry of Materials*, 30(10), 3199–3203.

86 Sibari, A., Marjaoui, A., Lakhal, M., Kerrami, Z., Kara, A., Benaissa, M., Ennaoui, A., Hamedoun, M., Benyoussef, A. and Mounkachi, O. (2018) Phosphorene as a promising anode material for (Li/Na/Mg)-ion batteries: a first-principle study. *Solar Energy Materials and Solar Cells*, 180, 253–257.

87 Mortazavi, B., Dianat, A., Rahaman, O., Cuniberti, G. and Rabczuk, T. (2016) Borophene as an anode material for Ca, Mg, Na or Li ion storage: a first-principle study. *Journal of Power Sources*, 329, 456–461.

88 Ikeda, S., Takahashi, M., Ishikawa, J. and Ito, K. (1986) Solid electrolytes with multivalent cation conduction. 1. Conducting species in Mg-Zr-PO_4 system. *Solid State Ionics*, 23, 125–129.

89 Ikeda, S., Kato, Y., Nomura, K., Rasai, A. and Ito, K. (1990) Solid electrolytes with multivalent cation conduction (2) zinc ion conduction in Zn-Zr-PO_4 system. *Solid State Ionics*, 40/41, 79–82.

90 Nomura, K., Ikeda, S., Ito, K. and Einaga, H. (1992) Framework structure, phase transition, and transport properties in $MIIZr_4(PO_4)_6$ compounds (MII = Mg, Ca, Sr, Ba, Mn, Co, Ni, Zn, Cd, and Pb). *Bulletin of the Chemical Society of Japan*, 65, 3221–3227.

91 Adamu, M. and Kale, G. M. (2016) Novel sol–gel synthesis of $MgZr_4P_6O_{24}$ composite solid electrolyte and newer insight into the Mg^{2+}-ion conducting properties using impedance spectroscopy. *The Journal of Physical Chemistry C*, 120, 17909–17915.

92 Higashi, S., Miwa, K., Aoki, M. and Takechi, K. (2014) A novel inorganic solid state ion conductor for rechargeable Mg batteries. *Chemical Communications*, 50, 1320–1322.

93 Roedern, E., Kühnel, R.-S., Remhof, A. and Battaglia, C. (2017) Magnesium ethylenediamine borohydride as solid-state electrolyte for magnesium batteries. *Scientific Reports*, 7, 46189.

94 Yamanaka, T., Hayashi, A., Yamauchi, A. and Tatsumisago, M. (2014) Preparation of magnesium ion conducting $MgS–P_2S_5–MgI_2$ glasses by a mechanochemical technique. *Solid State Ionics*, 262, 601–603.

95 Canepa, P., Bo, S.-H., Sai Gautam, G., Key, B., Richards, W. D., Shi, T., Tian, Y., Wang, Y., Li, J. and Ceder, G. (2017) High magnesium mobility in ternary spinel chalcogenides. *Nature Communications*, 8, 1759.

96 Wang, L.-P., Zhao-Karger, Z., Klein, F., Chable, J., Braun, T., Schür, A. R., Wang, C.-R., Guo, Y.-G. and Fichtner, M. (2019) $MgSc_2Se_4$—a magnesium solid ionic conductor for all-solid-state Mg batteries? *ChemSusChem*, 12, 2286–2293.

97 Aubrey, M. L., Ameloot, R., Wiers, B. M. and Long, J. R. (2014) Metal–organic frameworks as solid magnesium electrolytes. *Energy & Environmental Science*, 7, 667–671.

98 Shao, Y., Rajput, N. N., Hu, J., Hu, M., Liu, T., Wei, Z., Gu, M., Deng, X., Xu, S., Han, K. S., Wang, J., Nie, Z., Li, G., Zavadil, K. R., Xiao, J., Wang, C., Henderson, W. A., Zhang, J.-G., Wang, Y., Mueller, K. T., Persson, K. and Liu, J. (2015) Nanocomposite polymer electrolyte for rechargeable magnesium batteries. *Nano Energy*, 12, 750–759.

99 Dissanayake, M. A. K. L., Bandara, L. R. A. K., Karaliyadda, L. H., Jayathilaka, P. A. R. D. and Bokalawala, R. S. P. (2006) Thermal and electrical properties of solid polymer electrolyte PEO_9 $Mg(ClO_4)_2$ incorporating nano-porous Al_2O_3 filler. *Solid State Ionics*, 177, 343–346.

100 Oh, J.-S., Ko, J.-M. and Kim, D.-W. (2004) Preparation and characterization of gel polymer electrolytes for solid state magnesium batteries. *Electrochimica Acta*, 50, 903–906.

101 Pandey, G. P., Agrawal, R. C. and Hashmi, S. A. (2011) Magnesium ion-conducting gel polymer electrolytes dispersed with fumed silica for rechargeable magnesium battery application. *Journal of Solid State Electrochemistry*, 15, 2253–2264.

102 Pandey, G. P., Agrawal, R. C. and Hashmi, S. A. (2009) Magnesium ion-conducting gel polymer electrolytes dispersed with nanosized magnesium oxide. *Journal of Power Sources*, 190, 563–572.

103 Chusid, O., Gofer, Y., Gizbar, H., Vestfrid, Y., Levi, E., Aurbach, D. and Riech, I. (2003) Solid-state rechargeable magnesium batteries. *Advanced Materials*, 15, 627–630.

104 Manjuladevi, R., Thamilselvan, M., Selvasekarapandian, S., Mangalam, R., Premalatha, M. and Monisha, S. (2017) Mg-ion conducting blend polymer electrolyte based on poly(vinyl alcohol)-poly (acrylonitrile) with magnesium perchlorate. *Solid State Ionics*, 308, 90–100.

105 Kumar, G. G. and Munichandraiah, N. (1999) Reversibility of Mg/Mg^{2+} couple in a gel polymer electrolyte. *Electrochimica Acta*, 44, 2663–2666.

106 Kumar, G. G. and Munichandraiah, N. (2000) A gel polymer electrolyte of magnesium triflate. *Solid State Ionics*, 128, 203–210.

107 Yoshimoto, N., Yakushiji, S., Ishikawa, M. and Morita, M. (2002) Ionic conductance behavior of polymeric electrolytes containing magnesium salts and their application to rechargeable batteries. *Solid State Ionics*, 152–153, 259–266.

108 Yoshimoto, N., Yakushiji, S., Ishikawa, M. and Morita, M. (2003) Rechargeable magnesium batteries with polymeric gel electrolytes containing magnesium salts. *Electrochimica Acta*, 48, 2317–2322.

109 Yoshimoto, N., Shirai, T. and Morita, M. (2005) A novel polymeric gel electrolyte system containing magnesium salt with ionic liquid. *Electrochimica Acta*, 50, 3866–3871.

110 Morita, M., Shirai, T., Yoshimoto, N. and Ishikawa, M. (2005) Ionic conductance behavior of polymeric gel electrolyte containing ionic liquid mixed with magnesium salt. *Journal of Power Sources*, 139, 351–355.

111 Kumar, Y., Hashmi, S. A. and Pandey, G. P. (2011) Ionic liquid mediated magnesium ion conduction in poly(ethylene oxide) based polymer electrolyte. *Electrochimica Acta*, 56, 3864–3873.

112 Du, A., Zhang, H., Zhang, Z., Zhao, J., Cui, Z., Zhao, Y., Dong, S., Wang, L., Zhou, X. and Cui, G. (2019) A crosslinked polytetrahydrofuran-borate-based polymer electrolyte enabling wide-working-temperature-range rechargeable magnesium batteries. *Advanced Materials*, 31, 1805930.

113 Pandey, G. P., Agrawal, R. C. and Hashmi, S. A. (2010) Electrical and electrochemical properties of magnesium ion conducting composite gel polymer electrolytes. *Journal of Physics D: Applied Physics*, 43, 255501.

114 Deivanayagam, R., Cheng, M., Wang, M., Vasudevan, V., Foroozan, T., Medhekar, N. V. and Shahbazian-Yassar, R. (2019) Composite polymer electrolyte for highly cyclable room-temperature solid-state magnesium batteries. *ACS Applied Energy Materials*, 2, 7980–7990.

115 Song, S., Kotobuki, M., Zheng, F., Li, Q., Xu, C., Wang, Y., Li, W. D. Z., Hu, N. and Lu, L. (2017) Communication—a composite polymer electrolyte for safer Mg batteries. *Journal of The Electrochemical Society*, 164, A741.

8

Advanced Characterization and Electrochemical Testing

8.1 Characterization Techniques in Batteries

8.1.1 Scanning Electron Microscopy

8.1.1.1 Principle of Scanning Electron Microscopy

In recent years, scanning electron microscopy (SEM) analysis technology has become a primary method for characterizing material surface information. As an important branch of microscopy, SEM boasts advantages such as a wide magnification range, suitability for a broad range of samples, effective three-dimensional imaging, and strong comprehensive analysis capabilities [1]. It holds unique advantages in characterizing morphology, assisting mechanism studies, and analyzing micro-area elemental composition.

Scanning electron microscope is an electron microscope that uses a high-energy focused electron beam to scan the sample surface, thereby obtaining sample information. It uses an electron beam as the illumination source, which scans the sample surface. Imaging is based on the combination of information generated by the interaction of electrons with matter and electron optics principles (Figure 8.1). By detecting specific signals (mainly secondary electrons, backscattered electrons, and characteristic X-rays) with dedicated detectors, morphological or compositional information of the corresponding sample area can be obtained (the information carried, analysis depth, and generation principles of various signals are listed in Table 8.1). The performance of a scanning electron microscope is primarily judged based on its resolution and effective magnification. Resolution is the minimum distinguishable distance.

Resolution is typically defined using the Abbe's formula, given by (units in nm)

$$R = \frac{0.6\lambda}{n \sin \alpha} \tag{8.1}$$

In Equation (8.1), R is the resolution, λ denotes the wavelength, n is the refractive index, and α is the aperture radius.

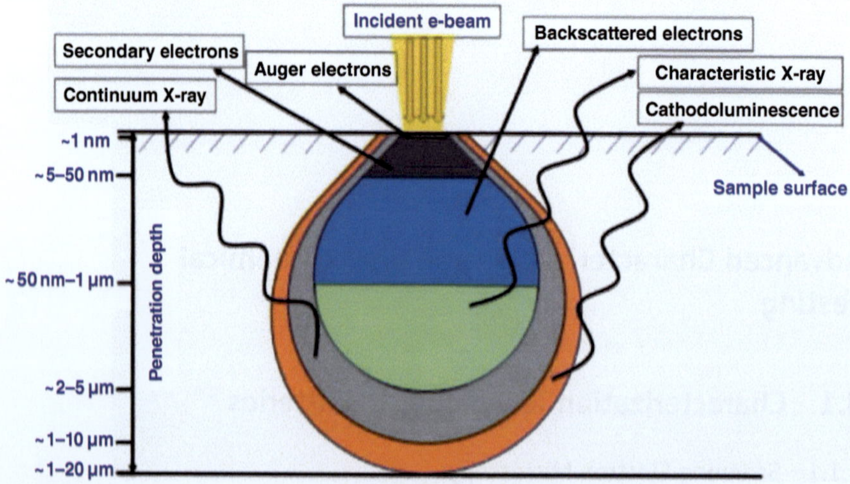

Figure 8.1 Various signals generated by the interaction of the electron beam with the sample.

Table 8.1 Information carried, analysis depth, and generation principles of electron signals in SEM.

Signal name	Information carried	Analysis depth	Generation principle
Secondary electron	Morphology	~10 nm	Electron beam ionizes outer shell electrons, forming free electrons that leave the sample surface becoming secondary electrons
Backscattered electron	Morphology and composition	0.1–1 μm	Some incident electrons undergo elastic scattering in various directions by atomic nuclei in the sample and escape the surface as backscattered electrons
Characteristic X-ray	Composition	1–several μm	Inner shell electrons are ionized forming an excited state; characteristic X-rays are emitted during the transition from the excited state back to the ground state

Effective magnification is defined as

$$M_e = \frac{R_p}{R_m} \quad (8.2)$$

In Equation (8.2), R_p is the minimum distance that the human eye can distinguish and R_m is the minimum distance that the machine can distinguish.

For example, if the human eye can distinguish 0.2 mm and the instrument can distinguish 1 nm, then the effective magnification of this instrument is 200,000 times. The scanning range also varies depending on the effective magnification.

8.1.1.2 Components of SEM

A scanning electron microscope mainly consists of three major parts: the electron optical system, the signal detection system, and the vacuum system.

The electron optical system comprises four parts: the electron gun, electromagnetic lenses, the objective aperture, and scanning coils. This system primarily provides the SEM with an electron beam having controllable energy, sufficient intensity, adjustable spot size, selectable scanning range, symmetrical shape, and stability.

In the signal detection system, the electron beam interacts with the sample. Depending on the material, different voltages and detectors are used, such as secondary electron detectors, backscattered electron detectors, X-ray energy-dispersive or wavelength-dispersive spectrometers, and electron backscatter diffraction detectors.

Different voltages result in different penetration depths and generate different electrons, which can be represented by the following formula (units in μm):

$$R = \frac{0.0276AE^{1.67}}{Z^{0.89}\rho} \quad (8.3)$$

In Equation (8.3), A is the mass number, Z is the atomic number, ρ denotes the material density, and E is the acceleration voltage.

From this equation, it can be seen that a higher acceleration voltage results in a greater penetration depth.

The vacuum system ensures that the entire SEM is in a vacuum environment, preventing air contact and contamination of the sample. Under standby conditions, the laboratory typically uses high-purity argon gas to protect the entire apparatus. The vacuum level in the column directly affects the service life of the tungsten filament.

8.1.1.3 SEM Application Examples

Bucur et al. [2] used polymer/inorganic clay nanofilms to encapsulate nanocarbon composite sulfur particles and prepared high-density sulfur cathode materials using a layered self-assembly method. The film structure included anchoring zones, conductive zones, and rigid clay zones, synergistically promoting uniform lithium deposition and suppressing dendrite growth, while effectively blocking polysulfide diffusion into the electrolyte. Experiments showed that this strategy achieved excellent cycling stability and high energy density (up to 1,500 Wh kg^{-1} and 1,000 Wh L^{-1}) under high sulfur loading, significantly improving sulfur material utilization and

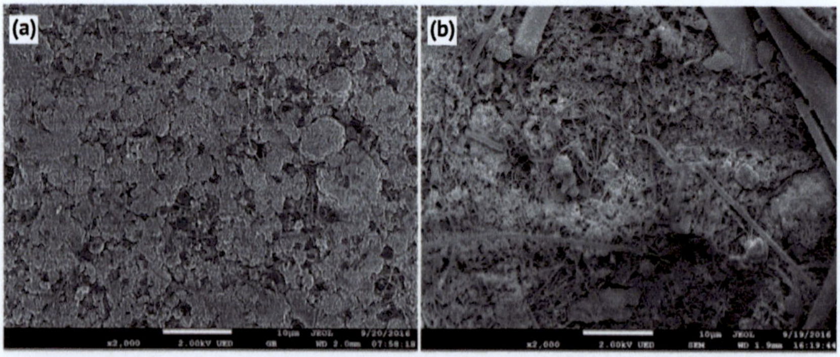

Figure 8.2 SEM images of the battery cathode (a) before assembly and (b) after the end of life. *Source*: [2] / with permission of the Royal Society of Chemistry.

the coulombic efficiency of the lithium anode. In this work, SEM was primarily used to observe the morphological changes of the sulfur cathode material before and after battery assembly (Figure 8.2). The results showed that after the end of battery life, the morphology of the sulfur particles was very similar to that before assembly; sulfur particles could still be clearly observed after cycling. This indicates that the sulfur particle structure was maintained during battery cycling without significant morphological changes or loss. Furthermore, SEM images also showed residual fibers on the separator, further proving the stability of the sulfur particle structure. These observations supported the authors' conclusion that polysulfide dissolution is not the main cause of capacity loss.

Xu et al. [3] used two-dimensional expanded vermiculite nanosheets, prepared via ion exchange and chemical exfoliation processes, to create a freestanding vermiculite separator rich in lithium ions. This separator had a dense structure, and its negatively charged surface could effectively inhibit polysulfide shuttling through electrostatic repulsion and spatial blocking, while providing fast transport channels for lithium ions. SEM was used to observe and analyze the structural morphology of the vermiculite separator (Figure 8.3). SEM results showed that after vacuum filtration, vermiculite nanosheets dispersed in solution could reassemble well into a layered structure. Cross-sectional observation of the separator clearly showed layered stacks of vermiculite silicate sheets, exhibiting an overall dense packing morphology with a thickness of about 10 μm. Additionally, SEM revealed that the thickness of the vermiculite nanosheets was typically several tens of nanometers, with an average size of about several tens of micrometers, assembling into a uniform and dense separator. These structural features provided the physical basis for rapid lithium-ion transport and suppression of polysulfide penetration through the separator, while also helping to inhibit lithium dendrite growth and improve the safety of lithium–sulfur (Li–S) batteries.

Pan et al. [4] selected an inexpensive, abundant natural fibrous clay mineral—sepiolite—as the cathode matrix material. Through simple mixing and heat treatment processes, sulfur was coated on the surface and within the pores of

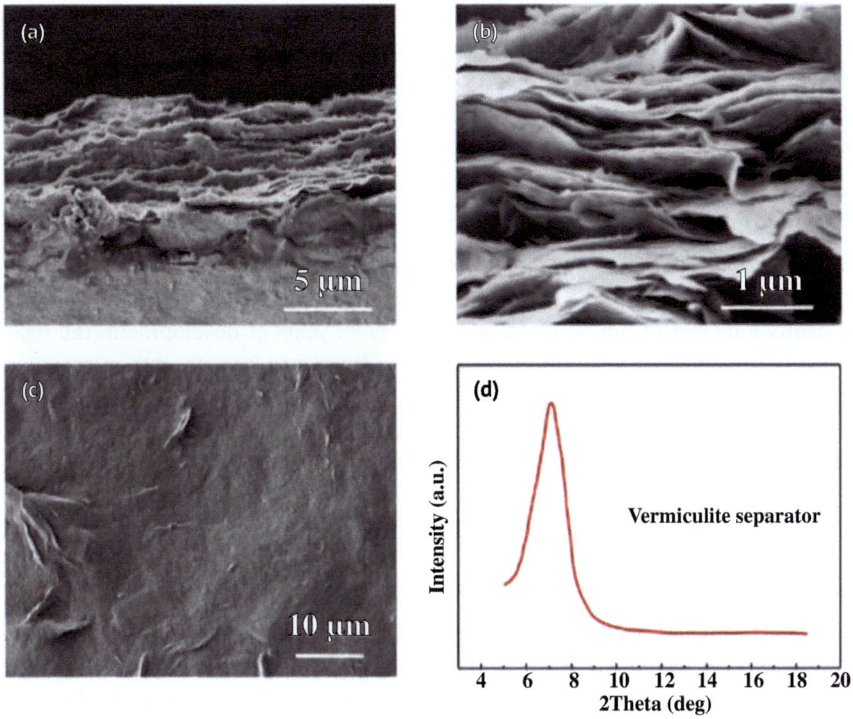

Figure 8.3 Morphological and structural characteristics of the vermiculite separator. (a, b) Cross-sectional SEM images of the vermiculite separator; (c) top-view SEM image of the vermiculite separator at different resolutions; and (d) XRD pattern of the vermiculite separator. *Source*: [3] / with permission of Elsevier.

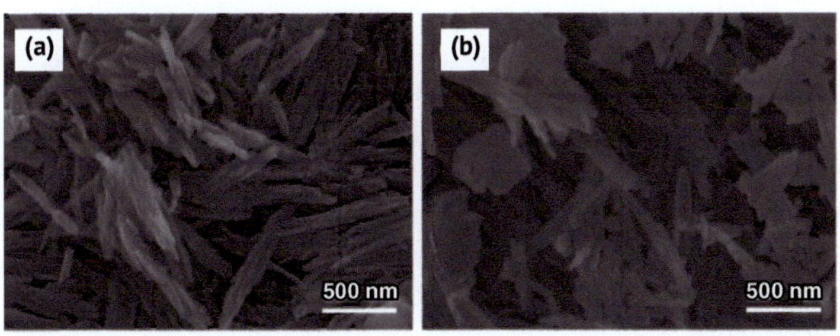

Figure 8.4 The morphologies of sepiolite and sepiolite/S composite: SEM images of (a) sepiolite and (b) sepiolite/S composite. *Source*: [3] / with permission of Elsevier.

sepiolite, forming a sepiolite/sulfur composite material. The SEM images before and after sulfur loading was observed. Figure 8.4a shows the microscopic morphology of pristine sepiolite powder, consisting of short microfiber bundles with lengths ranging from 200 nm to 2 μm and diameters from 50 to 200 nm; while Figure 8.4b shows the morphology of sepiolite after sulfur infusion. The surface of the sepiolite powder was covered with sulfur, and the average fiber diameter increased, but the high-aspect-ratio fibrous structure was maintained. This indicates that the basic structure of sepiolite did not change significantly during sulfur infusion, only that sulfur was attached to the surface, leading to fiber aggregation phenomena, further proving the physical properties and stability of sepiolite as a sulfur host.

8.1.2 Transmission Electron Microscopy

8.1.2.1 Principle of Transmission Electron Microscopy

In 1932, German Scientists Knoll and Ruska successfully developed the world's first transmission electron microscope. After nearly 90 years of development, the resolution of transmission electron microscope has improved from the sub-micrometer scale to the sub-nanometer scale at the atomic level, and its functions have become more diverse. Nowadays, it is widely used in materials science, life sciences, physics, and other fields, becoming one of the indispensable characterization tools in modern scientific research. Transmission electron microscope is an amplification device that images using electrons transmitted through the sample. The electron beam passing through the sample carries information about the sample. After processing and amplifying this information, the ultrastructure of the material can be displayed on a fluorescent screen. The resolution of transmission electron microscope can reach up to 1 Å, with a magnification ranging from several hundred to 800,000 times, continuously adjustable. It is mainly used for observing the internal ultrastructure of materials, composition analysis, and particle size determination.

The imaging principles of transmission electron microscopy (TEM) are divided into three types:

1. When the electron beam irradiates a sample surface with high density and thickness, areas with greater thickness scatter electrons at larger angles, resulting in fewer transmitted electrons and thus darker imaging brightness. This is the absorption contrast principle of early transmission electron microscope.
2. Due to varying diffraction capabilities of different parts of a crystal in the sample, after the sample diffracts the incident electron beam, the diffracted wave amplitude becomes unevenly distributed across the sample. When crystal defects are present, the diffraction capability of the defect region differs from that of the perfect region, leading to an uneven distribution of diffracted wave amplitude corresponding to the defect distribution. This is the diffraction contrast imaging principle.
3. When the sample thickness is below 100 Å, which is less than the wavelength of the electron beam, the beam can penetrate the sample directly. The amplitude change of the electron beam wave is negligible, allowing observation of differences based on phase changes. This is the phase contrast imaging principle.

Different imaging methods obtain information related to the sample by modifying the wave function of the electron beam exiting the sample. The observed image intensity depends on both the amplitude and phase of the electron wave. Phase objects affect the wave phase far more than the wave amplitude, thus requiring further complex analysis to relate the observed image intensity to the sample properties.

Currently, many manufacturers produce transmission electron microscopes, including Hitachi, JEOL, Philips (now FEI), Nion, etc. Today, TEM has become one of the most effective means for material characterization from the mesoscale down to the atomic level. The main factors driving these changes include improvements in the electron optical system, enhancements in the illumination system, and the application of new imaging principles.

8.1.2.2 TEM Application Examples

Wang et al. [5] addressed the challenges of low electronic conductivity and lithium-ion diffusion coefficient in $LiFePO_4$ as a cathode material for lithium-ion batteries (LIBs). Based on using PVDF as a fluorine source to prepare fluorine-doped carbon (FC) and coating it on $LiFePO_4$, they constructed a three-dimensional conductive network. Through systematic characterization and electrochemical testing, they confirmed that FC coating could introduce structural defects, increase surface area and porosity, and enhance electron and ion conduction, significantly improving the rate capability and cycling stability of the material. This work filled the theoretical and application research gap regarding the synergistic regulation of anion doping and carbon coating, providing new strategies and experimental basis for the design of high-performance LIB cathode materials. TEM images confirmed that the LFP@FC-II nanomaterial possessed a "core–shell" structure, where LFP (lithium iron phosphate) nanoparticles were uniformly coated with an amorphous carbon layer about 2.5 nm thick. This structure not only effectively prevented further particle growth but also formed a network conducive to electron conduction, contributing to the enhancement of the material's electrochemical performance. Furthermore, elemental mapping showed uniform distribution of Fe, P, and O, and successful doping of F element into the carbon layer. EDS results further verified the presence of Fe, P, O, C, and trace F elements in the material. In summary, TEM images (Figure 8.5) illustrated the uniform carbon coating and F element doping on the surface of the LFP@FC-II material, revealing its "core–shell" conductive network structure, which plays an important role in improving electrochemical performance.

Gao et al. [6] addressed the issues of poor conductivity and structural instability due to volume expansion during charge/discharge in anode materials for LIBs. They proposed a designable and controllable strategy for preparing multiwalled Sn/SnO_2@C hollow nanofibers with multi-structure control. Through steps such as electrospinning, polypyrrole coating, and carbonization reduction, they successfully synthesized composite nanofibers with a "wire–core/double-layer tube–shell" structure. This structure combined high volumetric energy density with excellent cycling stability, effectively alleviating electrode pulverization caused by volume expansion while enhancing lithium-ion and electron transport capabilities. Through TEM image analysis, the formation process and structural evolution of the multiwalled hollow nanofibers (such as WIDWT Sn/SnO_2@C nanofibers) were revealed. Specifically, TEM image in Figure 8.6 displays various stages from the precursor to the final multilayer structure, including solid precursor, wire-in-tube structure,

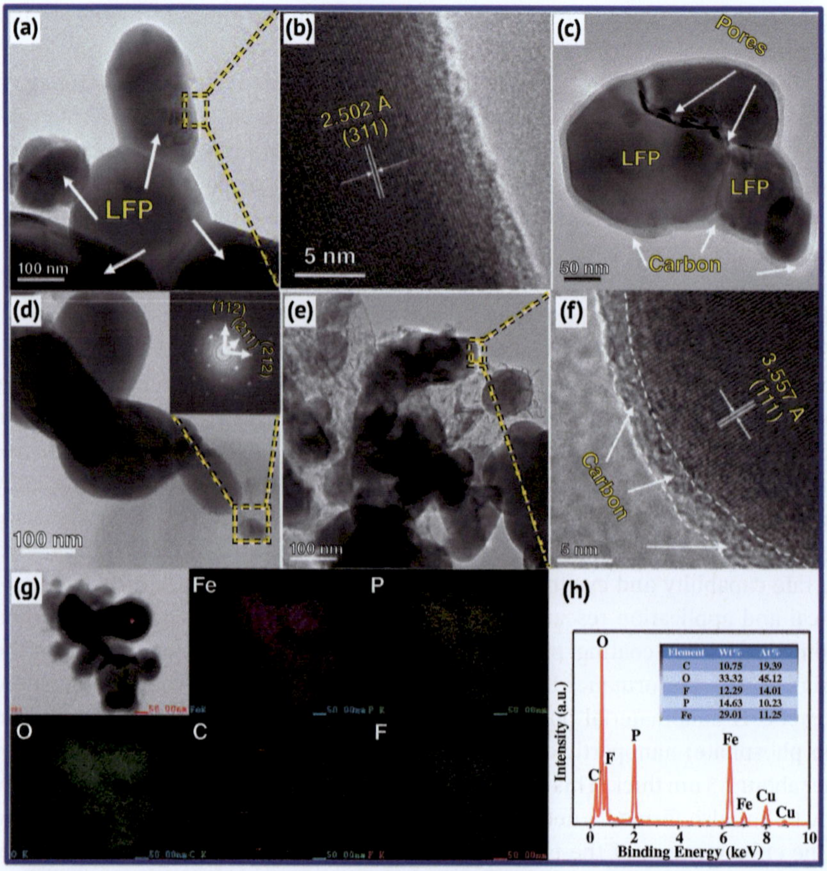

Figure 8.5 TEM images of (a) LFP and (c–e) LFP@FC-II. HRTEM images of (b) LFP and (f) LFP@FC-II corresponding to the yellow-boxed areas in panels (a) and (e), respectively. (g) TEM image and element mapping images of LFP@FC-II. (h) EDS image of LFP@FC-II corresponding to the red-cross dot in panel (g). The inset of panel (d) shows the SAED pattern of LFP@FC-II. *Source*: [5] / with permission of Elsevier.

polymer coating, and the final multiwalled hollow structure after carbonization. Line scanning analysis indicated that the distribution of elements (Sn, O, C, etc.) was closely related to structural changes.

Jiang et al. [7] addressed the problems of volume expansion, unstable SEI (solid electrolyte interface), and poor conductivity in silicon (Si)-based anode materials for LIBs caused by alloying/dealloying processes. Based on silicon–carbon composite theory, they designed and prepared a silicon anode material with a double carbon shell structure (DCS-Si), using low-temperature zinc thermal reduction and in-situ multilayer carbon coating technology. Systematic characterization of its structure and performance showed that DCS-Si significantly outperformed traditional materials in terms of cycling stability and rate capability, providing new ideas

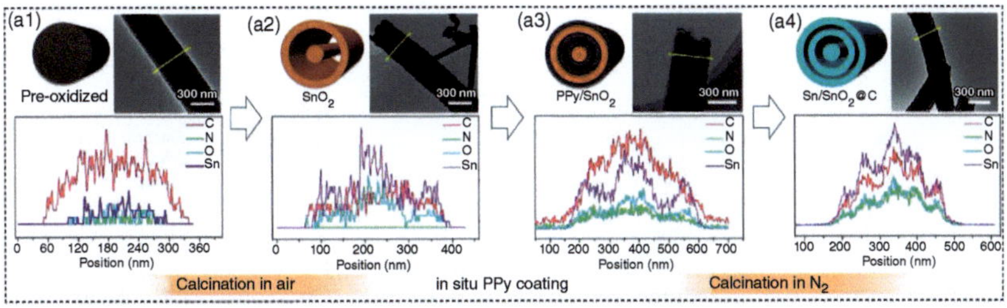

Figure 8.6 The transformation track of WIDWT $Sn/SnO_2@C$ nanofibers monitored by TEM and line scanning profiles. There are three key transitions: (a1) pre-oxidized nanofiber containing tin precursor and undecomposed polymer is calcinated to (a2) wire-in-tube SnO_2 nanofiber. Then the wire-in-tube nanofiber is immersed in pyrrole and Fe^{3+} solution and then (a3) PPy is polymerized in situ on both the outer and inner surfaces of wire-in-tube nanofiber. After the PPy/SnO_2 nanofiber is calcinated in N_2, PPy is carbonized to N-doped carbon. Furthermore, carbon further reduces part of SnO_2 to metal Sn and forms (a4) WIDWT $Sn/SnO_2@C$ nanofiber. *Source*: [6] / with permission of John Wiley & Sons.

and technical support for the development and application of high-performance silicon-based anodes. Through TEM image analysis (Figure 8.7), the microstructure of the material and the characteristics of the carbon coating layer were revealed. Specifically, TEM images showed the particle morphology and carbon coating thickness of Si/C-1 and DCS-Si samples. For Si/C-1, TEM images showed a particle size of 100–200 nm, with a clear carbon coating layer on the surface, about 3–5 nm thick. In contrast, TEM images of the DCS-Si sample indicated a thicker outer carbon coating, further confirming the double-carbon-layer coating structure of DCS-Si. These results directly proved the rationality of the material structure design and provided a microstructural basis for its electrochemical performance.

8.1.3 Atomic Force Microscopy

8.1.3.1 Principle of Atomic Force Microscopy

Atomic force microscope is a micro-area testing instrument that infers sample surface topography, mechanical, and electrical properties by measuring the interaction force between a probe and the sample surface. In 1986, Binning from IBM and Quate from Stanford University collaborated to invent the atomic force microscope [8]. Under ideal conditions, its imaging resolution can reach the atomic level [9]. Regarding the working environment, since the imaging principle does not involve an electron beam, atomic force microscope is one of the few scientific instruments that can perform microscopic imaging without a vacuum environment, making it suitable for in-situ experiments under variable temperature and pressure and in liquid phases. Functionally, atomic force microscope can obtain three-dimensional topography of the sample surface while simultaneously acquiring information on surface conductivity [10], surface potential distribution [11], and mechanical properties [12], demonstrating its multifunctionality.

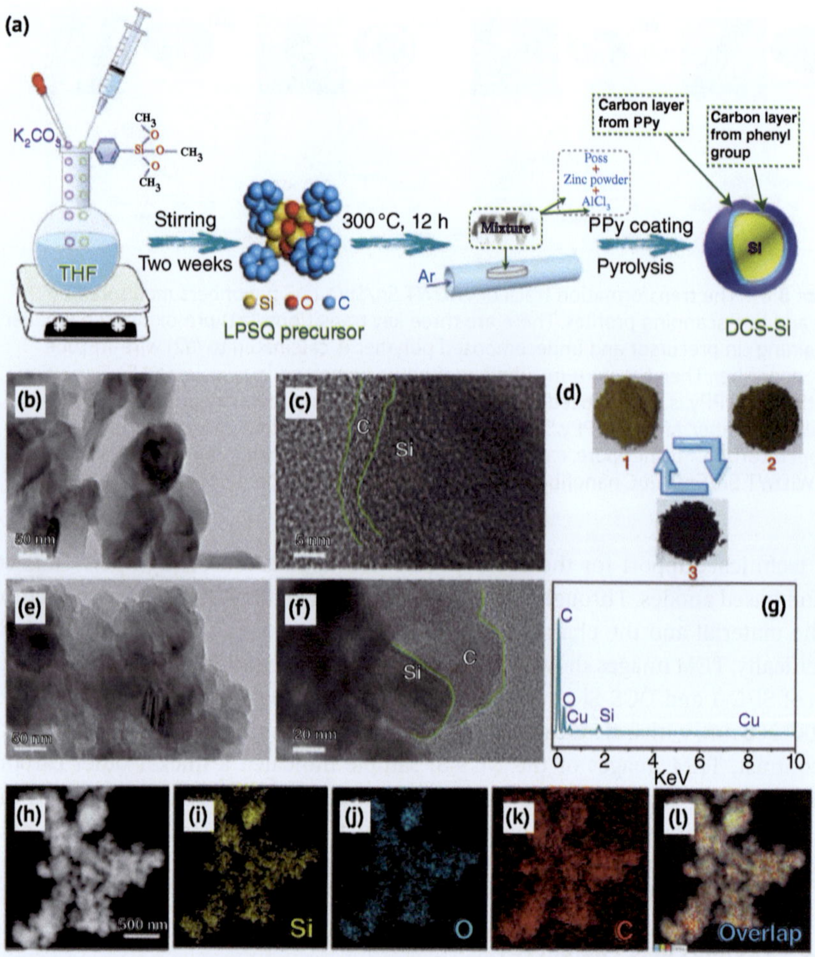

Figure 8.7 (a) Schematic illustration of the preparation steps for DCS-Si; (b) TEM and (c) HRTEM images of Si/C-1; (d) optical graphs of samples (1–3 represent the product after zinc thermal reduction, Si/C-1, and DCS-Si, respectively); (e) TEM and (f) HRTEM images of DCS-Si; (g) EDS spectrum of DCS-Si; and (h–l) the corresponding elemental mapping images of Si, O, and C, respectively. *Source*: [7] / with permission of Elsevier.

Currently, the mainstream imaging mode for atomic force microscopy (AFM) instruments is the constant force mode. Its basic principle involves converting the local attractive or repulsive force between the tip and the sample into a detectable deflection of the micro-cantilever. Then, via a feedback loop in the controller, voltage is applied to the scanner causing it to contract or extend, thereby adjusting the distance or force between the probe and the sample to keep it within a preset range. Finally, by recording the scanner's deformation over time in the *x*-, *y*-, and *z*-directions, combined with the set scanning rate, the three-dimensional surface topography of the sample can be reconstructed. In practice, atomic force microscope uses a fine probe mounted on a micro-cantilever to detect the sample surface. When the probe approaches the sample surface, the interaction force between the probe tip and the sample causes the cantilever to bend. This deflection is recorded by the position of a laser beam reflected from the back of the micro-cantilever [13], using a position-sensitive photodiode detector (PSPD). The detector is typically divided into four quadrants, and signals from these quadrants are combined to calculate the position information of the laser spot. Before testing begins, the operator needs to preset the signal threshold in the system, known as the set-point value in the operating software. During testing, the PSPD signal is fed into the controller in real time, which uses a feedback loop to control the probe to maintain a constant distance or interaction force with the sample surface. Early atomic force microscopes used the constant height mode for imaging. In this mode, the probe height is fixed, and variations in the sample surface height cause changes in the distance between the probe and the sample, leading to variations in the interaction force and thus the deflection of the probe, enabling 3D surface imaging. However, when testing samples with large surface roughness in constant height mode, the large variation in probe–sample distance can cause the interaction force to span both attractive and repulsive regimes, where the force–distance relationship is nonlinear, resulting in poor imaging accuracy. This mode is no longer used today.

AFM operating modes include contact mode, tapping mode, etc. (see Table 8.2). Contact mode operates in the repulsive force regime, where the force–distance relationship is approximately linear, simplifying the feedback loop and facilitating precise force control. Its disadvantage is that both the probe and sample are in repulsive contact, making lateral forces unavoidable. Imaging soft, sticky samples in contact mode may damage the surface. When using contact imaging mode in air, the water film adsorbed on the sample surface leads to capillary forces, resulting in relatively large interaction forces between the probe and sample, potentially damaging the sample. In tapping mode, the probe oscillates near its resonance frequency (tens to hundreds of kHz) and intermittently contacts the sample surface, almost eliminating lateral force effects. When imaging relatively soft surfaces, the interaction force range spans both attractive and repulsive regimes. As the probe approaches the sample surface, the interaction force causes a change in the probe's amplitude, which is used to infer the magnitude of the interaction force. Its drawback is that the relationship between amplitude change and interaction force is nonlinear, making

Table 8.2 Basic operating modes of atomic force microscope.

Feedback physical quantity P	Operating mode
Cantilever deflection D	Contact mode
Cantilever amplitude A	Tapping mode
Probe–sample interaction force F	Peak force tapping mode

the feedback loop more complex and force control less precise. Because vibration dissipation time is long in vacuum, leading to long feedback times and affecting imaging speed, tapping mode is not suitable under vacuum conditions [14]. Operation in liquid phase is somewhat limited because the vibration frequency falls within the sensitive range of human hearing, creating noise during testing.

In addition to the basic AFM operating modes mentioned above, by detecting different physical signals between the probe and sample, functionalized atomic force microscopes can be derived. These include mechanical AFM, conductive AFM (C-AFM), piezoresponse force microscopy (PFM), scanning capacitance microscopy (SCM), scanning spreading resistance microscopy (SSRM), electrostatic force microscopy (EFM), Kelvin probe force microscopy (KPFM), magnetic force microscopy (MFM), and scanning electrochemical microscopy (SECM), detailed in Table 8.3.

8.1.3.2 AFM Application Examples

Cresce et al. [15] used in-situ AFM to observe the SEI formation process on a graphite surface under electrochemical redox conditions in lithium bis(trifluoromethanesulfonyl)imide (LiTFSI)/ethylene carbonate (EC) electrolyte, revealing the bilayer structure characteristics of the SEI. This included an upper

Table 8.3 Types of functionalized AFM.

Detected physical quantity	Functionalized AFM type	Function
Peak force	Mechanical AFM	Force curve/elastic modulus measurement
Tip–sample current	C-AFM	Measure sample conductivity
Nonlinear force (piezoresponse)	PFM	Suitable for materials with piezoelectric effect
Tip–sample capacitance	SCM	Characterize carrier concentration distribution in nonuniformly doped samples
Tip–sample current (spreading resistance)	SSRM	Measure resistance between conductive probe and current collector
Tip–sample electrostatic force	EFM	Measure sample surface electric field distribution
Tip–sample electric field force	KPFM	Measure sample surface potential distribution
Tip–sample magnetic force	MFM	Measure sample surface magnetic force distribution
Tip–sample electrochemical micro-current	SECM	Describe local electrochemical behavior of materials

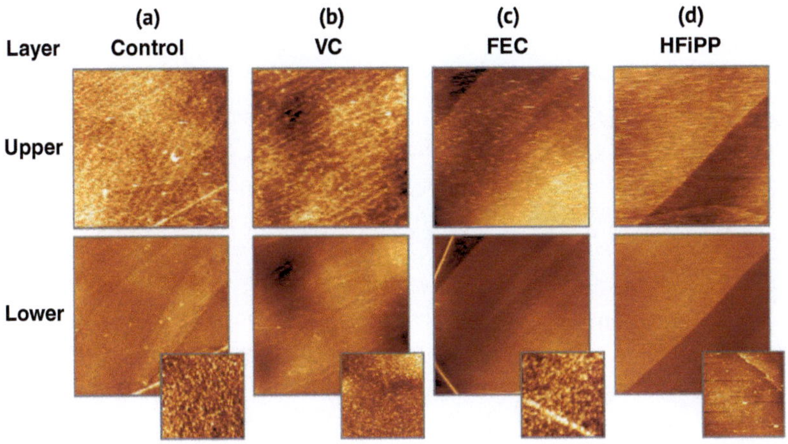

Figure 8.8 Representative AFM images comparing the upper and lower layers of the interphase formed on HOPG in electrolytes with or without additives at 0 V. Top: first image of a given region before removing the upper layer. Bottom: image of the same region as in the top row after scanning multiple times until the lower layer was exposed, and no further topography changes were observed. Insets: magnifications of the lower layers. Image parameters: 5 × 5 μm^2, insets 1 × 1 μm^2, all topography, and $F_c \approx$ 14–18 nN, with excursions up to 30 nN. *Source*: [15] / with permission of the American Chemical Society.

soft/polymeric layer and a lower hard/salt-like layer, along with their distribution and morphology on the graphite surface. AFM imaging results (Figure 8.8) revealed that the upper structure was relatively loose and easily disturbed by the probe, while the lower layer consisted of nanoscale nodular features with stronger adhesion, covering both the basal planes and edges of the graphite. These observations indicated that the SEI is not a single homogeneous film but has a layered, heterogeneous structure, which is important for understanding the function of LIB interfaces and optimizing battery performance.

Xu et al. [16] addressed the performance degradation of LIBs in extremely low-temperature environments by proposing and systematically evaluating a novel electrolyte system based on MDFA/MDFSA–TTE. The work screened soft solvents using density functional theory (DFT) to optimize solvent–lithium-ion binding energy and salt dissociation ability, achieving high ionic conductivity. Experiments showed that NMC811||graphite pouch cells delivered a capacity of 160.2 mAh·g^{-1} at 0.2C rate and −30 °C, maintaining 80% capacity after 360 cycles, with no significant capacity decay even with very little electrolyte. Interface characterization found that this electrolyte promoted the formation of a LiF-rich bilayer SEI and CEI, only about 2 nm thick, effectively suppressing lithium dendrites and side reactions, while maintaining high stability at elevated temperatures. Three-electrode tests showed well-matched cathode and anode impedances without lithium plating at low temperatures. This research significantly improved the low-temperature and high-temperature performances and safety of LIBs, holding great importance for the practical application of power batteries in extreme environments. AFM was

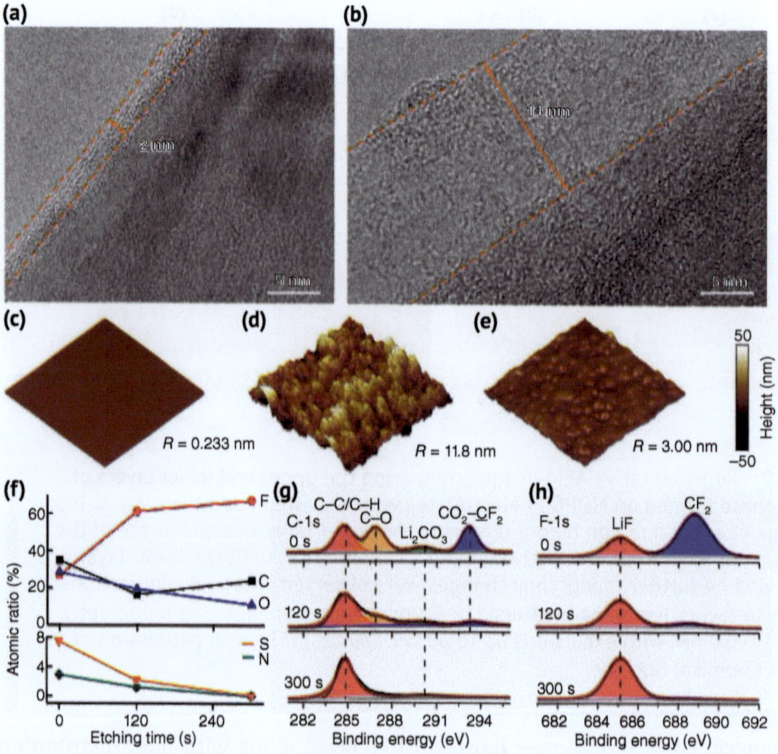

Figure 8.9 Characterization of SEI layers on graphite anode after cycling at −30 °C. HRTEM images with (a) 1 M LiTFSI MDFA/MDFSA–TTE electrolyte and (b) 1 M $LiPF_6$ EC/DEC electrolyte. The orange dashed lines are used to outline the SEI on the cycled graphite anode. Morphologies of the graphite electrode (c) before and (d) after cycling in 1 M $LiPF_6$ EC/DEC electrolyte, and in (e) 1 M LiTFSI MDFA/MDFSA–TTE electrolyte. AFM roughness (*R*) comparison on an area of 0.5 × 0.5 μm^2. XPS results of the SEI on cycled graphite in 1 M LiTFSI MDFA/MDFSA–TTE electrolyte: quantified atomic ratios of the elements in (f) SEI, and (g) C-1s and (h) F-1s spectra displayed in columns of the corresponding depth profiling results. *Source*: [16] / with permission of Springer Nature.

used to analyze the morphology and roughness of the SEI on the graphite anode surface. The results (Figure 8.9) showed that the SEI thickness and roughness on the graphite surface cycled with the MDFA/MDFSA-based electrolyte were significantly lower than those with the traditional EC/DEC system, indicating that the new electrolyte helps form a thinner, more uniform SEI layer, thereby improving battery performance. Specifically, the surface roughness of graphite cycled with MDFA/MDFSA electrolyte was about 3 nm, while it was as high as about 11.8 nm with the EC/DEC system, further validating the advantages of the new electrolyte in interface optimization.

Besides in-situ observation of SEI formation in various electrolyte systems, AFM can also characterize various properties of the SEI film based on its different modes, such as SEI thickness, modulus, viscoelasticity, and conductivity.

Shen and coworkers [17] investigated the effects of two electrolyte additives (LiDFOB and LiBOB) on the morphology, thickness, and mechanical properties of the SEI layer on highly oriented pyrolytic graphite (HOPG) electrodes in LIBs. The study used in-situ AFM, quantitative nanomechanical mapping (QNM), and X-ray photoelectron spectroscopy (XPS) to systematically analyze the SEI formation process and chemical composition. Results showed that the LiDFOB additive promoted the formation of a thin and hard SEI layer dominated by inorganic components (such as LiF and Li_2CO_3) at a higher potential (1.6 V), with an average Young's modulus of 15.14 GPa; whereas LiBOB formed a thick and soft SEI layer dominated by organic species (such as $ROCO_2Li$) at 1.8 V (modulus: 4.73 GPa). Both additives effectively passivated the electrode surface, inhibited solvent decomposition, and improved interface stability. The study presented the morphology and mechanical property changes of the SEI layer formed on the HOPG electrode surface after the first cycle in different electrolyte systems. Specifically, AFM images (Figure 8.10) showed differences in SEI layer thickness and surface roughness among the systems, and scratch tests revealed the mechanical strength and adhesion of the SEI layers. Additionally, the figure included line profile analysis of height changes in the scratched areas, aiding in the comparison of the effects of different additives (like LiDFOB and LiBOB) on SEI structure and performance.

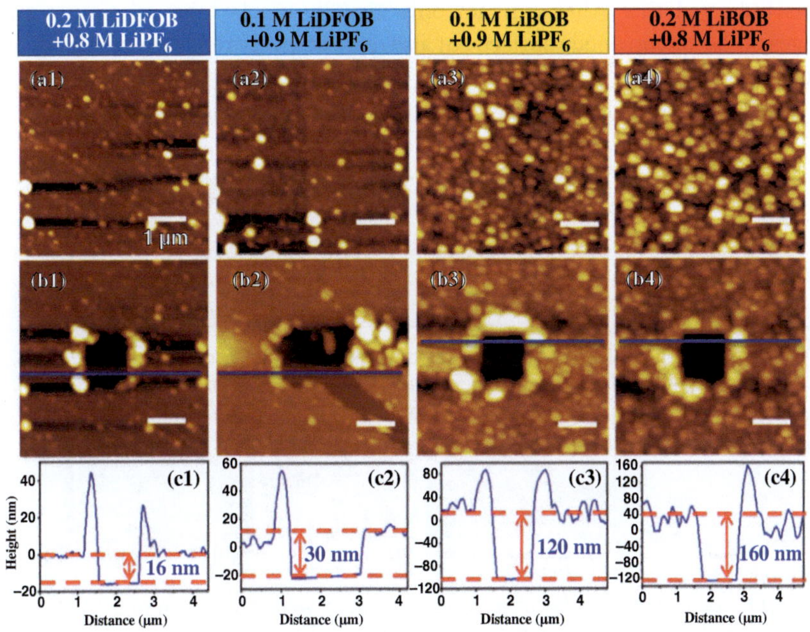

Figure 8.10 (a) AFM images of SEI layer formed on HOPG electrode after first CV cycle in various electrolytes. (b) AFM images of HOPG electrode following scratching of the SEI layer by repeating contact mode scanning. (c) Line profile analysis of the height of the location marked by blue line in image (b). *Source*: [17] / with permission of Elsevier.

Zheng et al. [18] focused on the SEI film formed on silicon anodes in nonaqueous LIBs, exploring the impact of its microstructure and mechanical properties on battery performance. The research team used scanning force spectroscopy to visually demonstrate, for the first time, the three-dimensional multilayer structure and mechanical characteristics of the SEI on silicon anodes, and quantitatively analyzed the SEI coverage via two-dimensional projection. Experiments found that the SEI structure was highly heterogeneous, including monolayers, multilayers, pores, and interlayer structures, with mechanical responses varying with layer number and depth. Temperature and electrolyte additives (like VC) significantly affected SEI thickness and stability. Adding VC effectively improved SEI stability and coverage, promoted its continuous distribution, and enhanced battery cycling performance. Furthermore, the study revealed the dynamic evolution mechanism of the SEI and proposed a formation and decomposition model for the multilayer SEI structure. The analysis focused on the surface morphology changes and roughness evolution of the silicon anode under different electrochemical states. Specifically, AFM imaging (Figure 8.11) revealed the evolution of the silicon anode surface structure during charge/discharge, from initial spherical particles gradually becoming blurred, developing cracks, and becoming smoother in the charged state, changes closely related to SEI deposition. Additionally, surface roughness (R_{RMS} values) measured from AFM images showed that the silicon anode surface roughness first increased and then decreased during charge/discharge. These results indicated that AFM can not only visually display the microstructural changes of the silicon anode surface but also quantitatively reflect the evolution of its surface roughness.

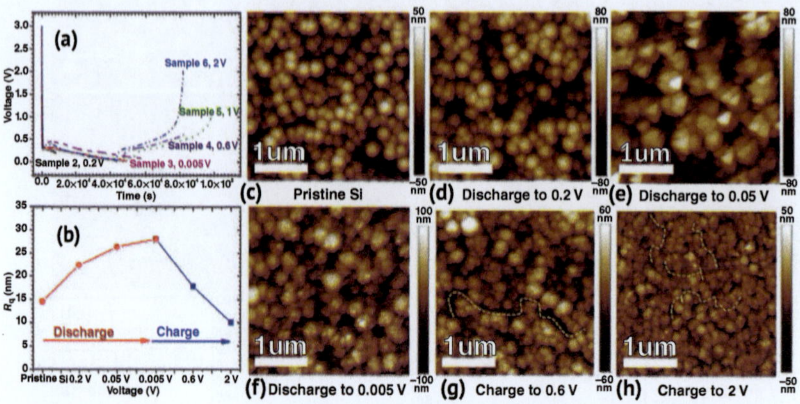

Figure 8.11 (a) Voltage profiles of samples 2–6, which were discharged to 0.2, 0.05, and 0.005 V and charged to 0.6 and 2 V, respectively. (b) The surface roughness values of the above samples. (c–h) AFM topographical images ($3 \times 3\ \mu m^2$) of samples 1–6, where sample 1 was pristine silicon thin film. *Source*: [18] / with permission of The Royal Society of Chemistry.

8.1.4 X-ray Photoelectron Spectroscopy

8.1.4.1 Introduction to XPS

XPS, also known as electron spectroscopy for chemical analysis (ESCA), was developed in the 1960s by Swedish scientist Professor Kai Siegbahn. For his major contributions to the theory and technology of photoelectron spectroscopy, Professor Kai Siegbahn was awarded the Nobel Prize in Physics in 1981. Today, XPS has made significant progress both theoretically and experimentally. It has evolved from initially focusing on qualitative analysis of chemical elements to an important technique for surface elemental qualitative analysis, semiquantitative analysis, and elemental chemical state analysis. The research fields of XPS are no longer limited to traditional chemical analysis but have expanded to the rapidly developing field of materials science. Currently, this analysis method accounts for about 50% of daily surface analysis work, making it one of the most important surface analysis tools.

XPS is based on the photoelectric effect. When a beam of photons irradiates the sample surface, photons can be absorbed by electrons in atomic orbitals of an element in the sample, causing the electrons to break free from the nucleus's binding and be emitted as free photoelectrons with a certain kinetic energy, while the atom itself becomes an excited ion. In conventional XPS instruments, MgK and AlK X-rays are typically used as excitation sources. The photon energy is sufficient to cause photoionization of all elements except hydrogen and helium. Thus, XPS is a method capable of performing a full analysis of all elements at once, which is very effective for qualitative analysis of unknown materials. After X-ray irradiation, the intensity of photoelectrons emitted from the sample surface is linearly related to the concentration of that atom in the sample, allowing for semiquantitative analysis.

However, since the photoelectron intensity depends not only on atomic concentration but also on the photoelectron mean free path, sample surface roughness, the chemical state of the element, X-ray source intensity, and instrument condition, XPS generally cannot provide the absolute content of the analyzed elements, only relative contents. Because the elemental sensitivity factor is related not only to the element type but also to its state in the material and the instrument condition, the relative content measured without calibration can also have significant errors. It should also be noted that XPS is a surface-sensitive analysis method with high surface detection sensitivity, reaching 10^{-3} atomic monolayers, but its bulk detection sensitivity is only about 0.1%. The XPS sampling depth is 2–5 nm; it provides elemental content from the surface, which can differ greatly from the bulk composition. The sampling depth also depends on material properties, photoelectron energy, and the angle between the sample surface and the analyzer.

8.1.4.2 XPS Application Examples

Zhang et al. [19] addressed the research gap in applying $CoMoP_2$ material as an anode material for LIBs. Based on heterostructure and doping engineering theory, they prepared a sandwich composite material of $CoMoP_2$–MoP nanoparticles supported on an N, P co-doped carbon matrix using solution evaporation and high-temperature calcination. Through systematic physical and electrochemical characterization, they revealed the mechanism by which it synergistically enhances lithium-ion and electron transport and interfacial reaction kinetics, significantly improving electrode performance. This provided theoretical and methodological innovation for the design of new high-performance LIB anode materials. From Figure 8.12a–c, the characterization data of XRD patterns, Raman spectra of different samples, N_2 adsorption–desorption isotherms and pore size distribution. XPS testing was used to analyze the surface elemental composition and bonding of the CMP-2@NPC material. Relevant data and spectra are shown in Figure 8.12d–i, including the presence and ratio of elements and their specific chemical states. For example, the XPS survey spectrum (Figure 8.12d) indicated the presence of C, N, P, Co, Mo, and other elements in the material, estimating mass ratios of N to C at 6.97% and P to C at 3.43%. High-resolution spectra further confirmed the co-doping characteristics of N and P in the carbon matrix and revealed the chemical bonding states of various elements, such as C—N, C—P, Co—P, and Mo—P. These analysis results help understand the material's structure and performance, providing a theoretical basis for its application in LIBs.

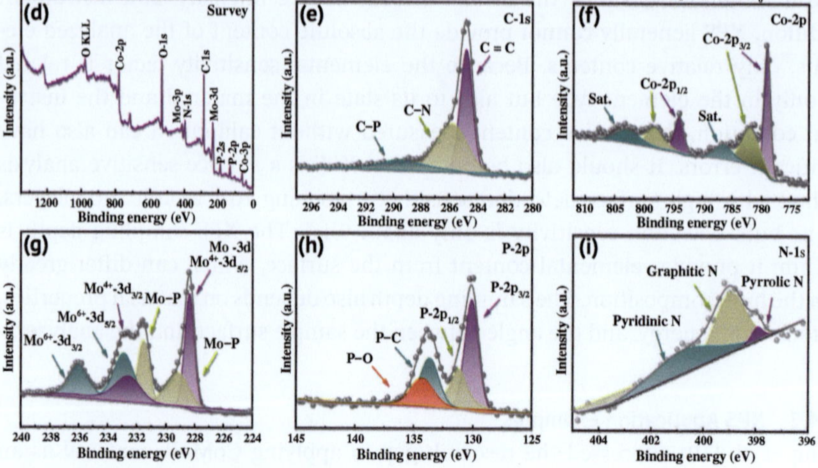

Figure 8.12 (a) XRD patterns and (b) Raman spectra of different samples; (c) N_2 adsorption–desorption isotherms and pore size distribution; (d) XPS survey; and (e–i) high-resolution XPS spectra of CMP-2@NPC. *Source*: [19] / with permission of Springer Nature.

Zhao et al. [20] addressed the inevitable contamination layer on the lithium metal surface and its negative impact on battery performance. They prepared different surface layers (lithium metal, Li_2O, and Li_3N) via in-situ plasma (Ar, H_2, O_2, N_2, and NH_3) treatment, combined with an integrated glovebox–ALD–XPS system and various characterization methods to systematically analyze the effects of each treatment on interface impedance, dendrite growth, and electrochemical performance. They found that the Li_3N protective layer generated by N_2 plasma effectively removed surface contamination, suppressed dendrites, reduced impedance, and significantly improved the cycling stability and capacity retention of lithium metal batteries. This provided a theoretical basis and new methods for interface regulation and surface protection of high-energy-density lithium metal batteries, filling the research gap previously ignoring the initial surface state of lithium and plasma mechanisms. The analysis of the surface "skin layer" was performed using XPS. Specifically, Figure 8.13 shows the characterization of lithium foil using XPS, revealing that the native lithium foil surface contains not only lithium but also a high proportion of carbon (57%) and oxygen (18%), primarily from surface-adsorbed carbon–oxygen compounds and unavoidable contaminants like Li_2CO_3, Li_2O, and LiOH. Detailed deconvolution of core levels such as C-1s, Li-1s, and O-1s determined the specific composition and distribution of these surface contaminants. This complex surface "skin layer" is inevitably formed during the production and transportation of lithium foil. The study also analyzed the effect of O_2 plasma treatment on the thickness and chemical composition of the lithium metal surface layer. According to Figure 8.14a, the lithium surface layer thickness changed with the number of O_2

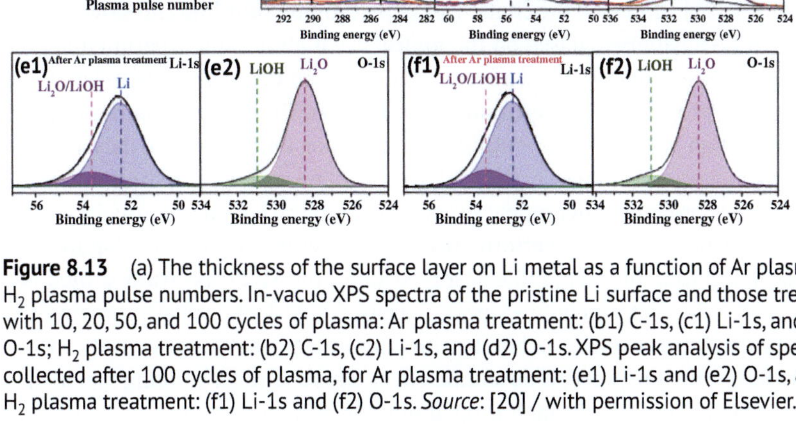

Figure 8.13 (a) The thickness of the surface layer on Li metal as a function of Ar plasma and H_2 plasma pulse numbers. In-vacuo XPS spectra of the pristine Li surface and those treated with 10, 20, 50, and 100 cycles of plasma: Ar plasma treatment: (b1) C-1s, (c1) Li-1s, and (d1) O-1s; H_2 plasma treatment: (b2) C-1s, (c2) Li-1s, and (d2) O-1s. XPS peak analysis of spectra collected after 100 cycles of plasma, for Ar plasma treatment: (e1) Li-1s and (e2) O-1s, and for H_2 plasma treatment: (f1) Li-1s and (f2) O-1s. *Source*: [20] / with permission of Elsevier.

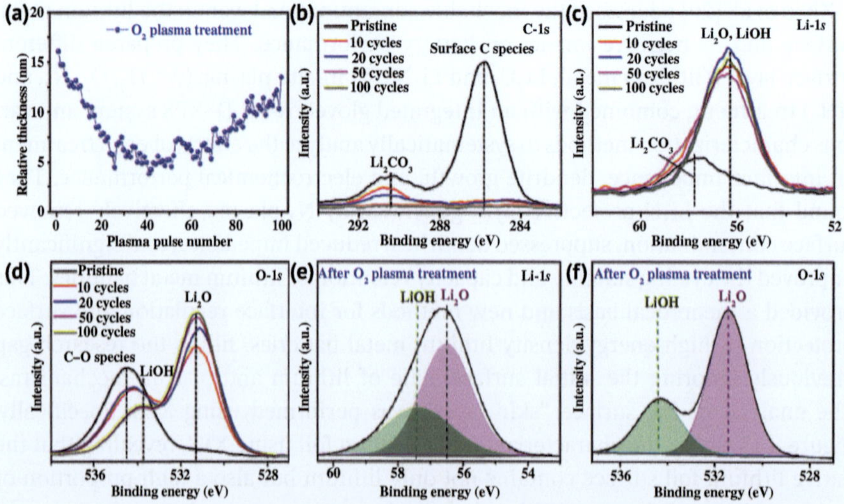

Figure 8.14 (a) The thickness of surface layer on Li metal as a function of O_2 plasma pulse number. In-vacuo XPS spectra of core levels of (b) C-1s, (c) Li-1s, and (d) O-1s of the pristine Li surface and those treated with 10, 20, 50, and 100 cycles of O_2 plasma. XPS peak analysis of (e) Li-1s and (f) O-1s after 100 cycles of O_2 plasma treatment. *Source*: [20] / with permission of Elsevier.

plasma pulses in three stages: the intrinsic skin was effectively removed within the first 40 pulses, the thickness remained stable afterward, and increased again after 60 pulses due to the formation of a lithium oxide layer. XPS analysis results showed that as the number of treatment cycles increased, the C-1s peak intensity dropped almost to zero, and the Li-1s signal strengthened, indicating the removal of surface contamination and exposure of more fresh lithium. The shoulder in the O-1s spectrum indicated the formation of lithium oxides. After 100 O_2 plasma treatment cycles, the surface layer consisted mainly of Li_2O (~61%) and LiOH (~39%), indicating that O_2 plasma can form a protective layer dominated by Li_2O on the lithium metal surface, as shown in Figure 8.14b–f.

Zeng et al. [21] addressed the challenges of polysulfide shuttling and slow reaction kinetics in Li–S batteries. Based on theoretical calculations and experimental analysis, they proposed and validated a separator design centered on a carbon–lepidolite-modified polypropylene (PP) separator (C–lepidolite@PP). Through enhanced Si—S bond adsorption and an ultralow lithium-ion diffusion energy barrier, they achieved high rate, low self-discharge, high areal capacity, and excellent cycling stability in Li–S batteries, providing theoretical support and application breakthroughs for separator optimization in high-performance Li–S systems. XPS results (Figure 8.15) analysis found that after lepidolite adsorbed Li_2S_6, the Si-2p spectrum shifted positively by 0.4 eV, and a new Si—S peak appeared (103.1 eV), clearly proving that lepidolite can form strong Si—S bonds with polysulfides, thereby effectively adsorbing and inhibiting the polysulfide shuttle effect. This chemical bonding mechanism further explains lepidolite's anchoring ability for polysulfides, helping to improve

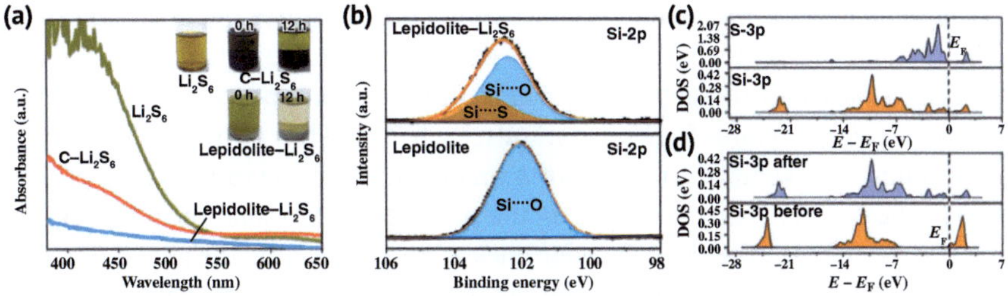

Figure 8.15 LiPS conversion electrochemistry. (a) UV–vis spectra of 0.01 M Li_2S_6 solution containing carbon or lepidolite. The insets are optical images showing the Li_2S_6 solution before and after soaking with carbon or lepidolite for 12 hours. (b) XPS spectra of Si of the lepidolite and the Li_2S_6-treated lepidolite, respectively. (c) The p-bands comparison of S-3p of Li_2S_8 and Si-3p of lepidolite. (d) The p-bands shapes of Si-3p of lepidolite before and after absorbing Li_2S_8. *Source*: [20] / with permission of Elsevier.

the performance of Li–S batteries. Additionally, the combination of theoretical calculations and XPS confirmed that the formation of Si—S bonds weakens the S—S bonds in polysulfides, promoting their rapid conversion.

8.1.5 X-ray Diffraction

8.1.5.1 Principle of X-ray Diffraction Technology

In 1912, Laue et al. [22], based on theoretical prediction and experimental confirmation, demonstrated that X-rays can produce diffraction phenomena when interacting with crystals, proving that X-rays have electromagnetic wave properties. When a monochromatic X-ray beam strikes a crystal, because the distance between these regularly arranged atoms is of the same order of magnitude as the incident X-ray wavelength, they can interfere with each other, producing strong X-ray diffraction (XRD) in certain specific directions. The spatial distribution and intensity of the diffracted rays are related to the crystal structure.

The relationship between the spatial orientation of diffracted rays and the crystal structure can be expressed by the Bragg's equation:

$$2d \sin \theta = n\lambda \tag{8.4}$$

where d is the interplanar spacing, θ is the grazing angle (Bragg angle), n denotes the order of reflection, and λ is the X-ray wavelength.

8.1.5.2 Main Applications of XRD Analysis

Identification of crystal varieties: each crystal has its own characteristic interplanar spacings. Therefore, for XRD with a certain wavelength and using a film of a certain size for photography, each crystal exhibits its own characteristic diffraction lines (powder lines). The relative intensity of the powder lines is also characteristic of the crystal phase.

Distinguishing crystalline from amorphous state: XRD is a characteristic of the crystalline state, requiring a periodic lattice structure to occur. The amorphous state lacks periodicity and thus cannot produce diffraction. On an X-ray film (regardless of the method used), no clear diffraction points or lines are obtained. Therefore, XRD can be used to distinguish between crystalline and amorphous states of matter.

Distinguishing mixtures from compounds: each crystal has its own characteristic powder lines. For example, a mixture of A and B will show the respective lines of A and B in the powder pattern, indicating the presence of two solid phases. If A and B react to form A_mB_n, then new powder lines appear, indicating the generation of a new phase. Based on this principle, one can determine whether the combination of two substances results in a mixture or a compound.

8.1.5.3 XRD Analysis Application Examples

Gangaja et al. [23] addressed the issues of complex processes and insufficient economic viability in the recycling of spent LFP battery materials. Based on the theoretical framework of material regeneration and chemical relithiation/sodiation, they used microwave treatment and chemical methods to achieve efficient reuse of spent LFP material. They systematically analyzed its structure and electrochemical performance, verified its feasibility as a cathode material for both LIBs and sodium-ion batteries, and proposed a simple, scalable closed-loop recycling scheme, providing a new technical approach and theoretical support for the green recycling of spent batteries. XRD was used to characterize the structural changes of the material at different processing stages. First, XRD analysis (Figure 8.16) confirmed

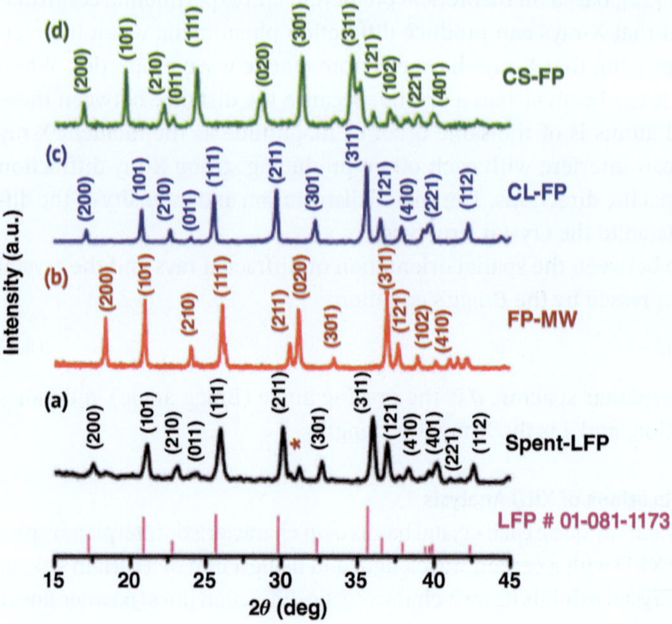

Figure 8.16 XRD patterns of (a) spent LFP, (b) FP-MW, (c) CL-FP, and (d) CS-FP. The standard JCPDS No. 01-081-1173 of the $LiFePO_4$ profile is displayed in the bottommost panel. *Source*: [22] / with permission of the American Chemical Society.

that the recycled spent LFP sample still maintained the olivine (orthorhombic) structure, but partial lithium loss led to the formation of $Li_xFePO_4(x < 1)$ and $FePO_4$ phases (see Figure 8.16a). After microwave treatment, XRD showed that the [200] and [211] peaks shifted to higher angles, proving complete lithium removal and the generation of $FePO_4$, while retaining the orthorhombic crystal structure (see Figure 8.16b). Subsequently, through chemical lithiation and sodiation, XRD verified the formation of $LiFePO_4$ and $NaFePO_4$, respectively. The sodiated product showed a high-intensity [301] peak at 33°, further confirming sodium insertion (see Figure 8.16c and d). These results indicated that XRD effectively tracked the material's structural transformation from spent LFP to $FePO_4$, and then to $LiFePO_4$ and $NaFePO_4$, confirming the effectiveness of chemical treatment and microwave technology in material regeneration.

Liu et al. [24], based on the performance bottlenecks of $LiFePO_4$ electrode material such as low electronic and lithium-ion conductivity, used theoretical analysis and literature review to prepare $LiFe_{1-x}Ni_xPO_4/C$ composite materials with different nickel doping levels via a hydrothermal method. They systematically studied the effect of nickel doping on its structure, microstructure, and electrochemical performance. They found that appropriate nickel doping could optimize the crystal structure, reduce particle size, and improve discharge specific capacity and cycling performance, providing new theoretical and experimental basis for the design of high-performance LIB cathode materials and filling the research gap on the structure–property relationship under nickel doping level control. XRD analysis (Figure 8.17) showed that appropriate Ni doping could enhance the crystallinity of the $LiFe_{1-x}Ni_xPO_4/C$ composite cathode material while maintaining its olivine-type $LiFePO_4$ structure, with no significant impurities detected, indicating high product purity. Furthermore, with increasing Ni doping, the unit cell volume and interlayer

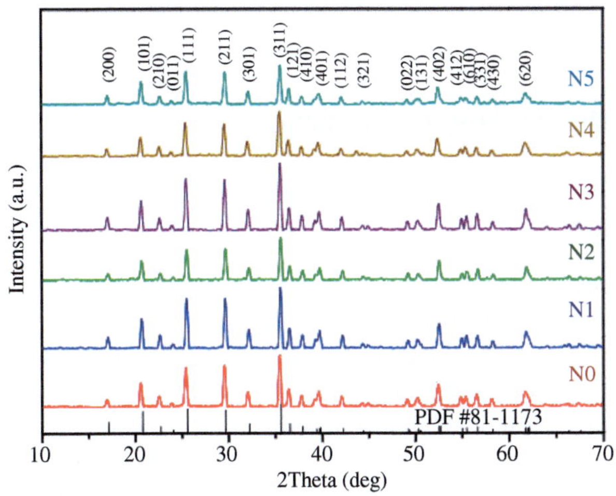

Figure 8.17 XRD powder patterns of the samples from N0 to N5. *Source:* [23] / with permission of Elsevier.

spacing slightly decreased, reflecting the successful substitution of smaller Ni^{2+} ions for Fe^{2+} and their incorporation into the lattice structure. XRD results also confirmed that a small amount of Ni doping helped promote nucleation, reduce grain size, thereby improving electrode–electrolyte contact and enhancing electrochemical performance. In summary, XRD analysis demonstrated the optimizing effect of Ni doping on the material's structure and performance.

Park et al. [25] addressed the shortcomings of existing oxide and sulfide solid-state electrolytes in terms of interface contact, mechanical plasticity, and high-voltage stability. They developed a new hybrid metal chloride solid electrolyte $Li_{3-x}M_{1-x}Zr_xCl_6(M = Er,Y)$. Through systematic structural analysis and Zr substitution experiments, they found that the material undergoes three structural phase transitions, eventually forming a unique structure with additional tetrahedral lithium sites at high Zr content (phase III), significantly enhancing lithium-ion conductivity (up to 1.4 mS cm^{-1}, with activation energy as low as 0.33 eV). Electrochemical tests showed that this chloride electrolyte exhibited excellent oxidation stability and could be directly paired with non-coated 4-V-class cathode materials (like $LiCoO_2$) without significant interface decomposition, achieving high capacity (>110 mAh·g^{-1}), high coulombic efficiency (> 96%), and excellent cycling stability. This achievement broke through the interface and stability bottlenecks in solid-state battery design, providing new ideas and material foundations for the practical application of high-energy-density ASSBs. XRD analysis (Figure 8.18) illustrated the crystal structure evolution of $Li_{3-x}Er_{1-x}Zr_xCl_6$ and $Li_{3-x}Y_{1-x}Zr_xCl_6$ at different Zr doping levels (x values). As the Zr^{4+} doping ratio increased, the material underwent various structural phase transitions, including from the initial trigonal structure (phase I) to a phase similar to Li_3LuCl_6 (phase II), and then to a new solid solution region (phase III), with significant differences in the XRD patterns of different structural stages. Additionally, XRD analysis results indicated that the average ionic radius of the transition-metal ions had an important influence on the crystal structure, and the structural changes were closely related to the improvement in the material's ionic conductivity.

8.2 Basic Introduction to Electrochemical Testing Methods

8.2.1 Cyclic Voltammetry

8.2.1.1 Introduction to Cyclic Voltammetry

Cyclic voltammetry (CV) involves applying a cyclically varying voltage to a closed circuit formed by the working electrode and counter-electrode, changing the potential at the working electrode at a certain rate, forcing the active material on the working electrode to undergo oxidation–reduction reactions, thereby obtaining the magnitude of the response current during electrochemistry. The corresponding response current and voltage are recorded, and then a cyclic voltammogram is plotted with current as the ordinate and voltage as the abscissa. Generally, CV uses

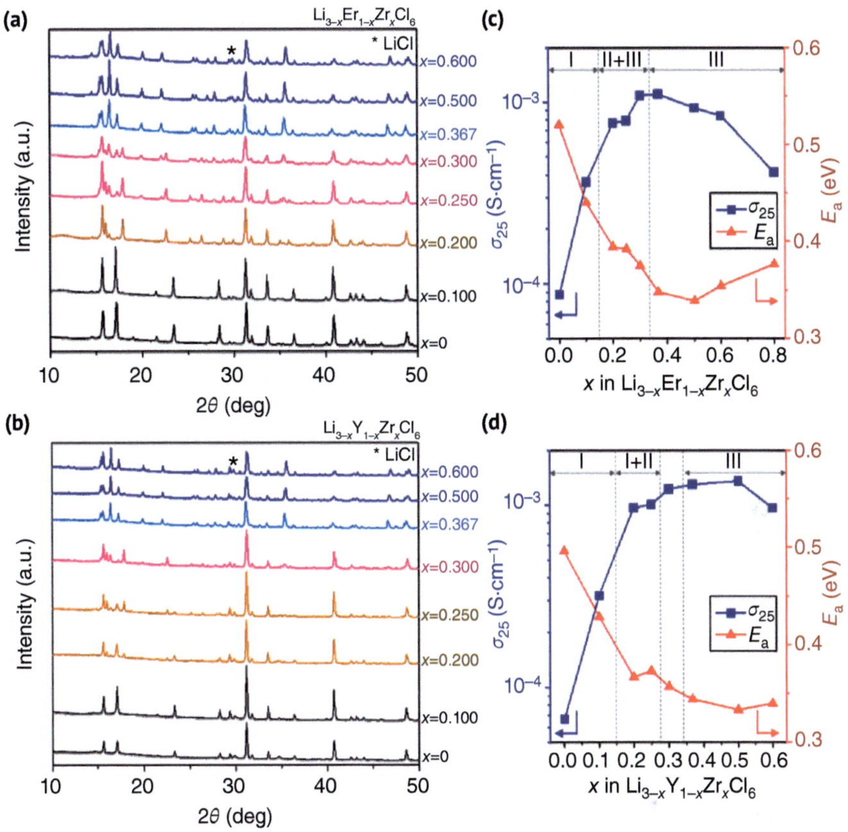

Figure 8.18 Structural evolution upon Zr^{4+} substitution. XRD patterns (CuKα radiation) of (a) $Li_{3-x}Er_{1-x}Zr_xCl_6$ and (b) $Li_{3-x}Y_{1-x}Zr_xCl_6$ ($0 \leq x \leq 0.6$). Ionic conductivities and activation energies of (c) $Li_{3-x}Er_{1-x}Zr_xCl_6$ and (d) $Li_{3-x}Y_{1-x}Zr_xCl_6$. *Source*: [24] / with permission of the American Chemical Society.

either a two-electrode system (working electrode and counter-electrode) or a three-electrode system (working electrode, reference electrode, and counter-electrode).

Processing the data obtained from CV testing yields a CV curve. Typically, one or more reaction peaks appear on the CV curve, generally divided into anodic peaks and cathodic peaks. The anodic peak refers to the first peak appearing in the anodic direction on the CV curve, usually representing an oxidation reaction. The appearance of the anodic peak is due to a chemical reaction at the interface generating electrons, and electrons flowing from the sample to the electrode. The height of the anodic peak is related to the reaction rate and the oxidizing ability of the sample. The cathodic peak refers to the first peak appearing in the cathodic direction on the CV curve. The cathodic peak usually represents a reduction reaction, and its electron flow direction is opposite to that of the anodic peak formation, with electrons flowing from the electrode to the sample. The height of the cathodic peak is also related to the electrochemical properties of the sample. By analyzing the height and potential of the cathodic peak, the reaction mechanism and rate can be inferred. Simultaneously, the potential difference between the anodic and cathodic peaks can be calculated to analyze the electron transfer rate of the reaction. However, some samples may not have both anodic and cathodic peaks; some samples only undergo oxidation or reduction reactions, so only one peak appears on the CV curve. CV has two important experimental parameters: one is the peak current (cathodic peak current denoted as i_{pc} and anodic peak current denoted as i_{pa}), and the other is the peak potential (cathodic peak potential denoted as E_{pc} and anodic peak potential denoted as E_{pa}). Often, the peak potential difference $\left(\Delta E_p = \left[\left|E_{pa} - E_{pc}\right|\right]\right)$ is also used. The magnitude of the peak current and peak potential is influenced by many factors, which is what we need to explore in experiments. Furthermore, the peak current and peak potential of the CV curve can indirectly analyze the rate and kinetics of electrochemical reactions on the electrode and at the interface.

8.2.1.2 CV Application Examples

Yuan et al. [26] addressed the insufficient cycling and rate performance of Li–S battery cathodes caused by the "shuttle effect" of lithium polysulfides (LiPSs) and the low conductivity of sulfur and Li_2S. Based on a composite stable network constructed from natural diatomite (Sep), carbon nanotubes (CNTs), and polyaniline (PANI), they prepared and systematically characterized the structure, composition, and electrochemical performance using chemical oxidation and vacuum heat treatment. They proposed for the first time that PANI coating improves sulfur active material utilization and cycling stability. By regulating the composition, they achieved high initial capacity (1,100 mAh·g^{-1}), excellent cycling (650 mAh·g^{-1} after 300 cycles), and high coulombic efficiency (above 93%), filling the research gap on the poor conductivity and low high-rate efficiency of natural diatomite, and providing new theories and methods for the design of Li–S battery cathode materials. In the study, the authors used CV analysis to illustrate the sulfur conversion mechanism and electrochemical performance of different electrode materials during charge/discharge. CV curves (Figure 8.19) showed that all four composite materials (Sep/CNT/S and

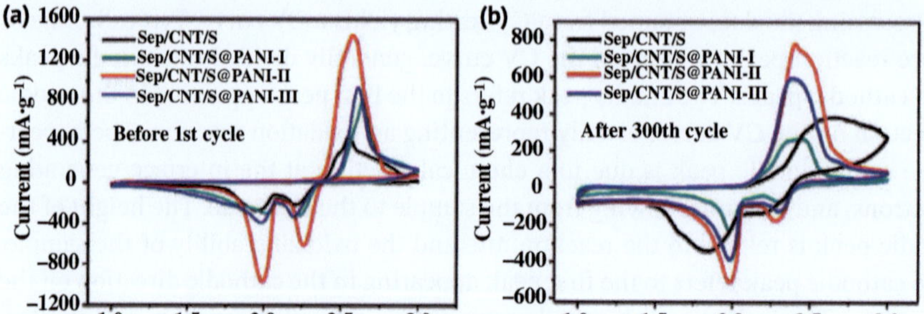

Figure 8.19 The CV curves of the Sep/CNT/S composite and Sep/CNT/S@PANI-I, II, and III composites (a) before the first cycle and (b) after 300 cycles with a scan rate of 0.1 mV s^{-1} in a voltage range between 3 and 1 V vs. Li metal reference electrode at 2C discharge rate. *Source*: [25] / with permission of the Royal Society of Chemistry.

Sep/CNT/S@PANI-I, II, and III) exhibited two reduction peaks (around 2.25 and 2 V) and one oxidation peak (around 2.60 V) in the first cycle, corresponding to the conversion of elemental sulfur to long-chain and short-chain LiPSs, and the reoxidation of polysulfides and lithium sulfide back to elemental sulfur, respectively. Furthermore, PANI-coated composites showed higher peak currents, indicating a larger amount of active material, especially Sep/CNT/S@PANI-II. After 300 cycles, the reduction and oxidation peaks of the PANI-coated materials remained clear, indicating better cycling stability.

Zhu et al. [27] addressed the issue of poor cycling stability of single-phase $LiMnO_2$ due to susceptibility to Jahn–Teller distortion. They proposed an in-situ electrochemical conversion method to transform Mn_3O_4 nanowall arrays into a spinel–layered $LiMnO_2$ (SPL-LMO) structure with self-assembled heterointerfaces. They detailed the mechanism of spinel-to-layered phase transformation driven by Mn^{2+} extraction and oxygen release, and the suppression of Jahn–Teller distortion by interfacial orbital ordering. Electrochemical test results showed that the SPL-LMO electrode exhibited high capacity (254.3 $mAh{\cdot}g^{-1}$) and excellent cycling stability (90.4% capacity retention after 2,000 cycles). This study revealed the important role of heterostructure and interface engineering in mitigating Jahn–Teller distortion and improving the performance of manganese-based cathode materials, providing new ideas for developing high-performance, low-cost, and sustainable manganese-based battery cathode materials. CV curves (Figure 8.20) revealed the redox processes of the SPL-LMO electrode, characterized by a pair of clear redox peaks at approximately 3.1 and 2.8 V (vs. Li/Li^+), corresponding to the cubic-to-tetragonal phase transition of the spinel structure from $LiMn_2O_4$ to $Li_2Mn_2O_4$. Additionally, two pairs of less distinct redox peaks appeared between 4 and 4.3 V, attributed to the insertion and extraction of lithium ions in the 8a tetrahedral sites of cubic spinel $LiMn_2O_4$; while small peaks detected at 3.9 and 3.8 V corresponded to lithium-ion insertion/extraction in the octahedral sites of monoclinic layered $LiMnO_2$, respectively. These features indicate the existence of multiple phase transitions and lithium-ion storage

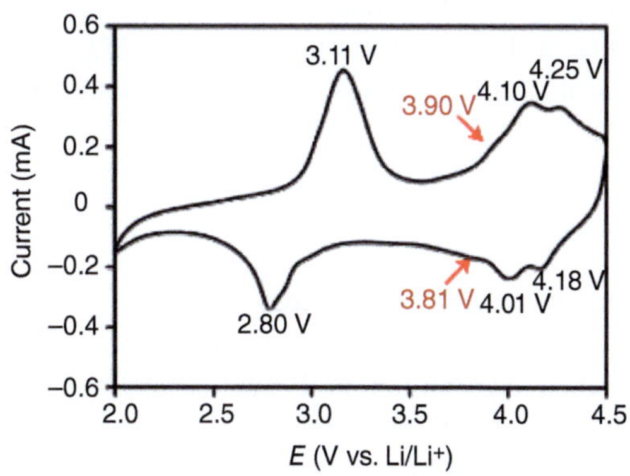

Figure 8.20 The typical CV curve of the SPL-LMO cathode at a scan rate of 0.1 $mV\,s^{-1}$. *Source*: [26] / with permission of Springer Nature.

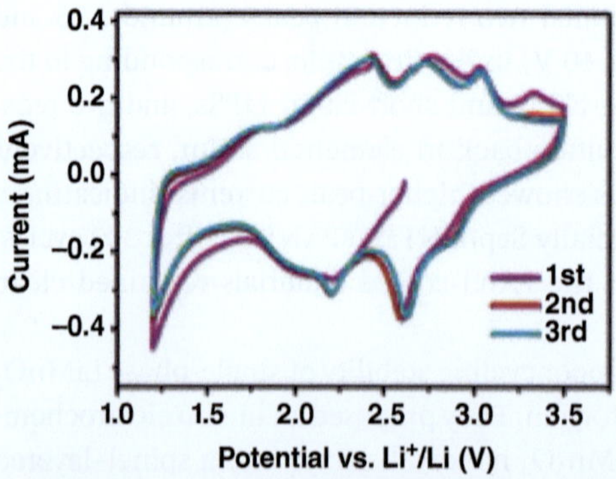

Figure 8.21 The initial three-cycle CV curves of BQ1-COF at a scan rate of 0.5 mV s^{-1}. *Source*: [27] / with permission of Elsevier.

mechanisms in different voltage ranges, further elucidating the electrochemical reaction process and structural evolution of the SPL-LMO composite material.

Wu et al. [28] designed and synthesized a two-dimensional covalent organic framework (COF) material with multiple redox sites (BQ1-COF), obtaining a high density of C=O and Π-conjugated C=N active groups through a simple condensation reaction. The material structure was confirmed by various characterization methods, exhibiting a stable two-dimensional honeycomb framework and good microporous channels. Electrochemical tests showed that BQ1-COF, as a cathode for LIBs, exhibited ultrahigh reversible capacity (502.4 mAh·g^{-1} at 0.05C), excellent capacity retention (still 81% after 1,000 cycles), and outstanding rate performance (170.7 mAh·g^{-1} at 10C), far exceeding existing COFs and polymer cathode materials. Its high energy density (up to 1,033 Wh kg^{-1}) and excellent cycling stability are attributed to the abundant redox groups and stable framework structure. CV test results (Figure 8.21) showed that the BQ1-COF electrode exhibited three distinct reduction peaks in the initial three cycles, located at 2.64, 2.20, and 1.2 V (vs. Li^+/Li), respectively, indicating multi-electron redox reactions during charge/discharge, reflecting the high efficiency and multiple reaction sites of the material. Furthermore, the good reproducibility of the CV curves indicated that BQ1-COF has a highly reversible electrochemical reaction process, making it suitable as a high-performance organic cathode material for LIBs.

8.2.2 Electrochemical Impedance Spectroscopy

8.2.2.1 Introduction to Electrochemical Impedance Spectroscopy

Electrochemical impedance spectroscopy (EIS) technology enables precise characterization of the internal impedance distribution and state-of-health (S_{OH}) assessment of lithium batteries through broad-frequency AC impedance testing. During implementation, the battery's steady state is first determined, typically tested at open-circuit potential or a specific state of charge (S_{OC}). The frequency scan range is set from 10^{-2} to 10^5 Hz, covering the characteristic frequencies of various electrochemical processes inside the battery. Then, the AC signal amplitude is set to 5–10 mV, ensuring the system is in the linear response region and avoiding disturbance of the battery's normal state. During testing, scanning is performed point by point from high frequency to low frequency. At each frequency point, a sinusoidal AC voltage signal is applied, and the corresponding current response is measured. The impedance value at each frequency point is obtained through Fourier transform processing, and the impedance data is expressed in complex form, as shown in Equation (8.5):

$$Z(\omega) = Z'(\omega) + jZ''(\omega) \tag{8.5}$$

where $Z(\omega)$ is the complex impedance (Ω), ω is the angular frequency (rad s^{-1}), $Z'(\omega)$ is the real part of impedance (resistive component) (Ω), j is the imaginary unit (dimensionless), and $Z''(\omega)$ is the imaginary part of impedance (reactive component) (Ω).

8.2.2.2 EIS Application Examples

Sun et al. [29] addressed the issues of capacity decay, low coulombic efficiency, and safety concerns in the practical application of Li–S batteries. They proposed using natural polymetallic oxide nanorods—attapulgite (ATP) compounded with conductive acetylene black (AB) to modify the separator. Through FTIR, XPS, and other characterizations, they confirmed that ATP has abundant surface polar sites and high porosity, which are conducive to strong adsorption of intermediate polysulfides (LiPSs), inhibiting the "shuttle effect." Electrochemical tests showed that Li–S batteries using the Attp40 separator exhibited excellent specific capacity and cycling stability at different rates, with an initial capacity of 1,215 mAh·g^{-1} and 792.5 mAh·g^{-1} after 200 cycles. EIS results (Figure 8.22) showed that before cycling, the impedance plots (Nyquist plots) of cells with various separators all exhibited a semicircle in the high-frequency region and a slanted line in the

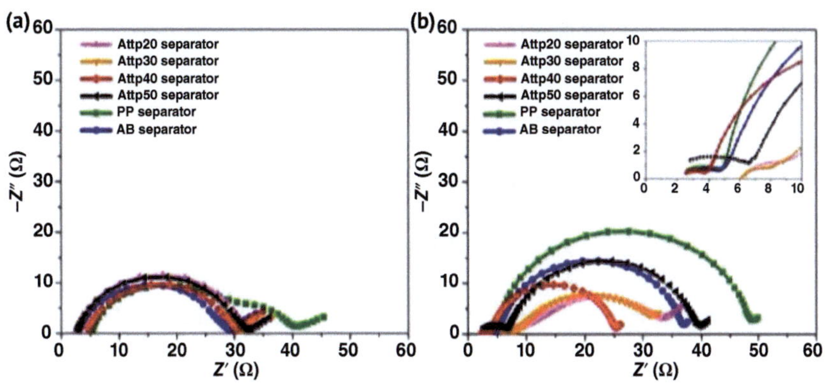

Figure 8.22 EIS plots of Li–S batteries with different separators (a) before and (b) after cycling. *Source*: [29] / with permission of Elsevier.

low-frequency region, corresponding to the charge-transfer resistance (R_{ct}) and the Warburg impedance (Z_W) related to lithium-ion diffusion, respectively. After cycling, a new semicircle appeared in the impedance plot, representing the deposition resistance (R_d) of Li_2S/Li_2S_2, reflecting the impact of deposition products on battery performance. Specifically, the AB separator had the smallest R_{ct} before cycling due to its better electronic conductivity, while R_{ct} increased with higher ATP content due to its nonconductive nature. After cycling, the Attp40 separator had a smaller R_d compared to other separators, indicating that ATP can effectively inhibit the polysulfide "shuttle effect" through chemical bonding, and AB helps activate the deposition products, improving battery cycling stability.

Li et al. [30] focused on graphene–sulfur composite cathode materials doped with different proportions of fly ash. Based on its unique quartz and mullite structure, they used it to adsorb polysulfide ions to effectively inhibit the shuttle effect in Li–S batteries, reduce polarization and impedance, and improve cycling performance. Using material characterization (FTIR, XRD, and SEM), electrochemical testing (CV, EIS, and charge/discharge), and quantum mechanical calculations, they filled the research gap on the weak interaction between nonpolar carbon materials and polysulfides, revealed the theoretical mechanism of fly ash ion adsorption and intermolecular interactions, and clarified the impact of fly ash doping amount on conductivity and specific capacity, providing new low-cost material ideas and theoretical basis for achieving high efficiency and stable performance in Li–S batteries. Through EIS testing, they analyzed the electrochemical performance and reaction kinetics of different composite cathode materials (G–S, G–5% F–S, and G–10% F–S) in Li–S batteries. Results (Figure 8.23) showed that the addition of fly ash effectively

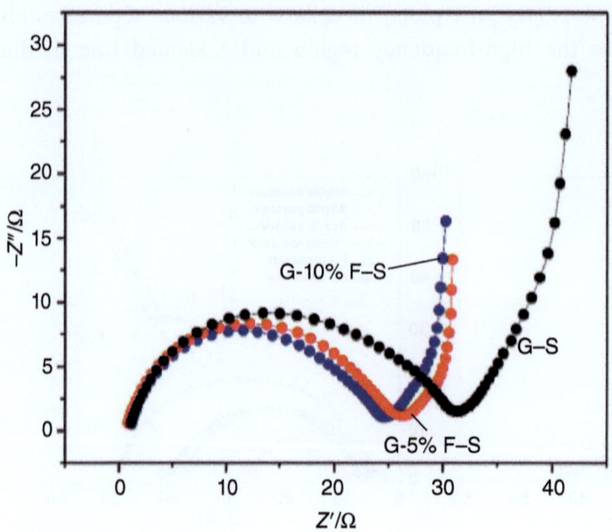

Figure 8.23 Impedance diagram of G–S, G–5% F–S, and G–10% F–S. *Source:* [29] / with permission of Elsevier.

reduced the charge-transfer resistance (R_{ct}) of the battery. G–10% F–S had the lowest impedance value (about 25 Ω), while G–S had the highest impedance (about 32 Ω), indicating that fly ash adsorption of polysulfides can promote lithium-ion diffusion, improving the charge/discharge efficiency and rate performance of the battery. Furthermore, the introduction of fly ash may also increase the porosity of the graphene–sulfur composite cathode surface, thereby improving electrolyte–electrode contact and further enhancing the utilization of active materials and battery performance.

Elibol et al. [31] addressed the research gaps in sulfur host materials for Li–S batteries, such as low electrical conductivity and difficulty in suppressing the LiPSs shuttle effect. They proposed using natural halloysite tubes as sulfur hosts and uniformly depositing NiO nanoparticles to enhance the adsorption capacity for LiPSs and electrochemical performance. Based on the theoretical framework of synergistic chemical adsorption by metal oxides and physical confinement, they synthesized NiO@halloysite composite materials via a hydrothermal method and assembled batteries for multiple electrochemical tests. Results showed that this composite cathode had higher discharge specific capacity and excellent cycling stability, indicating the potential of natural halloysite for low-cost, high-efficiency applications in energy storage systems like Li–S batteries. EIS was used to analyze the interfacial electrochemical characteristics of different cathode materials (S/etched-halloysite-550 and S/NiO@halloysite) in the Li–S battery system. EIS results (Figure 8.24) showed that the modified S/NiO@halloysite cathode surface was more uniform, with higher ionic permeability and lower impedance, indicating superior charge-transfer rate and better electrochemical reaction kinetics. Additionally, EIS test results also reflected the correlation between the overall system structure (including cathode, lithium foil anode, separator, etc.) and performance, helping to reveal the mechanism by which material modification enhances the overall battery performance.

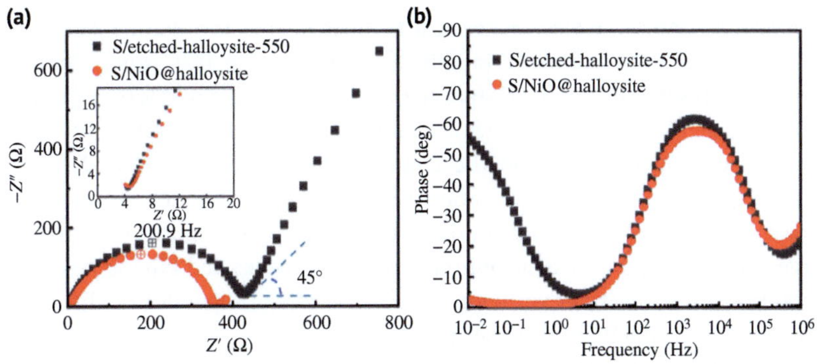

Figure 8.24 EIS data was recorded for Li–S half-cells based on S/etched-halloysite-550 and S/NiO@halloysite cathodes. (a) Complex plane plot and inset for enlarged spectra at the high-frequency range and (b) phase angle vs. frequency plots. *Source*: [30] / John Wiley & Sons / CC BY 4.0.

8.2.3 Galvanostatic Charge–Discharge

8.2.3.1 Introduction to Galvanostatic Charge–Discharge

Galvanostatic charge–discharge (GCD) is a testing method for evaluating the cycling stability performance of secondary batteries like LIBs. The process consists of two consecutive stages: first, during the constant current (CC) stage, a CC is applied for rapid charging, during which the battery voltage continuously rises until it reaches a preset upper cutoff voltage; then it automatically switches to the constant voltage (CV) stage, maintaining the voltage at this upper limit, while the charging current decays exponentially. Charging terminates when the current drops to a very small value, ensuring the battery is safely fully charged and preventing overcharging. A complete test typically combines the subsequent CC discharge process: after charging and a rest period, discharge with a CC to the lower cutoff voltage, constituting one full cycle. Analyzing the voltage–capacity curve of this cycle is key to evaluating the battery's actual capacity, coulombic efficiency, voltage plateau stability, and the polarization degree between charge and discharge curves; long-term repetition of this cycle is the core method for assessing battery cycle life and capacity fade. This test comprehensively reflects the battery's performance, safety, and durability by simulating scientific charging strategies.

8.2.3.2 GCD Application Examples

Cheng et al. [32] addressed issues such as the limited long-term cycling stability of Li–S battery separators. Based on the theoretical framework of activated biomass carbon, they prepared porous activated carbon (AC) material (AHC) using hazelnut shells as a precursor and a KOH high-temperature activation method, systematically characterizing its structure and performance. They ultimately confirmed that this material has a high specific surface area and a rich pore structure, which can effectively adsorb polysulfides, enhance ion migration and electrochemical performance, thereby significantly improving the cycling stability and capacity retention of Li–S batteries, filling the research gap in structural controllability and long-term performance of traditional separators and biomass carbon materials. GCD curves illustrated the electrochemical performance of the routine separator, HC-modified separator, and AHC-modified separator at a 0.2C rate. Charge–discharge curve results (Figure 8.25) showed that the modified separators (especially the AHC-modified separator) significantly improved the battery's initial discharge capacity and cycling stability, and the charge–discharge curves showed higher coincidence, proving their ability to effectively improve the cycling performance and electrochemical reaction rate of Li–S batteries. Furthermore, the charge–discharge curves reflected the distribution of oxidation peaks and kinetic processes of different separators, further verifying the positive role of separator modification in enhancing the overall battery performance.

Kuang et al. [33] addressed the limited cycling performance and capacity fade of $Li_4Ti_5O_{12}$ (LTO) anode material for LIBs. They proposed compounding Si and AC materials to construct a multilayer core–shell structure to enhance its electrochemical performance. The authors prepared modified samples such as

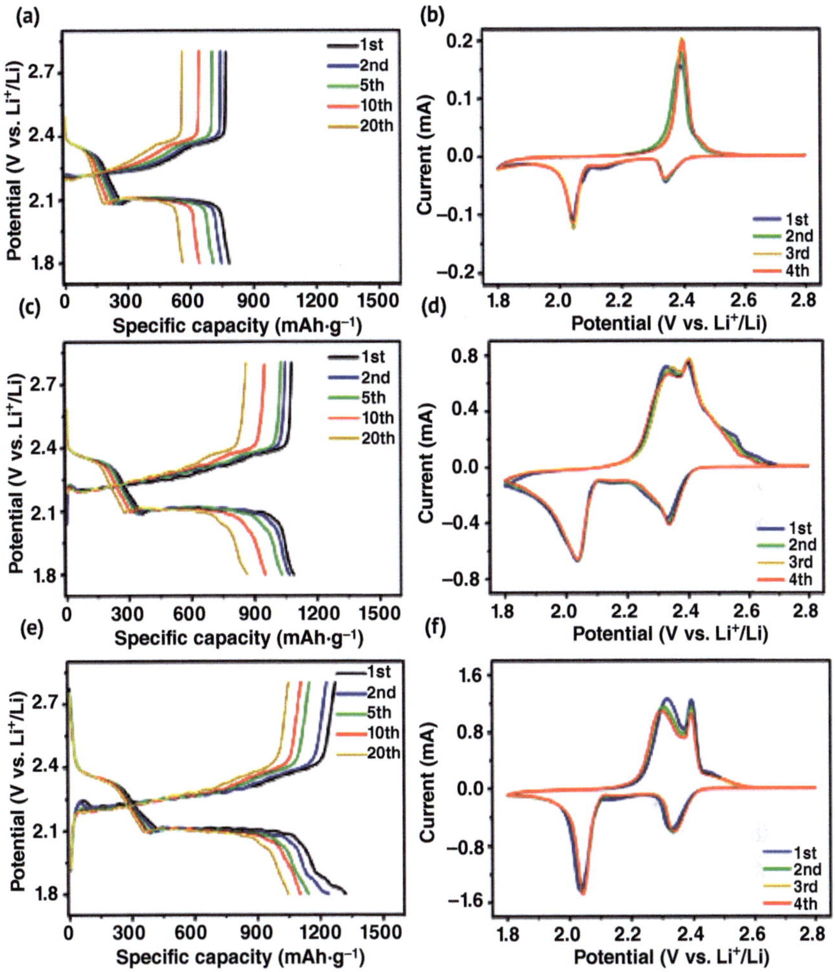

Figure 8.25 Charge–discharge diagrams of the (a) routine diaphragm, (c) HC-modified diaphragm, and (e) AHC-modified diaphragm at 0.2C and the CV curves of (b) routine diaphragm, (d) HC-modified diaphragm, and (f) AHC-modified diaphragm. *Source*: [31] / with permission of Elsevier.

LAC (LTO + AC), LSi (LTO + Si), and LSiAC (LTO + Si + AC) using solution mixing and high-temperature calcination processes, and conducted in-depth analysis of the material's structure and surface changes through various physical characterization methods including XRD, XPS, SEM, EDS, and HRTEM. Results showed that the LSiAC sample not only achieved substitution of part of Ti^{4+} by Si, forming Si—O bonds, but also effectively suppressed excessive SEI film growth and structural degradation due to the AC coating. In cycling tests, LSiAC maintained 89.37% capacity stability after 1,000 cycles at a high rate (5C), and the Li^+ diffusion coefficient was greatly improved, indicating excellent cycling life and kinetic performance of the material. This research provided important theoretical and practical references for

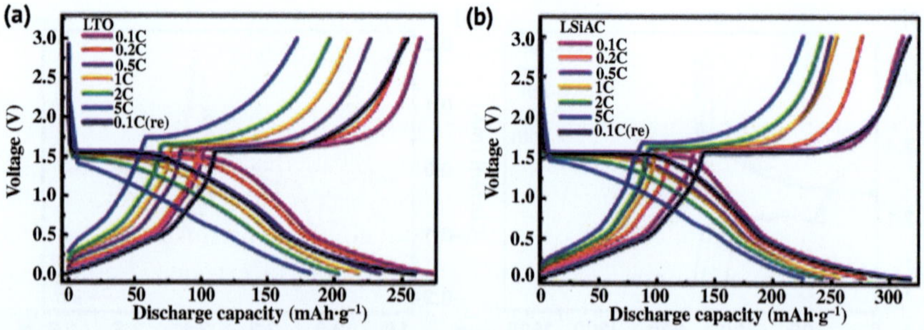

Figure 8.26 Charge–discharge curves at different rates of (a) bare LTO and (b) LSiAC. *Source*: [32] / with permission of Springer Nature.

the design and optimization of high-performance LIB anode materials, with broad application prospects. Charge–discharge curves (Figure 8.26) mainly displayed the electrochemical performance of bare LTO samples and modified LSiAC samples at different rates (from 0.1 to 5C). As the current increased, the polarization of the samples increased, and the capacity decreased somewhat, but the charge–discharge curves remained largely intact, indicating good rate capability and reversibility of the material. Furthermore, the specific capacity of the modified LSiAC sample was higher than that of the bare LTO sample at all rates, reflecting the significant capacity improvement and excellent cycling stability achieved through AC coating and silicon doping.

8.2.4 Rate Performance Test

8.2.4.1 Introduction to Rate Performance Test

Electrochemical rate testing is a systematic method to evaluate the performance evolution of batteries under different current intensities. Its core principle is based on the "C-rate" concept (a 1C current can complete charge or discharge in one hour) and the polarization effects in batteries under high currents (ohmic resistance, electrochemical polarization, and concentration polarization). These effects cause a decrease in the discharge voltage plateau and an increase in the charge voltage plateau, thereby reducing the actual releasable capacity. Testing follows a rigorous procedure: first, perform initial cycles at a low rate (e.g., 0.1C) to activate the electrode and accurately determine the rated capacity; then adopt a strategy of fixed charge rate and stepwise increasing discharge rates (e.g., from 0.2 to 5C). Each cycle includes CC–CV charging to the upper voltage limit, a rest period for equilibrium, and CC discharge to the cutoff voltage, completely recording capacity and voltage–time curves. The test often ends with low-rate recovery cycles to check the reversibility of the battery structure. Key data analysis includes two aspects: one is plotting the capacity retention rate vs. rate relationship to evaluate the capacity decay trend at high rates; the other is comparing voltage–capacity curves at different rates, quantifying the degree of polarization through the decrease in voltage plateau and changes in curve slope, thus comprehensively revealing the power output limit, reaction kinetics, and material structural stability of the battery, which is crucial for evaluating the comprehensive performance of power-type batteries.

8.2.4.2 Rate Performance Test Application Examples

Wu et al. [34] addressed the capacity fade in Li–S batteries caused by polysulfide dissolution and the shuttle effect. They proposed a new strategy of directly using natural expanded vermiculite as the sulfur cathode host. The study first analyzed the layered structure and surface cation characteristics of expanded vermiculite, finding that it can strongly adsorb polysulfide anions (S_n^{2-}), effectively inhibiting polysulfide dissolution. Vermiculite–sulfur composites with high sulfur content (60% and 80%) were prepared via a solution impregnation method. Combined with electrochemical performance tests, UV–vis absorption, ζ-potential measurements, and first-principles calculations, the adsorption mechanism between vermiculite surface cations and polysulfide anions and the electric double-layer effect formed by charge compensation were systematically elucidated. Experimental results showed that the vermiculite–sulfur composite cathode exhibited excellent rate performance and cycling stability (low decay rate of 0.04% per cycle after 1,000 cycles) at various rates, significantly outperforming conventional carbon-based hosts. This method utilizes cheap, abundant natural mineral materials, providing new ideas for high-performance, low-cost Li–S batteries and deepening the understanding of the interaction mechanism between natural layered minerals and polysulfides. Rate performance test results (Figure 8.27) showed that the expanded vermiculite–sulfur composite (V-S-80) exhibited excellent capacity and retention at different rates (C/10–2C). For example, its capacities at C/10, C/5, C/2, 1C, and 2C were approximately 1,185, 1,043, 920, 753, and 580 mAh·g^{-1}, respectively, and the capacity remained good when returning to low rates, indicating fast electrochemical reaction kinetics and stable structure. These data indicate that the material can not only charge and discharge rapidly at high rates but also effectively inhibit polysulfide dissolution and the shuttle effect, improving the long-term cycling stability

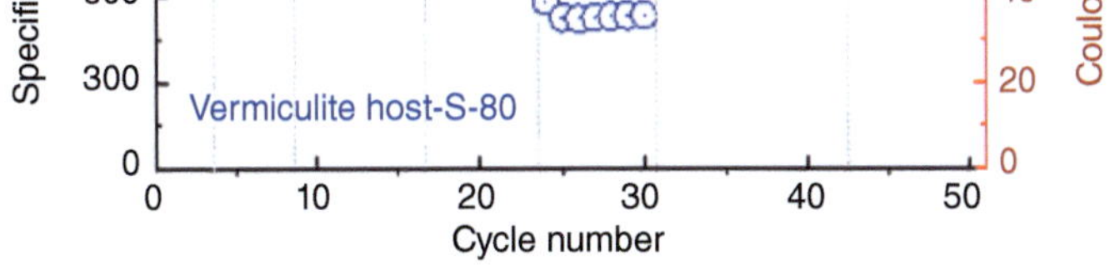

Figure 8.27 Discharge capacities at different C-rates. *Source*: [33] / with permission of John Wiley & Sons.

of the battery. Overall, expanded vermiculite as a sulfur host shows strong rate performance and cycling stability in Li–S batteries, with the potential to compete with traditional carbon-based composite materials.

Wang et al. [35] addressed the poor cycling stability of Li—S batteries caused by the polysulfide shuttle effect. They proposed a method of co-coating the Celgard separator with natural clay mineral illite/smectite (ISC) and carbon black (C) (ISC/C@ Celgard). The separator prepared by a simple blade-coating process exhibited good mechanical stability and super-electrolyte-philicity. The abundant —OH groups on ISC can rapidly adsorb polysulfides via Li—O and Li—S bonds. Combined with the conductivity provided by carbon black and the uniform nanoporous structure, it can physically block and chemically adsorb polysulfides, effectively inhibiting the shuttle effect, and promoting the reactivation of adsorbed polysulfides, improving sulfur utilization. Experimental results showed that this separator endowed Li–S batteries with extremely high initial reversible capacity (1,322 $mAh \cdot g^{-1}$ at 0.1C), excellent cycling stability (capacity decay rate only 0.054% per cycle after 500 cycles), and high areal capacity (5.9 $mAh \cdot cm^{-2}$ with sulfur loading of 8.9 mg cm^{-2}). The adsorption kinetics of polysulfides on ISC followed a pseudo-second-order model, superior to traditional materials. This work provided new ideas and technical paths for developing high-performance, low-cost, green Li–S battery separators using natural clay minerals. Experimental results (Figure 8.28) showed that batteries using the ISC/C@Celgard separator exhibited higher discharge capacity and more stable voltage plateaus at different rates (e.g., 0.1, 0.2, 0.5, 1, and 2C), especially at high rates, where its capacity significantly outperformed other separators like Celgard, C@Celgard, and CNT@Celgard. This indicates that the ISC/C@Celgard separator can effectively improve the reaction kinetics and rate performance of the battery. Therefore, the rate performance demonstrates that the ISC/C@Celgard separator can enhance the stability and practicality of Li–S batteries under high-power output conditions.

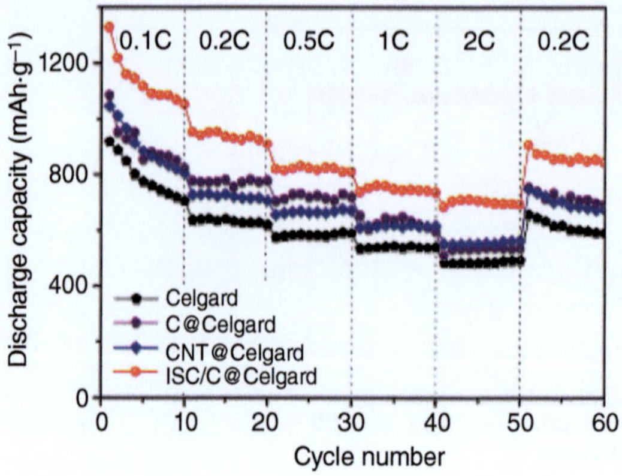

Figure 8.28 Charge/discharge profiles at 0.1C for measuring rate performance. *Source*: [34] / with permission of Elsevier.

8.2.5 Tafel Test

8.2.5.1 Introduction to Tafel Test

The electrochemical Tafel test is a classical method for quantitatively analyzing electrode reaction kinetics by measuring the steady-state polarization curves. Its core basis is the Tafel equation:

$$\eta = a + b\log(i) \tag{8.6}$$

where η is the overpotential, the difference between the actual electrode potential and the equilibrium potential $(\eta = E - E_{eq})$.

This equation describes the linear relationship between the overpotential (η) and the logarithm of the current density ($\log(i)$) under strong polarization conditions. During testing, the open-circuit potential of the electrode is first measured as a baseline. Then, a very slow potential scan rate is used to obtain the steady-state current-potential data. The key data processing step is converting the curve into a Tafel plot of overpotential vs. the logarithm of current density, and fitting the linear parts of the anodic and cathodic branches separately. Two core parameters can be extracted from the fitted lines: the Tafel slope is used to reveal the rate-determining step and mechanism (e.g., deducing the transfer coefficient and number of electrons transferred from the slope value), and the exchange current density is obtained by extrapolating the line to zero overpotential. This parameter directly measures the ease of the electrode reaction and the catalytic activity. This test is mainly used to evaluate the performance of electrocatalysts, estimate the corrosion rate of materials, and study the kinetic characteristics of electrode reactions in batteries. However, during testing, it is essential to ensure that steady state is reached and that effective solution I_R drop compensation is performed to guarantee data accuracy.

8.2.5.2 Tafel Test Application Examples

Zhang et al. [36] addressed the issue of delithiated Li_xCoO_2 cathode material from spent LIBs. They proposed and validated an electrochemical relithiation method to directly regenerate $LiCoO_2$. By performing Li^+ intercalation into the spent Li_xCoO_2 electrode in an aqueous solution, followed by high-temperature annealing, they successfully prepared regenerated Li_xCoO_2 powder suitable for new electrodes. The study systematically analyzed the structure, composition, and electrochemical performance of the relithiated product, finding that its crystal structure was similar to commercial Li_xCoO_2, with the main components being Li, Co, and O. Its charge/discharge capacity (136 mAh·g^{-1}) was close to that of the commercial product (140 mAh·g^{-1}), with no significant performance decay after 200 cycles. Kinetic analysis showed that the Li^+ intercalation process was controlled by charge transfer at high concentrations and by diffusion at low concentrations, with an activation energy of 22 kJ mol^{-1} and an equilibrium constant of 1.35×10^{-6} cm s^{-1}. Tafel curves (Figure 8.29) were used to analyze the kinetic parameters of the relithiation process of spent Li_xCoO_2 at different Li_2SO_4 concentrations. The slope of the curves reflected the electron transfer coefficient (α) of the reaction and indicated that at high lithium-ion concentrations (1, 0.8, and 0.5 M), the electrochemical

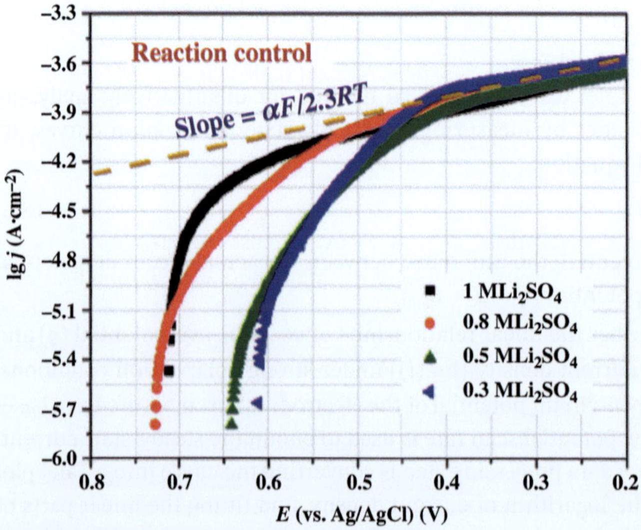

Figure 8.29 Tafel curves of the cathode with different Li^+ concentrations. *Source*: [35] / with permission of the American Chemical Society.

lithiation process was primarily limited by charge transfer, consistent with EIS analysis results. Furthermore, the intercept of the linear segment of the Tafel curve was used to calculate the exchange current density (j_0), thereby further determining the kinetic rate constant (k_0) for electrochemical lithiation, indicating that the reaction rate was related to electron transfer.

8.2.6 GITT Test

8.2.6.1 Introduction to GITT Test

GITT (galvanostatic intermittent titration technique) is a transient electrochemical testing method used for precise analysis of the thermodynamic and kinetic characteristics of battery materials. Its core principle involves decomposing continuous electrochemical reactions into discrete "titration" steps by alternately applying short CC pulses and long relaxation periods. During testing, a small CC pulse is first applied to the battery to cause intercalation/deintercalation of active material, leading to a sudden voltage change (including ohmic and electrochemical polarization). Then, the current is cut off for sufficient relaxation until the voltage stabilizes to a plateau (reflecting the thermodynamic equilibrium state). By analyzing the data from each pulse–relaxation cycle, three core applications can be achieved: calculating the solid-state diffusion coefficient of lithium ions in the electrode material based on Fick's second law and the voltage change model; plotting a precise equilibrium potential–composition curve based on the stabilized voltage values after relaxation to study the phase transition behavior; and evaluating the polarization overpotential of the battery at different states of charge by comparing the voltage difference at the end of the pulse and after relaxation. Although GITT can effectively separate kinetic and thermodynamic effects and provide high-resolution data, its main limitation is the extremely long testing period. This technique is a key means for investigating the intrinsic properties of battery materials (such as ion migration rate and phase equilibrium).

8.2.6.2 GITT Application Examples

Jin et al. [37] addressed key challenges in Li–S batteries, such as the polysulfide shuttle effect and lithium dendrite growth. They proposed using metal–organic framework (MOF) ZIF-67 as a precursor, and through chemical etching and high-temperature carbonization, successfully prepared a hollow porous-structured bimetallic carbide heterostructure material $Co_6Mo_6C_2$@Co@NC, used to modify the PP separator. Experimental results showed that the $Co_6Mo_6C_2$@Co@NC/PP separator had superior polysulfide adsorption and catalytic conversion capabilities, significantly enhancing lithium-ion diffusion rate and suppressing the shuttle effect, achieving high rate performance and excellent cycling stability (e.g., specific capacity of 782.6 $mAh{\cdot}g^{-1}$ after 300 cycles at 0.5C, with coulombic efficiency close to 100%). Under high sulfur loading, the battery also exhibited excellent capacity retention. GITT testing in this paper was used to evaluate the impact of different separators ($Co_6Mo_6C_2$@Co@NC/PP, Co@NC/PP, and PP) on the internal resistance and lithium-ion transport performance of Li–S batteries. Results (Figure 8.30) showed that the battery using the $Co_6Mo_6C_2$@Co@NC/PP-modified separator had the lowest internal resistance (I_R = 0.02946 V), significantly better than Co@NC/PP separator (I_R = 0.0312 V) and the traditional PP separator, proving that this material can effectively promote lithium-ion migration and improve battery reaction kinetics. Additionally, GITT data also verified that the $Co_6Mo_6C_2$ material enhances the adsorption and conversion of polysulfides, further inhibiting the shuttle effect and improving cycling stability and capacity retention. These results collectively indicated that the $Co_6Mo_6C_2$@Co@NC/PP separator significantly optimizes the electrochemical performance of Li–S batteries.

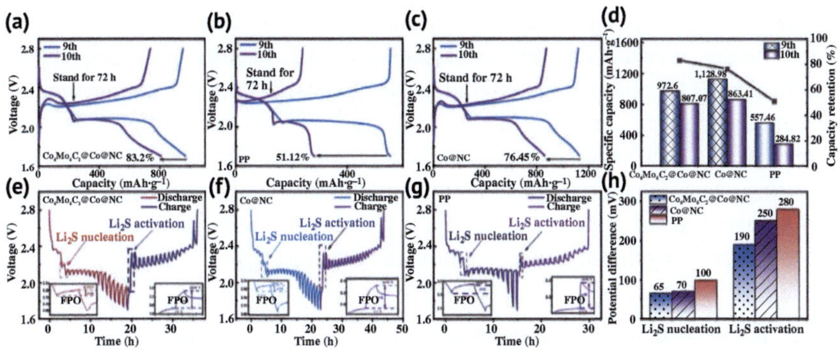

Figure 8.30 (a–d) The discharge–charge voltage profiles of self-discharge behavior with $Co_6Mo_6C_2$@Co@NC/PP, Co@NC/PP, and PP batteries. GITT voltage profiles of (e) $Co_6Mo_6C_2$@Co@NC/PP, (f) Co@NC/PP, and (g) PP batteries. (h) A columnar comparison of potential difference. *Source*: [36] / with permission of Elsevier.

8.3 Other Advanced Analysis Techniques

8.3.1 In-situ XRD

8.3.1.1 Introduction to In-situ XRD

The in situ XRD testing technique for LIBs is a powerful analytical method that combines the electrochemical charge/discharge process with real-time monitoring of crystal structure changes. Its core principle lies in establishing a direct correlation between the macroscopic electrochemical performance of the battery and the microscopic atomic structure changes of the electrode material. Technically, it relies on a specially designed in situ cell with an X-ray transparent window (typically beryllium or glassy carbon). During testing, while controlling the battery to perform CC charge/discharge, X-rays continuously irradiate the electrode material, and a series of XRD patterns varying with time or voltage are recorded in real time.

The physical basis for structure detection is the Bragg's equation. The intercalation and deintercalation of lithium ions in the electrode material lattice cause changes in the interplanar spacing (*d*-spacing), leading to dynamic evolution in the position, intensity, and even phase types of XRD peaks. By analyzing these evolutions, the battery's reaction mechanism can be precisely resolved: if the diffraction peak positions shift continuously, it indicates a solid-solution reaction with gradual lattice parameter changes; if old peaks disappear and new peaks generate, it signifies a two-phase reaction involving phase transition.

In practical applications, this technique is a key tool for revealing the internal secrets of batteries. It is primarily used to accurately analyze the charge/discharge mechanism of electrode materials, drawing precise phase diagrams corresponding to voltage, capacity, and phase, providing structural explanations for voltage plateaus. Second, it plays a prominent role in failure analysis, directly observing irreversible phase transitions, lattice collapse, and other structural degradations after long-term cycling, pinpointing the root cause of capacity fade. Furthermore, for next-generation high-performance materials (such as high-nickel cathodes, silicon-based anodes), in-situ XRD is an indispensable method for elucidating their unique lithium storage mechanisms and assessing structural stability, providing crucial theoretical basis for optimizing material design and battery process parameters.

In summary, electrochemical in-situ XRD, by real-time "capturing" a dynamic movie of the material's crystal structure during charge/discharge, serves as a bridge connecting electrochemical phenomena with atomic-scale structural changes, greatly deepening our understanding of the working mechanism of LIBs and continuously promoting the advancement of battery technology.

8.3.1.2 In-situ XRD Application Examples

Li et al. [38] focused on the thermal runaway mechanism of high-energy-density $LiNi_{0.8}Co_{0.1}Mn_{0.1}O_2$ (NCM811, ternary) LIBs, particularly the role of oxygen release from the cathode in triggering thermal runaway. The research team observed the thermal characteristics of the battery through full-cell ARC testing, combined with various characterization methods such as DSC, in-situ heating HRXRD–MS, and TEM, to systematically analyze the reaction processes of battery components. Results found that the cathode material releases active oxygen species (including O_2, O_2^-, O^-, etc.) at high temperatures, and these oxygen species undergo violent exothermic reactions with the flammable electrolyte, which is the direct trigger mechanism for thermal runaway. Further experiments showed that the electrolyte not only accelerates oxygen release but also promotes the phase transition of the cathode material. Control experiments capturing active oxygen confirmed that inhibiting oxygen release significantly improves battery safety. In-situ XRD revealed the structural changes of the NCM811 cathode during heating and its interaction with the electrolyte and TPFPB additive. Experimental results (Figure 8.31) showed that under three conditions—dry state, mixed with electrolyte, and with TPFPB added—the NCM811 cathode underwent phase transitions from a layered structure to a disordered spinel structure, and further to a rock-salt structure during heating. This process occurred between 100 and 600 °C, and the start and end temperatures of each phase transition stage are clearly marked in Figure 8.31. Furthermore, in-situ XRD analysis found that the presence of electrolyte accelerated the phase transition and oxygen release of the cathode, leading to decreased material structural stability, while adding TPFPB alleviated this process to some extent, improving the thermal stability of the cathode. These in-situ characterization results provided an important basis for understanding the battery thermal runaway mechanism and safety improvement strategies.

Zou et al. [39] addressed the slow kinetics of anode materials in lithium-ion capacitors (LICs) and proposed new intercalation-type anode materials to enhance their performance. The authors successfully synthesized orthorhombic phase $NaNbO_3$ nanocubes with large open tunnel structures allowing lithium-ion intercalation along the ⟨101⟩ and ⟨141⟩ directions, and studied their lithium storage mechanism and structural evolution in depth through in-situ XRD. To improve electron transport and prevent $NaNbO_3$ agglomeration, the study introduced rGO nanosheets to form a composite material with $NaNbO_3$. Experimental results showed that the $NaNbO_3$/rGO composite outperformed pure $NaNbO_3$ in reversible capacity, rate performance, and cycling stability, attributed to the synergistic effect and high lithium-ion diffusion coefficient. LIC devices based on this composite anode achieved high energy density (166.7 Wh kg^{-1}), excellent rate performance, and 90% capacity retention after 3,000 cycles. In situ XRD results (Figure 8.32) showed that no new phases formed in the $NaNbO_3$ electrode during charge/discharge. The main diffraction peaks ⟨101⟩ and ⟨141⟩ shifted to lower 2θ angles during discharge, indicating increased lattice spacing due to lithium-ion insertion; during charging, these peaks returned to their original positions, demonstrating the structural reversibility and intercalation behavior of the material during lithiation/delithiation. These results proved the performance advantages and excellent structural stability of $NaNbO_3$ as an intercalation-type material.

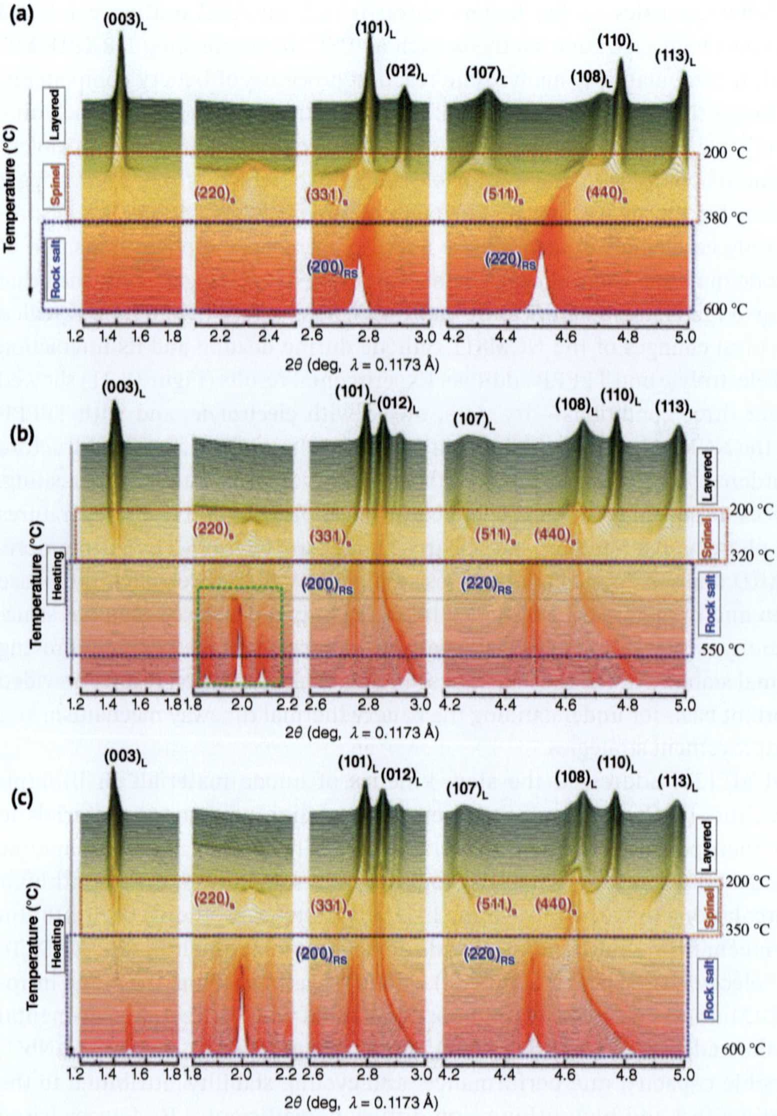

Figure 8.31 In-situ HEXRD characterization of (a) dry NCM811 cathode, (b) NCM811 cathode mixed with electrolyte (1 M $LiPF_6$ in EC/EMC), and (c) their mixture with TPFPB during heating. The temperatures on the right-hand side indicated when phase transformation started or finished. *Source*: [37] / with permission of Elsevier.

8.3.2 In-situ Raman Spectroscopy

8.3.2.1 Introduction to In-situ Raman Spectroscopy

The in-situ Raman spectroscopy electrochemical testing technique for LIBs is a real-time characterization method based on molecular vibrational spectroscopy, used to directly probe local chemical bonds, molecular structure, and phase transition dynamics of electrode materials during battery charge/discharge. Complementing in-situ XRD technology, which is sensitive to long-range ordered crystal structures,

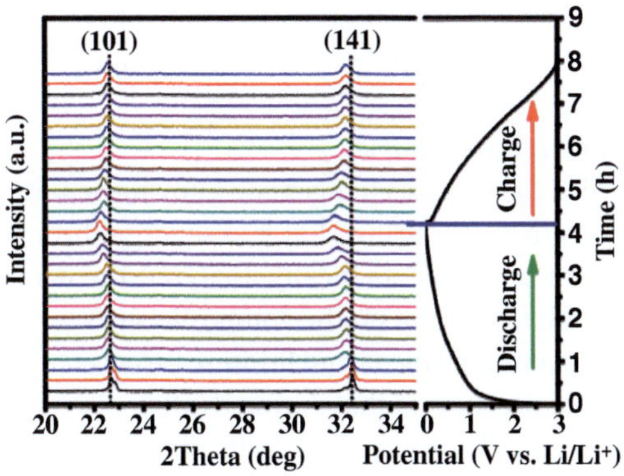

Figure 8.32 In-situ XRD patterns of the $NaNbO_3$ electrode during GCD at 0.1 A g^{-1}. *Source*: [38] / with permission of Elsevier.

in-situ Raman spectroscopy is extremely sensitive to short-range order, amorphous structures, and even surface/interface reactions, particularly adept at providing "fingerprint" features of chemical bonding changes.

Its basic principle is the Raman scattering effect. When a monochromatic laser beam irradiates the sample, it exchanges energy with the vibrational energy levels of the sample molecules, producing scattered light with shifted frequencies. This frequency shift (Raman shift) corresponds to specific chemical bond vibration modes of the molecules, like a molecular "fingerprint." During the operation of an LIB, the intercalation and deintercalation of lithium ions directly alter the bond lengths, angles, and charge distribution in the local environment of the electrode material, leading to changes in the position, intensity, or width of its characteristic Raman peaks, or even the appearance of Raman peaks representing new phases. By correlating these spectral changes with real-time electrochemical parameters (voltage, current), the reaction mechanism can be revealed.

Technically, it requires constructing a special in-situ cell with an optically transparent window (such as quartz glass) allowing laser incidence and collection of scattered signals. A confocal micro-Raman system is typically used to achieve micro-area probing and high spatial resolution.

The core applications of this technique are mainly reflected in the following aspects: first, in mechanism studies, it can accurately track the evolution of characteristic peaks of key materials during charge/discharge. For example, it can clearly monitor the formation and transformation of different lithium–graphite intercalation compounds during the staging process of graphite anodes, where the Raman G and D bands show regular shifts and intensity changes; for high-capacity silicon anodes, it can observe the amorphization process and the formation of Li–Si alloys in real time. Second, it has unique advantages in interface chemistry research, being able to identify in-situ the composition of the SEI film (including lithium carbonate, alkyl carbonates, etc.) and study its evolution with cycling, which is key to understanding battery coulombic efficiency and lifespan. Furthermore, for cathode materials, by monitoring the anionic (oxygen) redox activity in lithium-rich manganese-based materials, or observing the generation and conversion of polysulfides in sulfur cathodes (polysulfides have strong Raman signals), in-situ Raman spectroscopy can provide unique insights. Finally, it can also be used to study the decomposition behavior of electrolytes and their additives on the electrode surface, providing direct evidence for optimizing electrolyte formulations.

8.3.2.2 In-situ Raman Spectroscopy Application Examples

Nanda et al. [40] microscopically revealed the surface changes and alloying behavior of high-capacity Si/C composite anodes in LIBs during charge/discharge through in-situ Raman microscopy. The research team systematically analyzed the distribution of silicon particles, graphite, and hard carbon on the composite surface and real-time monitored the alloying process between silicon and lithium during charge/discharge cycles. Raman spectroscopy results showed that during the first discharge below 0.1 V, the silicon Raman signal completely disappeared, indicating that silicon had formed Li–Si alloys with lithium, while color changes in the graphite region and the presence of mixed phases were observed. During charging, no Raman peaks for crystalline Li–Si alloys were observed, suggesting the electrode consisted mainly of a mixture of amorphous silicon and Li–Si alloy phases. Additionally, accompanying ex-situ XRD experiments confirmed the gradual loss of crystallinity of the silicon structure. Through in-situ Raman experiments, the structural changes of silicon in the Si/C composite anode during the first charge/discharge process were detailed. Specific experimental results (Figure 8.33) showed that during the first discharge, as the voltage decreased below 0.2 V, the silicon Raman signal significantly weakened and eventually disappeared, indicating that silicon underwent an alloying reaction with lithium, forming lithium–silicon alloy phases, leading to

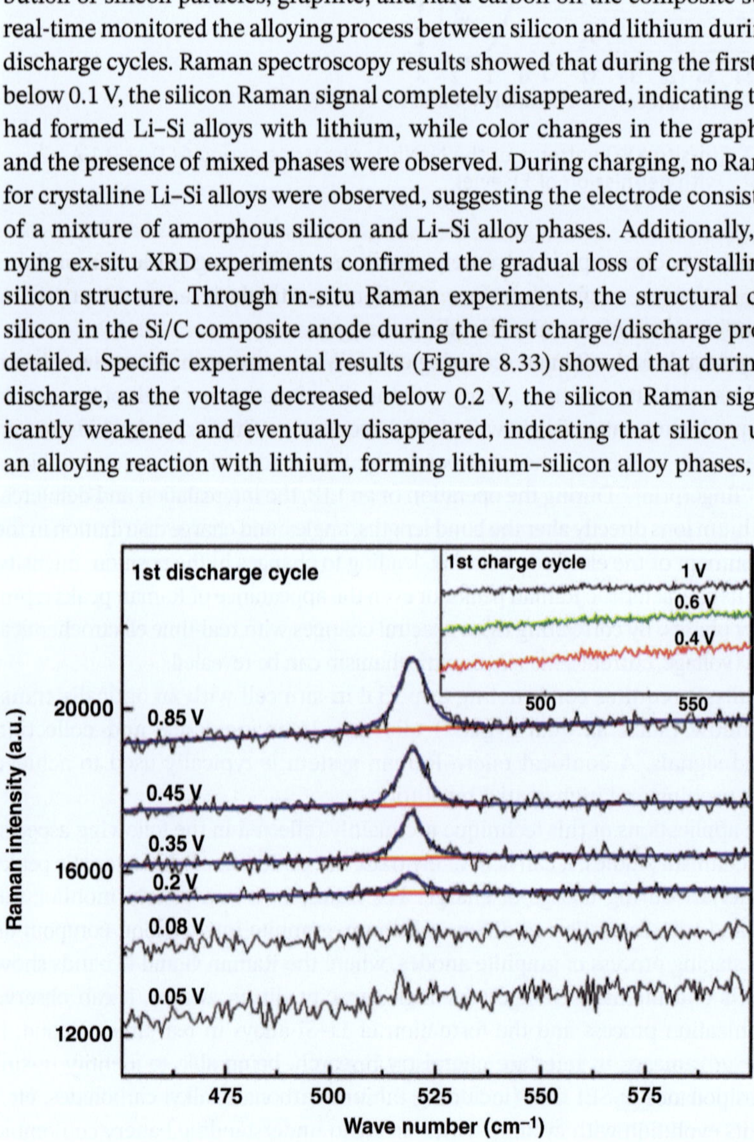

Figure 8.33 In-situ micro-Raman signals of silicon as a function of the first discharge and charge (inset) voltages. The solid blue lines are the fit to the experimental spectra. *Source:* [39] / with permission of Elsevier.

the destruction or transformation of the silicon crystal structure. During the first charge, no significant Raman active peaks for silicon or lithium–silicon alloys were observed, suggesting the electrode material may have transformed into a mixture of amorphous silicon and lithium–silicon alloy. These in-situ Raman observations revealed key phase transition and structural evolution processes of the silicon anode during electrochemical cycling, helping to deeply understand the performance evolution mechanism of silicon-based materials in LIBs.

8.3.3 In-situ Impedance Electrochemical Test

8.3.3.1 Introduction to In-situ Impedance Electrochemical Test

The in-situ impedance electrochemical testing technique for LIBs is a key method that uses small-amplitude AC perturbation signals to probe the evolution of kinetic processes inside the battery in real time. Unlike in-situ XRD and Raman techniques that provide structural information, in-situ impedance focuses on revealing the "kinetic" characteristics of processes such as charge transfer and ion transport, and can sensitively capture changes in the internal state of the battery during cycling and aging.

Its core principle is based on EIS. This technique applies a sinusoidal AC voltage (or current) signal with variable frequency (typically from high frequency to low frequency, e.g., from 100 kHz to 10 mHz) to the working battery, while measuring the battery's response to this perturbation (current or voltage signal). By analyzing the response at different frequencies, a complex plane plot (Nyquist plot) can be obtained. Features in different frequency regions of the plot (such as arcs, slanted lines) correspond to different physical–chemical processes inside the battery: the high-frequency region typically reflects lithium-ion transport through the SEI film on the electrode surface; the mid-frequency region is related to the charge-transfer process at the electrode/electrolyte interface; and the low-frequency region is dominated by the diffusion behavior of lithium ions in the solid electrode material. By establishing an equivalent circuit model to fit the spectrum, kinetic parameters of these processes can be quantified, such as interfacial charge-transfer resistance, SEI film resistance, and lithium-ion diffusion coefficient.

Technically, in-situ impedance testing is usually combined with charge/discharge cycles. It can be performed at a specific S_{OC} for fixed-point measurement, or continuous or intermittent scanning can be performed during the complete charge/discharge process, thus obtaining kinetic parameters as a function of voltage and time.

The power of this technique lies in its wide range of application scenarios. First, it is an indispensable tool in battery state monitoring and aging mechanism research. By tracking the growth trend of SEI film resistance and charge-transfer resistance during cycling, the main cause of capacity fade can be clearly distinguished, whether it is active lithium loss, active material failure, or kinetic performance degradation (e.g., increased impedance due to interfacial side reactions), providing quantitative basis for life prediction and failure analysis. Second, in electrode material performance evaluation, by comparing the charge-transfer resistance of different materials or electrodes prepared by different processes at different S_{OC}, their electrochemical activity level and polarization magnitude can be intuitively judged. Furthermore, it is extremely sensitive to interfacial side reactions and can monitor high-risk phenomena such as lithium plating on the anode in real time (usually manifested as abnormal changes in the mid-frequency charge-transfer resistance), providing key indicators for rapid assessment of battery safety (e.g., fast-charging performance). Finally, it can also be used to optimize battery operation strategies, for example, by analyzing impedance spectra at different temperatures to determine the optimal operating temperature range.

8.3.3.2 In-situ Impedance Electrochemical Test Application Examples

Xu et al. [41] addressed the issues of structural collapse and severe interfacial side reactions of high-nickel layered cathode material NCM811 under high voltage and long cycling conditions. They proposed a synergistic modification strategy of lanthanum (La) doping and Li_3BO_3 surface coating. Through a simple wet-chemical method and high-temperature sintering process, La atoms diffused into the surface lattice, inducing an amorphous domain, releasing mechanical stress, inhibiting microcrack generation, and enhancing crystal structure and particle integrity. The Li_3BO_3 coating layer consumed residual lithium compounds, accelerated interfacial ion transfer, blocked contact between the electrolyte and active material, and reduced side reactions and oxygen evolution. Various electrochemical and characterization tests showed that the modified NCM811 maintained a high specific capacity of 133.8 mAh·g^{-1} even at a high rate (4C), and achieved a capacity retention of 73.8% after 100 cycles at 0.5C between 2.8 and 4.8 V. Structural and interfacial stability were significantly improved, with almost unchanged interfacial impedance and increased lithium-ion diffusion coefficient. Through in-situ EIS testing, the impact of La doping and surface LBO coating on the interfacial lithium-ion transport kinetics and stability of NCM811 cathode material was systematically elucidated. Results (Figure 8.34) showed that the modified NCM–La–LBO sample had significantly lower total impedance than the unmodified sample, whether in charged or discharged states, indicating a more stable interface and smoother lithium-ion migration. Furthermore, in-situ EIS combined with DRT (distribution of relaxation times) analysis quantified the impedance changes of different processes such as lithium-ion diffusion, electron transfer, and CEI film growth. The impedance of the modified sample hardly increased after long cycling, indicating that La doping and LBO coating can effectively suppress direct contact between the electrolyte and cathode material, improve interfacial chemical stability, and maintain excellent lithium-ion transport performance. These results proved that the dual modification strategy can significantly improve the structural and interfacial stability of high-nickel cathode materials, thereby enhancing their long-term cycling performance and rate capability.

Hong et al. [42] addressed the interfacial stability and electrochemical performance bottlenecks of high-entropy oxides (HEOs) as anode materials for LIBs. They proposed synthesizing nanoporous high-entropy oxide (NHEO) using a NaCl hard template method and a new method of in-situ thermal treatment of polyacrylonitrile (PAN) to form cyclized PAN (cPAN) coating. Material characterization showed that NHEO possessed a uniform cubic structure and high specific surface area, significantly enhancing Li^+ diffusion kinetics. The cPAN coating not only introduced oxygen vacancies to improve material conductivity but also enhanced polarity and mechanical strength, effectively stabilizing the electrode structure and inhibiting

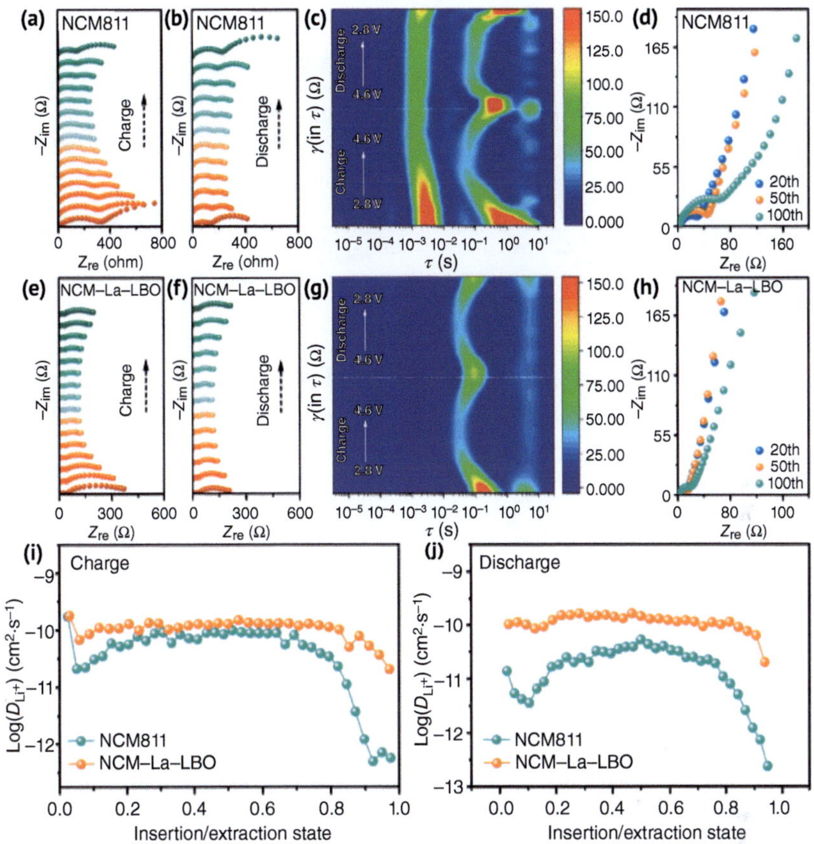

Figure 8.34 In-situ EIS and corresponding DRT results for (a–c) NCM811 and (e–g) NCM–La–LBO in the first charge/discharge states. The EIS plots of (d) NCM811 and (h) NCM–La–LBO at 0.5C at different cycles. (i, j) The calculated lithium diffusion coefficients of NCM811 and NCM–La–LBO based on GITT curves. *Source*: [40] / with permission of Elsevier.

volume expansion. Electrochemical test results showed that the NHEO-PAN500 electrode exhibited excellent specific capacity (up to 1,080 $mAh \cdot g^{-1}$) and cycling stability (86.5% capacity retention after 800 cycles) at high rates and long cycles. Interface analysis confirmed that the cPAN coating could form a high-quality SEI layer, reduce side reactions, and improve battery life. In-situ EIS testing revealed the reaction kinetics and interfacial changes of NHEO-PAN500 and NHEO-PVDF electrodes during charge/discharge. Specifically, in-situ EIS results (Figure 8.35) showed that as the battery discharged to 0.8 V, the semicircle radius in the high-frequency region gradually shrank, reflecting the gradual decrease in charge-transfer resistance (R_{ct}) with the reduction reaction; when discharging to 0.6 V, a new semicircle appeared in the high-frequency region, indicating the beginning of SEI film formation as Li^+ concentration increased, corresponding to the cathodic peak in the first cycle CV curve. Furthermore, the impedance gradually decreased during charging, suggesting

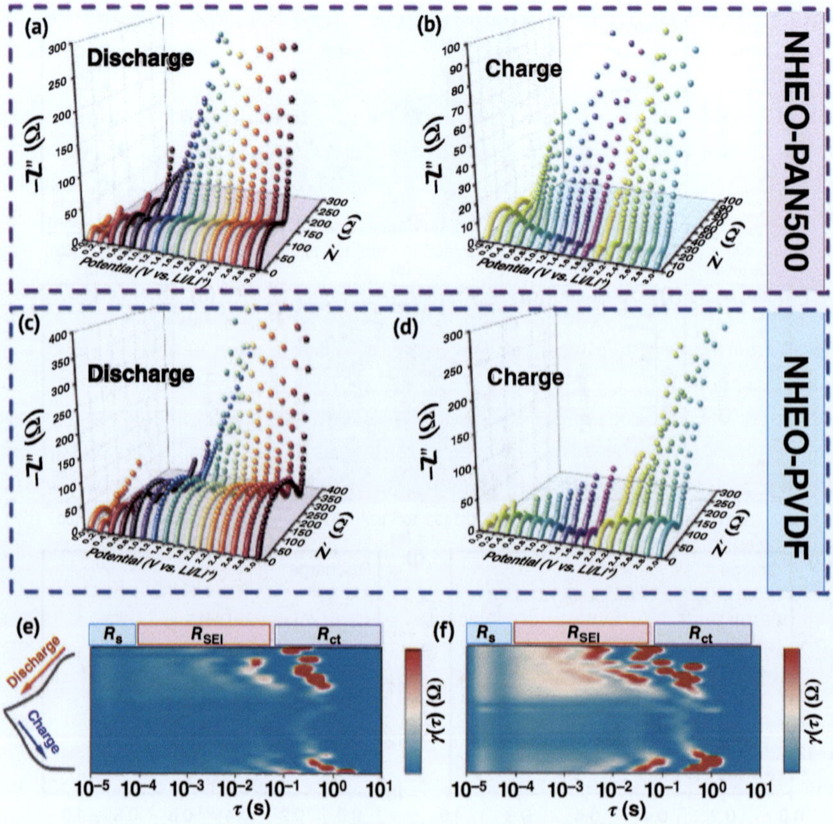

Figure 8.35 (a–d) In-situ EIS three-dimensional data, and DRT results calculated from the in-situ EIS analysis of (e) NHEO-PAN500 and (f) NHEO-PVDF. *Source*: [41] / with permission of John Wiley & Sons.

structural reorganization on the active material surface, forming an intermediate rock-salt phase, further promoting the electrochemical process. DRT analysis subdivided different impedance sources, including ohmic resistance (R_s), SEI film resistance (R_{SEI}), and charge-transfer resistance (R_{ct}), further proving that the cPAN coating on the NHEO-PAN500 surface can effectively stabilize the SEI film, reduce impedance and side reactions, and optimize charge/discharge performance.

8.3.4 In-situ AFM

8.3.4.1 Introduction to In-situ AFM

The in-situ AFM testing technique for LIBs is a cutting-edge characterization method that enables real-time observation of electrode surface topography evolution and measurement of its physical properties at the nano- or even atomic scale under operating battery conditions. Its unique advantage lies in directly correlating electrochemical processes with physical changes on the electrode surface (such as morphology, roughness, and modulus), providing intuitive surface/interface information that is difficult to achieve with other techniques.

Its basic principle involves scanning the electrode surface with an extremely fine probe. There is a very weak atomic force (including van der Waals forces, chemical bond forces, etc.) between the probe tip and the sample surface. By monitoring the bending or vibration state of the probe cantilever, the height variations of the sample surface are inferred, thus constructing a three-dimensional surface topography map. When performing in-situ electrochemical testing, a specially designed micro-electrochemical cell with an AFM probe is constructed, where the AFM probe acts simultaneously as a force sensor, and the electrode sample acts as the working electrode. While controlling the electrode to undergo charge/discharge (lithiation/delithiation), the AFM system continuously scans the same area, obtaining a series of dynamic topography images arranged in time or voltage sequence.

The core application value of this technique lies in its direct "visualization" of micro-processes on the electrode surface. First, in studying the structural stability of electrode materials, it can intuitively capture the entire process of electrode particle expansion, contraction, crack generation, and pulverization during charge/discharge. For example, for silicon anodes with huge volume effects, in-situ AFM can precisely measure their expansion rate at different lithiation degrees and directly observe crack initiation and propagation, providing the most direct evidence for understanding their capacity fade mechanism. Second, it plays an irreplaceable role in interface process research: first, it can observe the formation and growth of the SEI film in real time, seeing how the SEI film covers the electrode surface from scratch, from loose to dense, and can measure the mechanical properties (such as elastic modulus) of local areas through AFM's force spectroscopy mode, thereby correlating the stability of the SEI film with its mechanical properties; second, it can provide early warning of lithium dendrite precipitation, sensitively detecting nanoscale lithium metal protrusions on the anode surface, which is crucial for assessing battery safety. Additionally, it can be used to study the kinetic behavior of ion transport on the electrode surface during charge/discharge.

8.3.4.2 In-situ AFM Application Examples

Zhu et al. [43] focused on the formation and evolution mechanism of the SEI on graphite anodes in LIBs, specifically examining the influence of graphite surface defects on SEI structure and performance. Through in-situ electrochemical atomic force microscopy (EChem-AFM), ex-situ XPS, and DFT calculations, they systematically compared the SEI behavior of HOPG (defect-free) and graphite with multiple defects (GUITAR). The study found that the SEI formed on defective graphite surfaces at a more positive potential, was denser, more uniform, and thinner, contained no Li_2CO_3, and was rich in inorganic components like LiF and $Li_xPO_yF_z$, exhibiting stronger surface passivation and cycling stability. DFT calculations revealed that defects promote the adsorption and decomposition of EC–Li complexes on the graphite surface, facilitating early SEI formation. Through operando EChem-AFM, the formation and evolution processes of the SEI on HOPG and defective graphite electrode (GUITAR) surfaces were systematically monitored and compared. Results (Figures 8.36 and 8.37) showed that SEI preferentially formed at the step edges of HOPG at around 1.5 V and continuously expanded to the basal plane, while on the GUITAR surface, it presented a more uniform, dense, and thinner structure, with

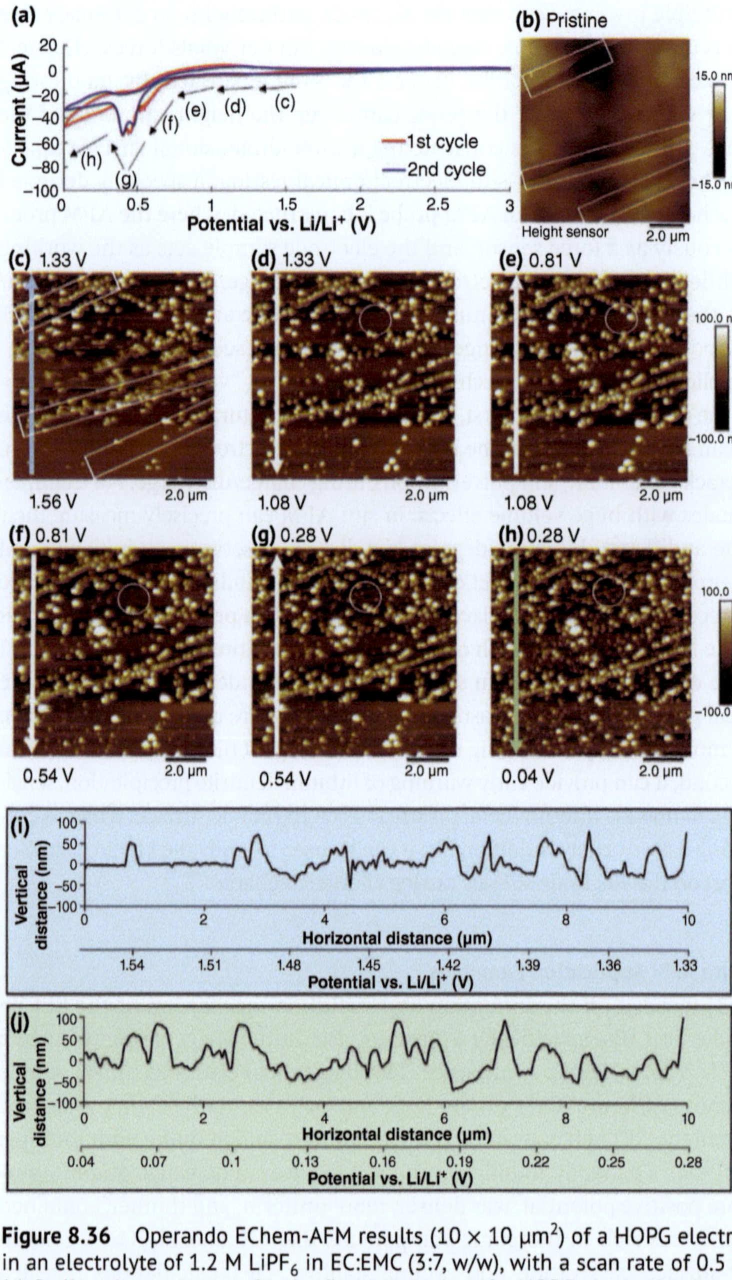

Figure 8.36 Operando EChem-AFM results (10 × 10 μm^2) of a HOPG electrode in an electrolyte of 1.2 M $LiPF_6$ in EC:EMC (3:7, w/w), with a scan rate of 0.5 mV s^{-1}. (a) Cyclic voltammograms for the first two cycles of a HOPG electrode; the arrows indicate the potential range in which the corresponding AFM images were collected. (b) AFM image of a pristine surface of HOPG, boxed areas indicate ≈10 layers of step edges. (c–h) Operando AFM images collected during the cathodic scan of the first cycle at different potential ranges, the circled areas highlight a small basal plane region. (i, j) Representative height profiles of panels (c) and (h) taken along the blue and green arrows, respectively. *Source*: [42] / with permission of John Wiley & Sons.

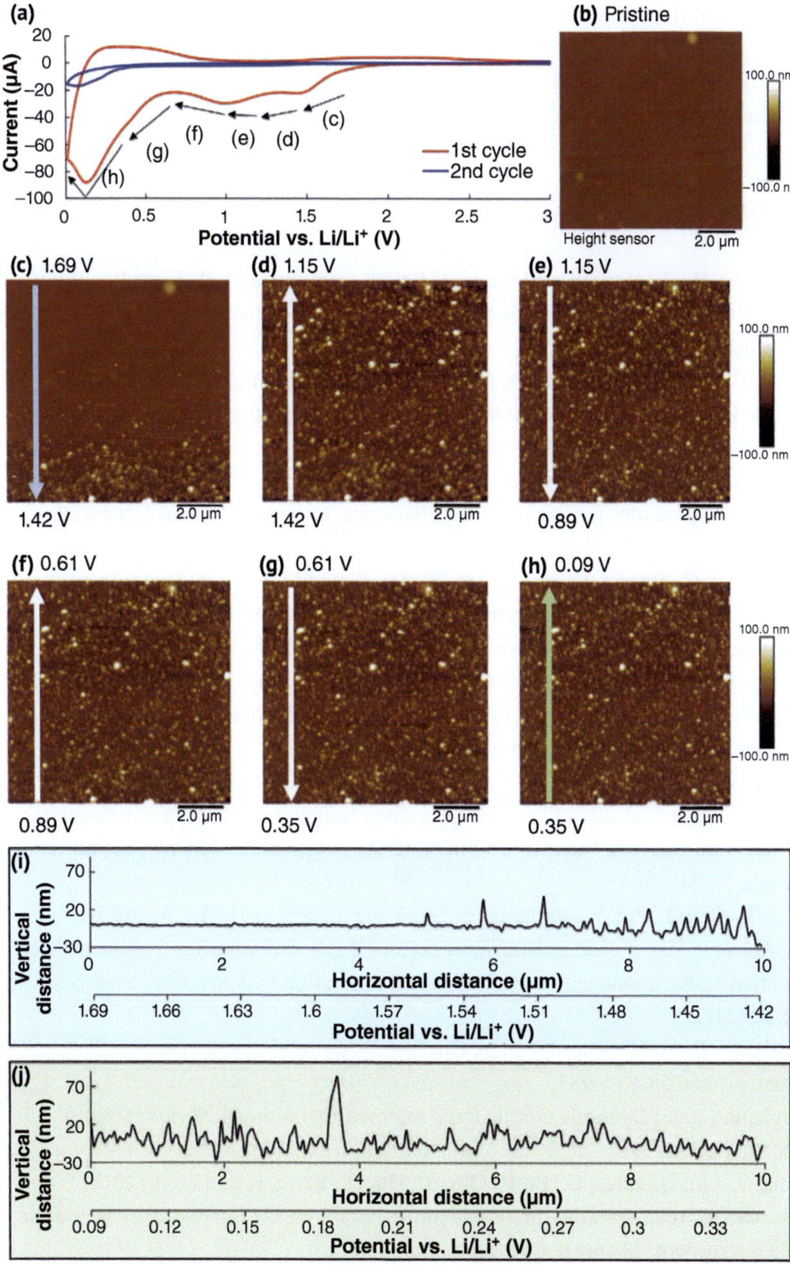

Figure 8.37 Operando EChem-AFM results (10 × 10 μm^2) of a GUITAR electrode in an electrolyte of 1.2 M $LiPF_6$ in EC:EMC (3:7, w/w), with a scan rate of 0.5 mV s^{-1}. (a) Cyclic voltammograms for the first two cycles of a GUITAR electrode. The arrows indicate the potential range in which the corresponding AFM images were collected. (b) AFM image of a pristine surface of GUITAR. (c–h) Operando AFM images collected during the cathodic scan of the first cycle at different potential ranges. (i, j) Representative height profiles of panels (c) and (h) taken along the blue and green arrows, respectively, plotted vs. the corresponding potentials. *Source*: [42] / with permission of John Wiley & Sons.

almost constant thickness during the first lithiation, demonstrating a higher passivation effect. This study revealed the influence of different graphite electrode defect degrees on SEI structure and performance, providing new insights for optimizing the interface engineering of LIB anode materials.

References

1. Zhang, N., He, J., Han, W. and Wang, Y. (2019) Composite solid electrolyte PEO/SN/$LiAlO_2$ for a solid-state lithium battery. *Journal of Materials Science*, 54, 9603–9612.
2. Bucur, C. B., Jones, M., Kopylov, M., Spear, J. and Muldoon, J. (2017) Inorganic–organic layer by layer hybrid membranes for lithium–sulfur batteries. *Energy & Environmental Science*, 10, 905–911.
3. Xu, R., Sun, Y., Wang, Y., Huang, J. and Zhang, Q. (2017) Two-dimensional vermiculite separator for lithium sulfur batteries. *Chinese Chemical Letters*, 28, 2235–2238.
4. Pan, J., Wu, C., Cheng, J., Pan, Y., Ma, Z., Xie, S. and Li, J. (2015) Sepiolite-sulfur as a high-capacity, high-rate performance, and low-cost cathode material for lithium–sulfur batteries. *Journal of Power Sources*, 293, 527–532.
5. Wang, X., Feng, Z., Hou, X., Liu, L., He, M., He, X., Huang, J. and Wen, Z. (2020) Fluorine doped carbon coating of $LiFePO_4$ as a cathode material for lithium-ion batteries. *Chemical Engineering Journal*, 379, 122371.
6. Gao, S., Wang, N., Li, S., Li, D., Cui, Z., Yue, G., Liu, J., Zhao, X., Jiang, L. and Zhao, Y. (2020) A multi-wall Sn/SnO_2@carbon hollow nanofiber anode material for high-rate and long-life lithium-ion batteries. *Angewandte Chemie International Edition*, 59, 2465–2472.
7. Jiang, X., Sun, L., Lu, Y., Wang, H., Shi, J., Yang, L., Zhang, L., Lv, R. and Jin, Z. (2024) Ladderlike polysilsesquioxanes derived dual-carbon-buffer-shell structural silicon as stable anode materials for lithium-ion batteries. *Journal of Power Sources*, 602, 234331.
8. Binnig, G., Quate, C. F. and Gerber, Ch. (1986) Atomic force microscope. *Physical Review Letters*, 56, 930–933.
9. García, R. (2002) Dynamic atomic force microscopy methods. *Surface Science Reports*, 47, 197–301.
10. Wang, S., Liu, Q., Zhao, C., Lv, F., Qin, X., Du, H., Kang, F. and Li, B. (2018) Advances in understanding materials for rechargeable lithium batteries by atomic force microscopy. *Energy & Environmental Materials*, 1, 28–40.
11. Danis, L., Gateman, S. M., Kuss, C., Schougaard, S. B. and Mauzeroll, J. (2017) Nanoscale measurements of lithium-ion-battery materials using scanning probe techniques. *ChemElectroChem*, 4, 6–19.
12. Butt, H.-J., Cappella, B. and Kappl, M. (2005) Force measurements with the atomic force microscope: technique, interpretation and applications. *Surface Science Reports*, 59, 1–152.

13 Eaton, P. and West, P. (2010) *Atomic Force Microscopy*. Oxford, UK: Oxford University Press. https://doi.org/10.1093/acprof:oso/9780199570454.001.0001

14 Xu, K., Sun, W., Shao, Y., Wei, F., Zhang, X., Wang, W. and Li, P. (2018) Recent development of PeakForce Tapping mode atomic force microscopy and its applications on nanoscience. *Nanotechnology Reviews*, 7, 605–621.

15 Cresce, A. V., Russell, S. M., Baker, D. R., Gaskell, K. J. and Xu, K. (2014) In situ and quantitative characterization of solid electrolyte interphases. *Nano Letters*, 14, 1405–1412.

16 Xu, J., Zhang, J., Pollard, T. P., Li, Q., Tan, S., Hou, S., Wan, H., Chen, F., He, H., Hu, E., Xu, K., Yang, X.-Q., Borodin, O. and Wang, C. (2023) Electrolyte design for Li-ion batteries under extreme operating conditions. *Nature*, 614, 694–700.

17 Huang, S., Wang, S., Hu, G., Cheong, L.-Z. and Shen, C. (2018) Modulation of solid electrolyte interphase of lithium-ion batteries by LiDFOB and LiBOB electrolyte additives. *Applied Surface Science*, 441, 265–271.

18 Zheng, J., Zheng, H., Wang, R., Ben, L., Lu, W., Chen, L., Chen, L. and Li, H. (2014) 3D visualization of inhomogeneous multi-layered structure and Young's modulus of the solid electrolyte interphase (SEI) on silicon anodes for lithium ion batteries. *Physical Chemistry Chemical Physics*, 16, 13229–13238.

19 Zhang, Y., Liu, L., Zhao, L., Hou, C., Huang, M., Algadi, H., Li, D., Xia, Q., Wang, J., Zhou, Z., Han, X., Long, Y., Li, Y., Zhang, Z. and Liu, Y. (2022) Sandwich-like $CoMoP_2$/MoP heterostructures coupling N, P co-doped carbon nanosheets as advanced anodes for high-performance lithium-ion batteries. *Advanced Composites and Hybrid Materials*, 5, 2601–2610.

20 Zhao, B., Li, J., Guillaume, M., Dendooven, J. and Detavernier, C. (2022) In vacuo XPS investigation of surface engineering for lithium metal anodes with plasma treatment. *Journal of Energy Chemistry*, 66, 295–305.

21 Zeng, G., Liu, Y., Chen, D., Zhen, C., Han, Y. and He, W. (2021) Natural lepidolite enables fast polysulfide redox for high-rate lithium sulfur batteries. *Advanced Energy Materials*, 11, 2102058.

22 Friedrich, W., Knipping, P. & Laue, M. (1913) Interferenzerscheinungen bei Röntgenstrahlen. *Annalen der Physik*, 346, 971.

23 Gangaja, B., Nair, S. and Santhanagopalan, D. (2021) Reuse, recycle, and regeneration of $LiFePO_4$ cathode from spent lithium-ion batteries for rechargeable lithium- and sodium-ion batteries. *ACS Sustainable Chemistry & Engineering*, 9, 4711–4721.

24 Liu, Y., Gu, Y.-J., Luo, G.-Y., Chen, Z.-L., Wu, F.-Z., Dai, X.-Y., Mai, Y. and Li, J.-Q. (2020) Ni-doped $LiFePO_4$/C as high-performance cathode composites for Li-ion batteries. *Ceramics International*, 46, 14857–14863.

25 Park, K.-H., Kaup, K., Assoud, A., Zhang, Q., Wu, X. and Nazar, L. F. (2020) High-voltage superionic halide solid electrolytes for all-solid-state Li-ion batteries. *ACS Energy Letters*, 5, 533–539.

26 Yuan, G., Pan, J., Zhang, Y., Yu, J., He, Y., Su, Y., Zhou, Q., Jin, H. and Xie, S. (2018) Sepiolite/CNT/S@PANI composite with stable network structure for high performance lithium sulfur batteries. *RSC Advances*, 8, 17950–17957.

27 Zhu, X., Meng, F., Zhang, Q., Xue, L., Zhu, H., Lan, S., Liu, Q., Zhao, J., Zhuang, Y., Guo, Q., Liu, B., Gu, L., Lu, X., Ren, Y. and Xia, H. (2021) $LiMnO_2$ cathode stabilized by interfacial orbital ordering for sustainable lithium-ion batteries. *Nature Sustainability*, 4, 392–401.

28 Wu, M., Zhao, Y., Sun, B., Sun, Z., Li, C., Han, Y., Xu, L., Ge, Z., Ren, Y., Zhang, M., Zhang, Q., Lu, Y., Wang, W., Ma, Y. and Chen, Y. (2020) A 2D covalent organic framework as a high-performance cathode material for lithium-ion batteries. *Nano Energy*, 70, 104498.

29 Sun, W., Sun, X., Akhtar, N., Li, C., Wang, W., Wang, A., Wang, K. and Huang, Y. (2020) Attapulgite nanorods assisted surface engineering for separator to achieve high-performance lithium–sulfur batteries. *Journal of Energy Chemistry*, 48, 364–374.

30 Li, R., Xie, Z., Zhou, Y., Wang, W. and Gui, X. (2025) Fly ash as a high-capacity, high-rate performance, and low-cost cathode material for lithium-sulfur batteries. *Journal of Power Sources*, 640, 236738.

31 Elibol, M. K., Jiang, L., Xie, D., Cao, S., Pan, X., Härk, E. and Lu, Y. (2023) Nickel oxide decorated halloysite nanotubes as sulfur host materials for lithium–sulfur batteries. *Global Challenges*, 7, 2300005.

32 Cheng, H., Yang, W., Ni, Z., Wang, Y., Hou, J., Deng, B., Zhu, Z., Zhang, Y., Li, X. and Zhang, Y. (2023) Hazel shell-based biomass-derived carbon modified diaphragm for high-performance lithium-sulfur batteries. *Ceramics International*, 49, 41248–41258.

33 Kuang, H., Xiao, L., Lai, Y., Shen, L., Zhou, A., Wu, J. and Zhu, Y. (2024) Effect of Si doping and active carbon surface modifications on the structure and electrical performance of $Li_4Ti_5O_{12}$ anode material for lithium-ion batteries. *Ionics*, 30, 1307–1317.

34 Wu, F., Lv, H., Chen, S., Lorger, S., Srot, V., Oschatz, M., van Aken, P. A., Wu, X., Maier, J. and Yu, Y. (2019) Natural vermiculite enables high-performance in lithium–sulfur batteries via electrical double layer effects. *Advanced Functional Materials*, 29, 1902820.

35 Wang, W., Yang, Y., Luo, H., Li, S. and Zhang, J. (2020) A separator based on natural illite/smectite clay for highly stable lithium-sulfur batteries. *Journal of Colloid and Interface Science*, 576, 404–411.

36 Zhang, L., Xu, Z. and He, Z. (2020) Electrochemical relithiation for direct regeneration of $LiCoO_2$ materials from spent lithium-ion battery electrodes. *ACS Sustainable Chemistry & Engineering*, 8, 11596–11605.

37 Jin, X., Guo, X., Dong, S., Li, S., Jin, S., Xia, P., Lu, S., Zhang, Y. and Fan, H. (2025) Synergistic regulation of polysulfides shuttle effect and lithium dendrites from cobalt-molybdenum bimetallic carbides (Co-Mo-C) heterostructure for robust Li-S batteries. *Chinese Chemical Letters*, 36, 110604.

38 Li, Y., Liu, X., Wang, L., Feng, X., Ren, D., Wu, Y., Xu, G., Lu, L., Hou, J., Zhang, W., Wang, Y., Xu, W., Ren, Y., Wang, Z., Huang, J., Meng, X., Han, X. and Wang, H. (2021) Thermal runaway mechanism of lithium-ion battery with $LiNi_{0.8}Mn_{0.1}Co_{0.1}O_2$ cathode materials. *Nano Energy*, 85, 105878.

39 Zou, B., Wang, T., Li, S., Kang, R., Li, G., El-Khodary, S. A., Ng, D. H. L., Liu, X., Qiu, J., Zhao, Y., Lian, J. and Li, H. (2021) In situ XRD and electrochemical investigation on a new intercalation-type anode for high-rate lithium ion capacitor. *Journal of Energy Chemistry*, 57, 109–117.

40 Nanda, J., Datta, M. K., Remillard, J. T., O'Neill, A. and Kumta, P. N. (2009) In situ Raman microscopy during discharge of a high capacity silicon–carbon composite Li-ion battery negative electrode. *Electrochemistry Communications*, 11, 235–237.

41 Xu, M., Lu, J., Sun, Z., Yang, M., Sheng, B., Chen, M., Chen, J., Zhang, Q. and Han, X. (2024) Lanthanum doping and surface Li_3BO_3 passivating layer enabling 4.8 V nickel-rich layered oxide cathodes toward high energy lithium-ion batteries. *Journal of Colloid and Interface Science*, 673, 386–394.

42 Hong, C., Tao, R., Tan, S., Pressley, L. A., Bridges, C. A., Li, H., Liu, X., Li, H., Li, J., Yuan, H., Sun, X. and Liang, J. (2025) In situ cyclized polyacrylonitrile coating: key to stabilizing porous high-entropy oxide anodes for high-performance lithium-ion batteries. *Advanced Functional Materials*, 35, 2412177.

43 Zhu, H., Russell, J. A., Fang, Z., Barnes, P., Li, L., Efaw, C. M., Muenzer, A., May, J., Hamal, K., Cheng, I. F., Davis, P. H., Dufek, E. J. and Xiong, H. (2021) A comparison of solid electrolyte interphase formation and evolution on highly oriented pyrolytic and disordered graphite negative electrodes in lithium-ion batteries. *Small*, 17, 2105292.

9

Advanced Assembly Technology of Full Battery

9.1 Preparation Method of Electrode Materials

The pursuit of higher energy density, enhanced safety, and longer cycle life in energy storage systems has propelled the development of all-solid-state lithium batteries (ASSLBs). Unlike conventional lithium-ion batteries (LIBs) that employ liquid electrolytes, ASSLBs utilize solid-state electrolytes, fundamentally altering the requirements and methodologies for electrode preparation. The electrode, a composite mixture of active materials, conductive additives, and solid electrolytes, must facilitate efficient ion and electron transport through solid–solid interfaces, a significant departure from the solid–liquid interfaces in LIBs [1]. Consequently, the preparation methods of electrode materials are critical in defining the microstructure, interfacial contact, and, ultimately, the electrochemical performance of ASSLBs.

Traditional electrode manufacturing for LIBs predominantly relies on the wet slurry casting method. This process involves dispersing active materials (e.g., $LiNi_xMn_yCo_zO_2$ [NMC], $LiFePO_4$ [LFP]), conductive carbon, and a polymeric binder in a volatile solvent to form a slurry [2]. This slurry is then coated onto a current collector (e.g., Al foil for cathode, Cu foil for anode) and dried to evaporate the solvent, leaving a porous electrode film. Binders like polyvinylidene fluoride (PVDF) are crucial in this process, providing adhesion that binds the active particles together and to the current collector. However, the transition to ASSLBs introduces severe complications to this established process. Many high-performance SEs, particularly sulfides (e.g., Li_6PS_5Cl, $Li_{10}GeP_2S_{12}$), are highly sensitive to moisture and react with polar aprotic solvents commonly used in slurry processing (e.g., *N*-methyl-2-pyrrolidone [NMP] for PVDF) [3]. This chemical incompatibility can lead to the decomposition of the SEs, generation of toxic H_2S gas, and the formation of insulating interfaces, drastically reducing ionic conductivity and overall battery performance.

To circumvent these issues, significant research has focused on developing solvent-free or dry processing techniques. Dry electrode manufacturing eliminates the need for solvents altogether, thereby avoiding compatibility issues and simplifying the production process [4]. One prominent method involves the use of fibrillizable binders like polytetrafluoroethylene (PTFE). In this process, active materials, SEs, conductive carbon, and a small amount of PTFE binder are mixed mechanically. Subsequently, the mixture is calendared under shear force, causing the PTFE fibrils to elongate and form a fibrous network that entangles the solid particles, creating a freestanding electrode film. This film can then be laminated onto a current collector. This method allows for the production of thicker electrodes with higher active material loadings (>20 mg cm−2) and lower tortuosity, which is beneficial for high-energy-density applications. Moreover, it is more environmentally friendly and potentially cheaper due to the absence of solvent recovery systems. However, challenges remain, such as achieving uniform distribution of the binder at very low concentrations (often <1 wt.%), ensuring sufficient adhesion strength, and the fact that PTFE itself is electrochemically unstable at low potentials against lithium metal [5].

Another approach within wet processing is the careful selection of nonpolar solvents and compatible binders. Researchers have explored using binders with specific functional groups that dissolve in solvents with low Gutmann donor numbers, which are less likely to attack sulfide SEs. Examples include using polystyrene-*block*-polybutadiene-*block*-polystyrene (SBS) rubber modified with carboxylic acid groups in *p*-xylene [6], ethyl cellulose in toluene [7]. These binders can provide adequate adhesion while maintaining the chemical integrity of the SEs. Furthermore, innovative binder design strategies, such as in situ polymerization or the use of protected functional groups, have emerged. For instance, a binder with tert-butyl-protected carboxyl groups can be processed in a low-polarity solvent. During a subsequent heat treatment step, the protective groups are removed, unveiling highly polar carboxyl groups that significantly enhance adhesion to active materials and SEs [8].

The choice of preparation method also profoundly impacts the electrode's microstructure. Wet-processed electrodes often result in a microstructure where binder and SEs coating are concentrated at the points of contact between active material particles, potentially leading to incomplete ionic conduction pathways. In contrast, dry-processed electrodes can achieve a more homogeneous distribution of SEs and a more continuous ionic conduction network due to the intense shear forces during calendaring, leading to improved rate capability. Therefore, the evolution of electrode preparation methods is intrinsically linked to the development of novel binders and processing techniques, all aimed at optimizing the complex solid–solid interfaces within ASSLB electrodes.

9.2 Development of Advanced Assembly Technology for Solid-state Batteries

The assembly of ASSLBs is a multifaceted challenge that extends far beyond simply combining individually prepared components. It requires innovative technologies to integrate anodes, solid electrolytes, and cathodes into a stable, high-performance, and durable cell. The core challenges stem from the rigid nature of solid–solid interfaces, which are prone to contact loss, high impedance, and mechanical degradation during cycling, unlike the self-wetting liquid electrolytes in conventional LIBs. Advanced assembly technologies are therefore pivotal in ensuring intimate interfacial contact, mitigating parasitic reactions, and accommodating volume changes [9].

A critical aspect of ASSLBs assembly is the management of stacking pressure. Applying external pressure is commonly employed to maintain intimate contact between the solid electrolyte and the electrodes, reducing interfacial resistance and preventing delamination. Laboratory-scale cells (e.g., coin cells) often use high static pressures (tens of MPa), which is impractical for commercial pouch or cylindrical cells. Therefore, a key development focus is designing cell configurations and assembly processes that can maintain effective contact at much lower pressures (<few MPa). Strategies include the use of soft and compliant interfacial layers. For instance, incorporating a ductile lithium metal foil or a lithiophilic interlayer (e.g., Ag–C composite, lithiated graphite) at the anode–SEs interface can improve contact. Similarly, at the cathode side, binders with excellent elastic recovery (e.g., Spandex) can help buffer the volume changes of high-capacity cathodes like NMC or sulfur, maintaining contact without requiring excessively high external pressure.

The integration of the electrolyte separator with the electrodes is another crucial assembly step. Techniques like hot pressing are widely used [10]. This involves stacking the anode, SEs membrane, and cathode together and applying heat and pressure. The heat softens polymer-based binders or SEs, allowing them to flow and fill the voids, thereby creating a more integrated and low-impedance interface [11]. For sulfide SEs, which are often ductile, cold pressing can sometimes suffice. Furthermore, bipolar stacking is an advanced assembly architecture for creating high-voltage battery packs. In this design, multiple unit cells are connected in series within a single cell package by sharing a common current collector between the cathode of one cell and the anode of the next. This significantly reduces the weight and volume of inert packaging materials compared to connecting individual cells externally, leading to a substantial increase in the overall energy density at the pack level [12].

In-situ polymerization represents a cutting-edge assembly technology. Instead of handling a prefabricated solid polymer electrolyte (SPE) membrane, liquid precursor solutions containing monomers, initiators, and lithium salts are injected into the preassembled cell (containing the anode and cathode) [13]. The cell is then sealed and heated, triggering the polymerization of the liquid precursor into a solid polymer matrix in situ. This process perfectly conforms to and wets all electrode surfaces, even penetrating deep into porous electrodes, resulting in exceptional interfacial contact and low impedance. It also simplifies the assembly process by avoiding the handling of fragile SPE films.

For sulfide-based ASSLBs, which are air-sensitive, assembly must often be conducted in inert environments (e.g., argon-filled gloveboxes), which complicates and increases the cost of manufacturing. Developing air-stable SEs or protective coatings that allow for brief air exposure during handling is a critical area of research aimed at simplifying assembly. Moreover, quality control and inspection during assembly are paramount. Techniques like ultrasonic imaging or X-ray computed tomography (CT) can be used nondestructively to detect internal defects such as voids, cracks, or misalignments between layers before the cell is sealed, helping to improve yield and reliability [14].

Finally, the assembly of different cell formats—pouch, cylindrical, or prismatic—presents unique challenges. Pouch cells offer high energy density but require robust sealing to prevent air and moisture ingress. The stack inside a pouch cell must be carefully designed to accommodate swelling during cycling. Cylindrical cells provide excellent mechanical stability but pose challenges in uniformly applying pressure to the jelly roll structure. Advanced assembly technologies are thus tailored to the specific cell format, aiming to achieve a perfect marriage of materials, interfaces, and mechanics for optimal performance.

9.3 Summary and Prospect

The advancement of all-solid-state batteries represents a paradigm shift in energy storage technology, promising unparalleled safety and energy density. This chapter has detailed the critical pillars enabling this transition: the evolution of electrode material preparation methods and the development of sophisticated assembly technologies. The shift from solvent-based wet processing to dry electrode manufacturing addresses the fundamental chemical incompatibility between sulfide solid electrolytes and polar solvents, enabling the fabrication of thick, high-loading electrodes. Concurrently, advanced assembly techniques—such as optimized stacking pressure management, hot pressing, in-situ polymerization, and bipolar stacking—are crucial for integrating these components into cells with intimate interfacial contact, low impedance, and mechanical resilience, all while aiming for practicality in terms of required operating pressure.

Despite significant progress, the path to widespread commercialization of ASSLBs remains fraught with challenges. A primary hurdle is the scalability of the proposed advanced methods. Dry processing, while promising, requires specialized equipment for calendaring and fibrillation that must be scaled up cost-effectively. The synthesis and processing of novel functional binders need to be compatible with large-scale industrial production. Furthermore, the stringent requirement for inert atmospheres during the assembly of sulfide-based cells adds complexity and cost. Cost itself is a significant barrier; many of the materials involved, such as high-performance sulfide SEs (e.g., containing germanium), lithium metal foil, and specialized binders, are currently expensive. Reducing cost through material innovation and process optimization is essential.

Looking forward, the future development of ASSLBs technology will likely focus on several key areas. First, the design of multifunctional binders will intensify. Ideal future binders will seamlessly combine strong adhesion, high ionic conductivity (perhaps as single-ion conductors), elasticity to accommodate volume change, and the ability to form stable interphases (SEI/CEI). Self-healing binders that can autonomously repair microcracks during cycling could dramatically extend battery life. Second, the integration of artificial intelligence and machine learning into material discovery and process optimization holds great promise. These tools can accelerate the screening of novel binder molecules, solvent systems, and processing parameters by establishing structure–property relationships from existing data, guiding experimental efforts more efficiently.

Third, the pursuit of truly dry and solvent-free manufacturing processes will continue. This includes not only dry electrode processing but also the development of dry techniques for applying SE layers. Advances in this area will enhance environmental sustainability and reduce energy consumption associated with solvent drying and recovery. Fourth, advanced in-situ and operando characterization techniques will become increasingly important. Tools like in-situ electrochemical transmission electron microscopy (TEM), X-ray tomography, and neutron depth profiling (NDP) are vital for directly observing the evolution of interfaces and the failure mechanisms of binders and electrodes during operation, providing invaluable feedback for material and cell design.

Finally, there is a growing recognition that success requires a holistic approach that co-optimizes materials, interfaces, and manufacturing processes simultaneously. Collaboration between academia and industry is crucial to bridge the gap between laboratory breakthroughs and commercial products. The standardization of testing protocols, especially regarding the application of stack pressure, will also be necessary for fair and comparable performance evaluations across the field.

In conclusion, the assembly of high-performance all-solid-state batteries is a complex but surmountable challenge. Through continued innovation in electrode preparation, binder science, and assembly technology, ASSLBs are poised to transition from laboratory curiosities to power the next generation of electric vehicles and grid storage systems, marking a new chapter in electrochemical energy storage.

References

1 Cao, D., Zhao, Y., Sun, X., Natan, A., Wang, Y., Xiang, P., Wang, W. and Zhu, H. (2020) Processing strategies to improve cell-level energy density of metal sulfide electrolyte-based all-solid-state Li metal batteries and beyond. *ACS Energy Letters*, 5(11), 3468–3489.

2 Liang, Y., Liu, H., Wang, G., Wang, C., Ni, Y., Nan, C. W. and Fan, L. Z. (2022) Challenges, interface engineering, and processing strategies toward practical sulfide-based all-solid-state lithium batteries. *InfoMat*, 4(5), e12292.

3 Nikodimos, Y., Huang, C.-J., Taklu, B. W., Su, W.-N. and Hwang, B. J. (2022) Chemical stability of sulfide solid-state electrolytes: stability toward humid air and compatibility with solvents and binders. *Energy & Environmental Science*, 15(3), 991–1033.

4 Hippauf, F., Schumm, B., Doerfler, S., Althues, H., Fujiki, S., Shiratsuchi, T., Tsujimura, T., Aihara, Y. and Kaskel, S. (2019) Overcoming binder limitations of sheet-type solid-state cathodes using a solvent-free dry-film approach. *Energy Storage Materials*, 21, 390–398.

5 Li, Y., Wu, Y., Wang, Z., Xu, J., Ma, T., Chen, L., Li, H. and Wu, F. (2022) Progress in solvent-free dry-film technology for batteries and supercapacitors. *Materials Today*, 55, 92–109.

6 Lee, K., Lee, J., Choi, S., Char, K. and Choi, J. W. (2018) Thiol–ene click reaction for fine polarity tuning of polymeric binders in solution-processed all-solid-state batteries. *ACS Energy Letters*, 4(1), 94–101.

7 Cao, D., Li, Q., Sun, X., Wang, Y., Zhao, X., Cakmak, E., Liang, W., Anderson, A., Ozcan, S. and Zhu, H. (2021) Amphipathic binder integrating ultrathin and highly ion-conductive sulfide membrane for cell-level high-energy-density all-solid-state batteries. *Advanced Materials*, 33(52), 2105505.

8 Lee, J., Lee, K., Lee, T., Kim, H., Kim, K., Cho, W., Coskun, A., Char, K. and Choi, J. W. (2020) In situ deprotection of polymeric binders for solution-processible sulfide-based all-solid-state batteries. *Advanced Materials*, 32(37), e2001702.

9 Xu, L., Lu, Y., Zhao, C. Z., Yuan, H., Zhu, G. L., Hou, L. P., Zhang, Q. and Huang, J. Q. (2020) Toward the scale-up of solid-state lithium metal batteries: the gaps between lab-level cells and practical large-format batteries. *Advanced Energy Materials*, 11(4), 2002360.

10 Liu, S., Zhou, L., Han, J., Wen, K., Guan, S., Xue, C., Zhang, Z., Xu, B., Lin, Y., Shen, Y., Li, L. and Nan, C. W. (2022) Super long-cycling all-solid-state battery with thin Li_6PS_5Cl-based electrolyte. *Advanced Energy Materials*, 12(25), 2200660.

11 Whiteley, J. M., Taynton, P., Zhang, W. and Lee, S.-H. (2015) Ultra-thin solid-state Li-ion electrolyte membrane facilitated by a self-healing polymer matrix. *Advanced Materials*, 27(43), 6922–6927.

12 Cao, D., Sun, X., Wang, Y. and Zhu, H. (2022) Bipolar stackings high voltage and high cell level energy density sulfide based all-solid-state batteries. *Energy Storage Materials*, 48, 458–465.

13 Ding, P., Lin, Z., Guo, X., Wu, L., Wang, Y., Guo, H., Li, L. and Yu, H. (2021) Polymer electrolytes and interfaces in solid-state lithium metal batteries. *Materials Today*, 51, 449–474.

14 Mao, C., Dong, J., Li, J., Zhai, X., Ma, J., Luan, S., Shen, X., Wang, Y., Zhang, P., Sun, H., Bie, X., Gao, X. and Song, J. (2025) Toward practical all-solid-state batteries: current status of functional binders. *Advanced Materials*, 37(16), 2500079.

Index

Note: Page numbers in *italics* and **bold** refers to figures and tables respectively.